T5-AQT-728

RESIDENTIAL AND LIGHT BUILDING CONSTRUCTION
Principles and Practices

Stephens W. Nunnally
Joan A. Nunnally

PRENTICE HALL
Englewood Cliffs, New Jersey 07632

Library of Congress Cataloging-in-Publication Data

Nunnally, S. W.
Residential and light building construction : principles and practices / S.W. Nunnally, J.A. Nunnally.
p. cm.
Bibliography: p.
Includes index.
ISBN 0-13-775156-7
1. Building. I. Nunnally, J. A. . II. Title.
TH145.N863 1990
690--dc20 89-35454
CIP

Editorial/production supervision: Tally Morgan
Cover design: Bruce Kenselaar
Manufacturing buyer: David Dickey
Cover photo: J. Zehrt/FPG

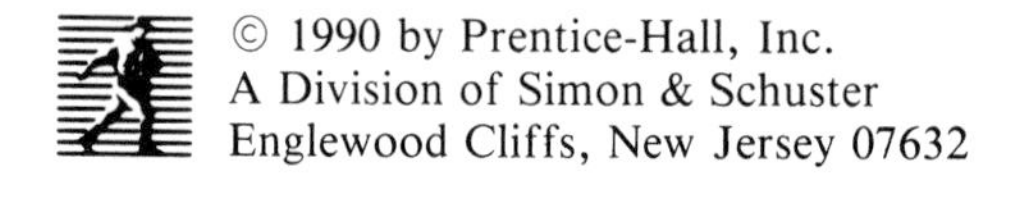

Printed in the United States of America
10 9 8 7 6 5 4 3 2 1

ISBN 0-13-775156-7

Prentice-Hall International (UK) Limited, *London*
Prentice-Hall of Australia Pty. Limited, *Sydney*
Prentice-Hall Canada Inc., *Toronto*
Prentice-Hall Hispanoamericana, S.A., *Mexico*
Prentice-Hall of India Private Limited, *New Delhi*
Prentice-Hall of Japan, Inc., *Tokyo*
Simon & Schuster Asia Pte. Ltd., *Singapore*
Editora Prentice-Hall do Brasil, Ltda., *Rio de Janeiro*

Dedicated to the men and women of the construction industry.

Contents

Preface

Construction is an exciting, dynamic process and is one of the largest industries in the United States. Therefore, it provides many satisfying career opportunities for both men and women. This book is a comprehensive introduction to the construction of small residential, commercial, and industrial buildings, a category of construction called "light building construction."

The material in this book is suitable for use as a text for a variety of college construction courses, as well as for reference and self-study by construction professionals. Chapter 1 provides an introduction to the construction industry, its contracting procedures, and governmental regulation. Chapter 2 describes the basic properties, behavior, and use of common construction materials. Chapter 3 discusses site investigation, soil characteristics, the estimation of cut and fill, and site preparation. Chapters 4 through 13 discuss common building construction procedures from the foundation through final finish. Chapters 14 and 15 discuss the construction aspects of electrical, mechanical, and plumbing systems. Lastly, Chapter 16 describes some of the most common hand tools, power tools, and construction equipment used by builders.

Throughout the book, a special effort has been made to explain the "why" as well as the "how" of many common construction procedures. The construction procedures described reflect generally accepted U.S. construction practices. However, the requirements of the project plans and specifications and of the local building code(s) must always be observed. Readers should consult the latest adopted version of the applicable codes when determining code requirements.

It would not have been possible to produce this book without the cooperation of many individuals and organizations within the construction industry. The assistance of construction industry associations, building material producers, and tool and equipment manufacturers in providing information and in permitting the reproduction of their material is gratefully acknowledged. Where possible, appropriate credit has been provided.

Readers are encouraged to advise the authors of any errors noted and to suggest improvements for future editions.

S. W. Nunnally
J. A. Nunnally

chapter 1

Introduction

1-1 THE CONSTRUCTION INDUSTRY

Construction is one of the largest industries in the United States, accounting in recent years for about 10 percent of the nation's gross national product and providing employment for over 5 million Americans. Construction is an exciting, dynamic process that often provides high income for workers and contractors. However, its seasonal nature and widely fluctuating volume of work often results in reduced annual income. When combined with the industry's highly competitive nature, these factors often result in a high bankruptcy rate for the industry. In spite of these and other problems, construction provides an exciting career opportunity with many challenges and rewards. This book focuses on the principles and practices of residential and light building construction, and it also provides a sound background for a professional career in construction.

The number of construction contractors in the United States has been estimated to exceed 800,000, but some 60 percent of these firms employ three or less workers. Only a few thousand firms employ 100 or more workers. Nevertheless, these few large firms account for more than 30 percent of the value of work performed. The trend in recent years has been for these large firms to capture an ever-increasing share of the total U.S. construction market.

Individuals and companies engaged in the business of construction are commonly referred to as *construction contractors* because they operate under a contract arrangement with the owner. Construction contractors are further classified as general contractors or specialty contractors. *General contractors* engage in a wide variety of construction work and carry out most major construction projects. *Specialty contractors* operate in only a limited field of construction, such as plumbing, electrical, heating and ventilating, or earthmoving. When a specialty contractor is employed by a general contractor to accomplish some portion of a project, the specialty contractor is referred to as a *subcontractor*; that is, subcontractors operate under a subcontract between themselves and the general or *prime contractor*. As you see, "prime contractor" and "subcontractor" are defined by the contract arrangement involved, not the work classification of the contractors themselves. Therefore, a specialty contractor employed by an owner to carry out and manage a particular project might employ a general contractor to execute some phase of the project. In this situation, the specialty contractor becomes the prime contractor for the project, and the general contractor becomes a subcontractor. The reverse situation is much more common. Contract construction is discussed in more detail in section 1-2 and in reference 3.

The major divisions of the construction industry include building construction (also called "vertical construction") and heavy construction (also called "horizontal construction"). *Building construction*, as the name implies, involves the construction of buildings. This category is often further subdivided into residential and nonresidential building construction. *Heavy construction* includes highways, airports, railroads, bridges, canals, harbors, dams, and other major public works and is therefore often referred to as "highway and heavy construction." Other specialty divisions of the construction industry include industrial construction, process plant construction, utility construction, and marine construction.

In 1988, building construction accounted for over 77 percent of total new construction in the United States (see Figure 1-1). Residential building construction, the largest single component of the building construction industry, accounted for 61 percent of building construction and 47 percent of total new construction. Residential housing construction is primarily carried out by local construction firms, both in the United States and abroad.

There are a number of professional organizations representing builders and contractors. Of these, the major organization representing the U.S. residential construction industry is the National Association of Home Builders (NAHB). This organization consists of over 800 state and local chapters with a membership of more than 140,000 builders and associates. In addition to representing the industry in its relations with the public, government, and financial institutions, the NAHB is active in fostering construction education and research. A large number of educational programs and seminars are sponsored each year by the Home Builders Institute, the educational arm of the NAHB. Research on home buildings materials, methods, standards, and equipment is sponsored by the NAHB Research Foundation, Inc.

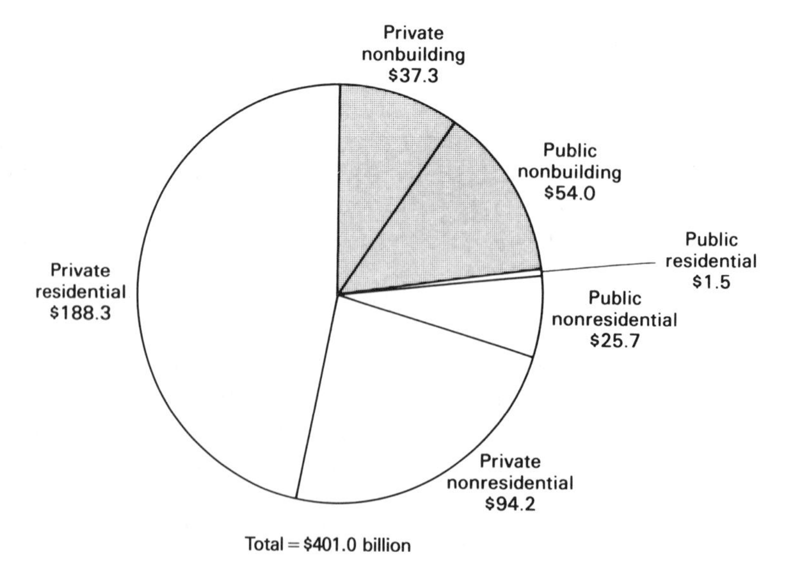

Figure 1-1 U.S. new construction volume. Annual rate in billions of dollars (July 1988) (Source: Bureau of the Census).

1-2 CONSTRUCTION CONTRACTING

Construction contracting is a highly competitive business with a high rate of bankruptcy. Therefore, knowledge of the business aspects of construction is as essential for success in the industry as is knowledge of the technical aspects. Because this book is primarily concerned with the technology of building, the following discussion simply provides a basic understanding of the environment in which the builder must operate.

Construction Contracts

Few builders are also qualified lawyers and must call on lawyers for legal assistance when needed. However, all construction professionals should have an understanding of the customary practices and underlying legal principles so that they know when to consult a lawyer. A brief overview of the construction contracting system practiced in the United States follows.

Virtually all construction is accomplished under a *contract* between an owner and a builder. The essential elements of a contract include an offer, an acceptance, and a consideration. The offer customarily consists of a *bid* or *proposal* by a builder to construct a facility in accordance with the plans, specifications, and conditions supplied by the owner. The most common form of acceptance is a *notice of award* from the owner notifying the builder that the bid was accepted. The legal consideration usually consists of a specified payment for the work, but it may be anything of value.

The major steps in the construction contracting process consist of bid solicitation, bid preparation, bid submission, contract award, and contract administration. However, before the bidding process can take place, the owner must determine the requirements for the building and have the necessary plans, specifications, and other documents prepared. For major projects, steps in the project development process include:

- Recognition of the need for a building.
- Determination of the technical and financial feasibility of the project.
- Preparation of detailed plans, specifications, and cost estimates for the building.
- Approval by regulatory agencies. This involves ascertaining compliance with zoning regulations, building codes, environmental regulations, etc.

Competitively Bid Projects

After obtaining approval of regulatory agencies and ascertaining that financing is available, the owner initiates the contracting process. Almost all public construction contracts and most large private building contracts are awarded on a *competitive bid* basis; that is, bids are solicited by either public advertising or by sending bid requests to several builders. Interested builders are required to submit their bids prior to a designated bid-closing time. After reviewing the bids to determine that each bid is responsible (from a qualified bidder) and responsive (complies with bid requirements), the winning bidder (normally the low bidder) is designated. The selected contractor is then notified of the bid award.

The most critical phase of a contractor's bid occurs in bid preparation. In the limited time available, the contractor must accurately estimate the cost for carrying out the project and determine the final bid price, which includes overhead and profit. If a bid is too low, there will be little profit or even a loss in carrying out the project. If the bid is too high, there is little chance that the contractor will be awarded the contract.

Negotiated Contracts

Construction contracts may also be awarded by *negotiation* between the owner and a builder. Here the price, time, and other terms of the contract are negotiated directly between the owner and the builder to their mutual satisfaction. Clearly there is less risk to the builder in such a contract. However, an owner is not likely to use a negotiated contract unless the project is complex or urgently needed.

Fixed-Price and Cost-Type Contracts

The major categories of contract by method of pricing are fixed-price and cost-type contracts. Most routine building contracts are executed on a *fixed-price* basis: the contract states a fixed price is to be paid upon completion of the project as specified. *Cost-type* (or cost-plus) contracts provide that the contractor be reimbursed for all costs directly associated with the project plus a designated fee or profit. To avoid legal disputes, cost-type contracts should clearly define the items of contractor cost that will be reimbursed as well as the criteria for determining the acceptability of costs. Again, cost-type contracts provide little risk to the builder, but they also give the builder little incentive to minimize project costs. Therefore, this type of contract is not likely to be used unless the project is unusually large or complex.

Contract Administration

It is common procedure for a construction contract to require the contractor to submit a proposed construction schedule to the owner shortly after contract award. After approval by the owner, this schedule becomes the basis for determining the contractor's progress. The owner's agent, usually an architect or engineer, verifies that work is being accomplished in accordance with the contract and also monitors project progress.

Most projects expected to extend longer than a few months provide for *progress payments*, payments made at periodic intervals to the contractor according to the stage of project completion. This procedure lessens the contractor's financial burden during construction.

Changes, Delays, and Disputes

Most construction contracts provide that the owner or his or her representative may direct changes "within the general scope of the contract" during construction. The document directing such a change is a *change order*. Ideally, a change order should specify a time and price adjustment satisfactory to the contractor for performing the work in the change order. In actual practice, time often does not permit such mutual agreement prior to implementing a change. This frequently results in the later submission of a contractor claim for the cost of work involved in carrying out the change.

A *claim* is simply a request by the contractor for additional time or money in compensation for an event under the control of the owner. Such events include changes directed by the owner (without an appropriate time and price adjustment), delay in delivery of owner-provided property, and changed site conditions.

Disputes are disagreements between the owner and contractor over some aspect of contract performance and must be settled by either arbitration or court action. *Arbitration*, such as that carried out under the procedures of the American

Arbitration Association, provides a faster, less costly resolution of disputes than does court action. However, both the owner and contractor must agree to the use of arbitration in lieu of court action, and normally no appeal of the arbitration decision is permitted.

Contract Completion or Termination

Owner acceptance of a completed project is commonly based on the results of a final inspection conducted by the owner's representative. Acceptance is also conditioned on correction of deficiencies noted during the final inspection. This list of deficiencies is referred to as the *punch list* of record. When inspection reveals that the building is substantially complete and suitable for occupancy, the owner's representative may execute a *certificate of substantial completion*. This document authorizes payment of the total contract amount less an amount reserved to ensure correction of punch list deficiencies. Final payment is made after correction of these deficiencies. Any deficiencies later discovered are corrected under the warranty provisions of the contract.

Contracts may also be *terminated* prior to completion of work. The usual basis for such termination is *breach of contract* by either the owner or builder. Breach of contract by the contractor most often involves contractor default or failure to make reasonable progress. Breach of contract by the owner is most often due to unreasonable delays by the owner or failure of the owner to make specified progress payments.

1-3 BUILDING REGULATIONS AND CODES

Buildings constructed in most areas of the United States must comply with a number of governmental regulations. *Building codes*, which are primarily concerned with public safety, provide minimum design and construction standards for structural and fire safety. *Zoning* regulations, which control land use, limit the type, size, and density of structures that may be erected at a particular location. *Environmental regulations* protect the public and environment by controlling such factors as vehicular traffic, water usage, precipitation runoff, and waste disposal. The profession of construction is also regulated by governmental licensing and certification procedures.

Building Codes

The danger to the public from the possible collapse of structures has been a public concern since early civilization. An early approach, contained in the Code of Hammurabi dating from about 1800 BC, simply condemned a builder to death if the structure collapsed and caused someone's death. In the United States, early building codes were primarily concerned with the threat of fire. In 1905 the Board of Fire Underwriters published a Recommended National Building Code, which provided minimum standards for fire protection as well as structural safety. This code, now known as the *Basic/National Building Code*, published by the Building Officials and Code Administrators International, was the only nationally recognized building code until 1927. Other major building codes now in use include the *Uniform Building Code* published by the International Conference of Building Officials and the *Standard Building Code* published by the Southern Building Code Congress International. In 1971, all of these model code groups cooperated to publish the *CABO One and Two Family Dwelling Code* (hereafter referred to as the CABO code) to provide a simplified standard for the construction of detached one- and two-family dwellings not more than three stories high. Most of these building codes include provisions for plumbing and mechanical systems. However, electrical work is commonly governed by the *National Electric Code*, published by the National Fire Protection Association under the auspices of the American National Standards Institute (ANSI).

The national model codes are purely advisory and must be put into force by local ordinance. Most local building codes are in fact based on one of the model codes. However, most local codes also contain modifications to the model code on which they are based. Such modifications, as well as local delays in code updating, result in code provisions that are often unnecessarily restrictive and result in increasing building costs. Another problem associated with building codes at the local level is the quality of code administration. Local jurisdictions often lack an adequate number of technically qualified building officials. This leads to rather cursory building inspections using a checklist approach and discourages builders from utilizing new materials and procedures.

A *building permit* must be obtained before construction can begin. After the permit is issued, the local building department inspects the project at certain stages of completion. While procedures vary widely, a typical inspection schedule for a single-family residence might call for the following inspections during the course of construction:

Foundation Inspection Performed prior to placing concrete or to check special requirements for wood foundations.

Inspection of Rough Plumbing, Electrical, and Mechanical Performed before these items are concealed and prior to setting fixtures, and prior to framing inspection.

Frame and Masonry Inspection Performed after the roof and all framing and bracing are in place. Follows the rough plumbing, electrical, and mechanical inspection.

Lath and/or Drywall Inspection Performed after all interior drywall or lathing is in place but before drywall joints are taped or plaster is applied.

Final Inspection Performed after construction is complete so that the building can be certified ready for occupancy.

The scheduling of these inspections often poses problems for the builder and may result in construction delays.

Regulations and Zoning

There may be a number of other regulations that must be complied with before a building permit can be obtained. Zoning regulations limit the type of structure that may be erected at a specific location. Typical zoning classifications include residential (with specified density), commercial, office, industrial, recreational, and agricultural. These zoning classifications are usually designated by a combination of letters and numbers. For example, the R-4 zoning classification may represent residential housing with a maximum density of 4 units per acre. In order to construct a building not conforming to the current zoning, it would be necessary to obtain a change in zoning or an administrative exception.

Large projects, such as large housing developments, major shopping centers, large industrial plants, or athletic stadiums, may require preparation and approval of an Environmental Impact Statement (EIS), describing and quantifying the effect the project will have on the environment. Preparation of the EIS is a complex, time-consuming, and expensive task that should only be undertaken by a professional with experience in such matters.

When municipal utility services are not available at the project site, additional permits may be required for water wells, water treatment plants, septic tanks or sewage treatment plants, and similar facilities.

Contractor Licensing

Most communities with building departments also require construction contractors to have their professional qualifications verified by licensing or certification. This may be done at the local level or by the state. State certification or licensing usually requires satisfactory completion of a comprehensive written examination plus proof of financial capacity and verification of character. A business or occupational license is also usually required of all builders.

1-4 REDUCING THE COST OF CONSTRUCTION

Construction Productivity and Construction Management

Several recent studies have identified major problems in U.S. construction industry productivity and competitiveness. The Construction Industry Cost Effectiveness Study, also known as the CICE Study, sponsored by the Business Roundtable, attempted to determine reasons for the industry's declining productivity and rising costs during the 1960s and 1970s. This study, completed in 1982, was probably the most comprehensive ever made of the U.S. construction industry. It identified a number of industry problems and suggested improvements in the areas of project management, construction technology, supply, training, and use of labor, and governmental regulation. It concluded that while much of the blame for industry problems should be shared by owners, contractors, labor, and government, many of the problems could be overcome by improved construction management.

The term *construction management* can be confusing because it is sometimes used to refer to a contractual arrangement under which a firm supplies construction management services to an owner while other firms accomplish the actual design and construction. However, in its more general use, "construction management" refers to the control of construction's basic resources of workers, material, machinery, money, and time. Every construction supervisor from crew chief to company president is part of construction management. While the principal objectives of a construction manager should be to complete every project on time and within budget, he or she has a number of other important responsibilities. Some of these responsibilities include innovation, improvement of technology, productivity improvement, safety, morale, and public relations. Some of the avenues available for reducing the cost of a specific building are discussed in the following paragraphs.

Reducing Construction Costs

Construction cost savings can be made in the design process even before construction begins. Some of the design factors that can reduce costs include the use of modular dimensions, incorporation of prefabricated components and assemblies, grouping plumbing fixture locations to minimize piping runs, use of economical materials, and utilization of new technology. One of the advantages claimed for the construction contract system that employs a professional construction manager is that the construction manager can inject constructability considerations into the design process.

During the actual construction phase, construction managers can increase productivity and reduce costs by such things as:

- careful job planning
- effective organization of work
- efficient scheduling of labor, equipment, and materials
- incorporation of labor-saving techniques (such as on-site or plant prefabrication)
- timely quality control to minimize rework
- minimizing accidents through good safety procedures

While on the subject of quality control, it is important for everyone in construction to realize that the *construction contractor* is primarily responsible for quality. Inspections by an owner's representative or government agency provide little more than spot checks to verify that some particular aspect of the project meets minimum standards. The extra costs associated with rework or accidents are either directly or ultimately borne by the contractor, even on cost-type contracts. Poor performance in these areas also results in the contractor gaining a reputation for poor work. The combined effect of added cost and poor reputation may well make the contractor noncompetitive and lead to company failure.

The Future

Industry problems of high cost and low productivity during recent years have served to reduce the demand for construction in the United States. In addition, studies of international competition in design and construction have found that the U.S. share of the world's construction market declined sharply in the decade from 1975–85. During the same period, foreign construction firms increased their portion of domestic U.S. construction work from less than $1 billion to nearly $10 billion. However, many observers are confident that the U.S. construction industry will rise to the challenge and in time regain its predominant position in the world construction market. In any event, the U.S. construction industry will continue to provide many opportunities and rewards to the innovative, professionally competent, and conscientious construction professional.

1-5 CONSTRUCTION SAFETY AND HEALTH

General

Building construction is an inherently dangerous process. Historically, the construction industry has had one of the highest accident rates among all industries. In 1986, the U.S. construction industry suffered 15.1 injuries and deaths per 100 workers, the highest rate of all industries. Concern over industrial accidents and occupational diseases in the United States led to the passage of the *Occupational Health and Safety Act of 1970* (OSHA). This act established health and safety standards for almost all U.S. industries, including construction.

OSHA has produced a comprehensive set of safety and health regulations, inspection guidelines, and record-keeping requirements. The act establishes both civil and criminal penalties for violations as shown in Table 1-1. Civil penalties are imposed administratively by OSHA officials, whereas criminal penalties must be adjudged by a federal court. When a job site condition presents imminent danger

TABLE 1-1
Maximum OSHA Penalties

Administrative penalties Offense	Maximum penalty
Willful or repeated, serious	$10,000/violation
Routine; failure to post citation	$1,000/violation
Failure to correct cited violation	$1,000/day

Criminal penalties Offense	Maximum fine	Maximum imprisonment
Assaulting, resisting or killing OSHA official	$10,000	Life
Willful violation resulting in death, second conviction	$20,000	1 year
Willful violation resulting in death, first conviction; falsifying required records	$10,000	6 months
Unauthorized prior notice of inspection	$1,000	6 months

of death or serious injury, OSHA officials may request a restraining order from a U.S. District Court stopping construction until the hazard has been eliminated.

Construction managers are understandably concerned about OSHA regulations and penalties. However, the financial consequences of accidents are often more serious than OSHA penalties. Premiums for workers compensation, public liability, property damage, and equipment insurance are all based on a firm's accident and loss rates. A poor safety record results in high insurance premiums. Studies have shown that a construction firm can be rendered noncompetitive by a poor safety record. In addition to insurance costs, accidents result in lost project time, loss of skilled workers, and other losses not covered by insurance.

Safety Considerations

To minimize accidents, every construction firm needs a realistic and comprehensive safety program. The most important element of any safety program is the enthusiastic support of top management. Supervisors must be convinced that safety is as important as production. The *Manual of Accident Prevention in Construction* (see references) contains a number of suggestions for establishing an effective construction safety program. Among the important elements in a safety program are:

- Formal safety training for all workers and supervisors.
- Periodic refresher training.
- A company safety officer responsible for identifying and correcting job hazards at the construction site.
- Necessary personal protective equipment.
- Adequate first aid equipment, trained emergency personnel, and procedures for emergency evaluation.
- Safety records and reports required by OSHA.

Most serious construction accidents involve falls from high places, collapse of concrete formwork or other temporary structures, structural failure of buildings under construction, or the operation of construction equipment. Therefore, particular attention should be paid to the safety of these operations. A number of safety precautions for specific construction operations are described in other sections of this book as well as in references 5 and 7.

Environmental Health

Although accidents have received the greatest OSHA attention, occupational health hazards to workers are also covered by the OSHA act. In recent years, increased attention has been directed toward environmental health hazards, both on and off the job. Some major environmental health hazards that may be encountered in construction include noise, dust, radiation, toxic materials, heat, and cold.

Under OSHA regulations, the maximum noise level to which a worker may be exposed depends on the length of exposure. When noise cannot be reduced to an acceptable level, personal ear protection must be provided to the workers.

Inhaling certain types of dust can produce lung disease. Asbestos and silica dust are particularly hazardous. OSHA health standards establish limits for the exposure of workers to dust. In addition, the reduced visibility produced by dust in the air may present a safety hazard.

There are two types of radiation that pose health hazards. The ionizing radiation produced by X-ray equipment and radioactive material causes tissue damage in humans. Regulations of the Nuclear Regulatory Commission control the use of ionizing radiation. The nonionizing radiation produced by lasers and microwave equipment may pose an eye hazard. Limits on nonionizing radiation are contained in OSHA regulations. Construction workers who may be exposed to laser light energy of more than 5 mW are required to be equipped with antilaser eye protection.

The toxic materials most likely to be encountered in construction include electricity, petroleum fuels, natural and synthetic gas, and carbon monoxide. An oxygen deficiency may result when the air in a confined area is displaced by a heavy gas. OSHA regulations require specific safety procedures and protective equipment when hazardous materials are likely to be encountered.

Construction workers exposed to extreme heat or cold may suffer temperature-related illness. While the human body acclimates itself to extreme temperature over a period of time, care must be taken to minimize the effect of such exposure. To reduce the effect of extreme heat on workers, provide hot-weather clothing, sun shades, cool rest areas, and an ample supply of drinking water. For work in extreme cold, provide cold-weather clothing, warm rest areas, and limit exposure time.

PROBLEMS

1. Describe at least three major characteristics of the U.S. construction industry.
2. Explain the meaning of "construction contractor."
3. Identify the principal divisions of the construction industry.
4. Define and briefly explain the effect of a change order.
5. Briefly discuss the advantages and disadvantages to the contractor of the use of arbitration in settling a dispute with an owner.
6. What is the purpose of a building code?
7. How does zoning affect the construction of a specific building at a particular location?
8. What party is responsible for construction quality control on a building project?
9. What construction operations account for the majority of serious injuries?
10. List the major environmental health hazards likely to be encountered in construction.

REFERENCES

1. Building Officials and Code Administrators International, Inc. *Basic/National Building Code*. Country Club Hills, IL.
2. Council of American Building Officials. *CABO One and Two Family Dwelling Code*. Falls Church, VA.
3. Clough, Richard H. *Construction Contracting*. 4th ed. New York: Wiley, 1981.
4. The National Association of Home Builders. *Land Development Manual*. Washington, DC, 1969.
5. The Associated General Contractors of America, Inc. *Manual of Accident Prevention in Construction*. Washington, DC, 1977.
6. The Business Roundtable. *More Construction for the Money*. New York, 1983.
7. Nunnally, S.W. *Construction Methods and Management*. 2nd ed. Englewood Cliffs, NJ: Prentice-Hall, 1987.
8. U.S. Department of Labor. *OSHA Safety and Health Standards: Construction Industry* (OSHA 2207). Washington, DC, 1983.
9. Southern Building Code Congress International, Inc. *Standard Building Code*. Birmingham, AL.
10. International Conference of Building Officials. *Uniform Building Code*. Whittier, CA.

chapter 2

Construction Materials

2-1 WOOD

Man has employed wood as a construction material since ancient times. Wood is the principal material employed in the construction of small buildings except in those areas of the world where timber is scarce or its use is inhibited by local conditions. In the United States, some 90 percent of all homes are constructed of wood.

Wood Classification and Properties

The two principal classes of wood by origin are softwood and hardwood. *Softwood* is the wood obtained from conifers, trees having needlelike or scalelike leaves, primarily evergreen trees. *Hardwood* is obtained from deciduous or leaf-shedding trees. The terms "softwood" and "hardwood" can be misleading since they indicate only the wood species involved and not the actual wood hardness. In fact, some softwoods are harder than some hardwoods.

The *moisture content* of wood is defined as the weight of water in the wood divided by the oven-dry weight of the wood and is expressed as a percentage. Moisture content is an important factor in determining wood strength. Wood is essentially in its natural state at moisture contents above 30 percent. As the moisture content falls below 30 percent, wood shrinks and gains strength. However, lumber also frequently warps as the wood dries out and shrinks. Warping can be minimized by shaping the lumber after it has been dried to within a few percent of the moisture content at which it will be used. Grading rules define lumber having a moisture content over 19 percent as *green lumber*; lumber having a moisture content of 19 percent or less but over 15 percent as *dry lumber*; and lumber having a moisture content of 15 percent or less as *kiln-dried lumber*.

Structural lumber is further classified according to its size and intended use and is graded to determine its allowable stresses (force per unit area). By use, structural lumber is divided into the four classifications shown below.

Board Pieces less than 2 in. (5 cm) thick and at least 2 in. (5 cm) wide.

Dimension Pieces at least 2 in. (5 cm) but less than 5 in. (12.7 cm) thick and 2 in. (5 cm) or more wide.

Beam and Stringer Pieces at least 5 in. (12.7 cm) thick and 8 in. (20 cm) wide, graded for their strength in bending when a load is applied to the narrow face.

Post and Timber Pieces approximately square in cross-section, at least 5 in. (12.7 cm) in thickness and width, and intended primarily for use as posts or columns where bending strength is not important.

Rough lumber has been sawn on all four sides but not surfaced. *Surfaced* or *dressed* lumber has been planed smooth on one or more sides or edges. Classifications of dressed lumber include surfaced one side (S1S), surfaced two sides (S2S), surfaced all four sides (S4S), surfaced one edge (S1E), surfaced two edges (S2E), and surfaced on a combination of edges and sides (S1S1E, S1S2E, and S2S1E). Section dimensions for common sizes of U.S. structural lumber are given in Table 2-1. U.S. structural lumber varies in length from 10 ft. (3 m) to 20 ft. (6 m) in 2-ft. (0.6 m) increments. Longer lengths may be available by special order.

Strength

As noted earlier, moisture content is a major factor in determining the strength of structural lumber. As an example of the increase in strength of lumber resulting from drying, the bending strength of common softwood increases by a factor of approximately 2 1/2 times as it dries from 30 percent to 19 percent moisture. In the United States, a national Grading Rules Committee established by the U.S. Department of Commerce sets lumber grading rules for the industry. The maximum allowable unit stresses and modulus of elasticity that may be used in design and construction are determined by the wood species, lumber grade, and moisture content. The principal U.S. grades of commercial lumber are identified in Table 2-2.

Maximum allowable lumber stresses are given in Reference 6. Note that the values given in the various tables must be adjusted for certain conditions as explained in the tables and footnotes. The major adjustment factors pertain to the duration of the loading and service under wet conditions. These principal adjustment factors are summarized in Table 2-3.

Plywood

Plywood is produced by gluing three or more thin layers of wood (veneers) together so that the grain of alternate layers runs perpendicular to each other. This produces a wood product having a high strength/weight ratio. U.S. plywood is produced in a wide range of strength and appearance grades established under grading rules of the American Plywood Association. Surface appearance grades include N, A, B, C, and D, with Grade N having the best appearance.

Strength grading rules for plywood divide wood veneers into Groups 1 through 5 according to strength and stiffness, with Group 1 having the highest strength. *Exterior* grade plywood is manufactured with waterproof glue and uses higher grade veneers than does *interior* grade plywood. In addition to exterior and interior grades, grading rules establish *engineered* grades and an *identification index*, which indicates the maximum allowable span of a plywood sheet under standard loads.

Plyform, a grade of plywood intended for use in concrete formwork, is manufactured in Class I, Class II, and Structural I grades, with Class II most readily available. Note that the maximum allowable stresses for Plyform set forth in Reference 2 have already been adjusted for a load duration of 7 days under wet condi-

TABLE 2-1
Dimensions of U.S. Structural Lumber (in.)

Nominal size	Standard dressed size (S4S)
1 × 3	3/4 × 2 1/2
1 × 4	3/4 × 3 1/2
1 × 6	3/4 × 5 1/2
1 × 8	3/4 × 7 1/4
1 × 10	3/4 × 9 1/4
1 × 12	3/4 × 11 1/4
2 × 3	1 1/2 × 2 1/2
2 × 4	1 1/2 × 3 1/2
2 × 6	1 1/2 × 5 1/2
2 × 8	1 1/2 × 7 1/4
2 × 10	1 1/2 × 9 1/4
2 × 12	1 1/2 × 11 1/4
2 × 14	1 1/2 × 13 1/4
3 × 4	2 1/2 × 3 1/2
3 × 6	2 1/2 × 5 1/2
3 × 8	2 1/2 × 7 1/4
3 × 10	2 1/2 × 9 1/4
3 × 12	2 1/2 × 11 1/4
3 × 14	2 1/2 × 13 1/4
3 × 16	2 1/2 × 15 1/4
4 × 4	3 1/2 × 3 1/2
4 × 6	3 1/2 × 5 1/2
4 × 8	3 1/2 × 7 1/4
4 × 10	3 1/2 × 9 1/4
4 × 12	3 1/2 × 11 1/4
4 × 14	3 1/2 × 13 1/4
4 × 16	3 1/2 × 15 1/4
6 × 6	5 1/2 × 5 1/2
6 × 8	5 1/2 × 7 1/2
6 × 10	5 1/2 × 9 1/2
6 × 12	5 1/2 × 11 1/2
6 × 14	5 1/2 × 13 1/2
6 × 16	5 1/2 × 15 1/2
6 × 18	5 1/2 × 17 1/2
6 × 20	5 1/2 × 19 1/2
6 × 22	5 1/2 × 21 1/2
6 × 24	5 1/2 × 23 1/2
8 × 8	7 1/2 × 7 1/2
8 × 10	7 1/2 × 9 1/2
8 × 12	7 1/2 × 11 1/2
8 × 14	7 1/2 × 13 1/2
8 × 16	7 1/2 × 15 1/2
8 × 18	7 1/2 × 17 1/2
8 × 20	7 1/2 × 19 1/2
8 × 22	7 1/2 × 21 1/2
8 × 24	7 1/2 × 23 1/2
10 × 10	9 1/2 × 9 1/2
10 × 12	9 1/2 × 11 1/2
10 × 14	9 1/2 × 13 1/2
10 × 16	9 1/2 × 15 1/2
10 × 18	9 1/2 × 17 1/2
10 × 20	9 1/2 × 19 1/2
10 × 22	9 1/2 × 21 1/2
10 × 24	9 1/2 × 23 1/2
12 × 12	11 1/2 × 11 1/2
12 × 14	11 1/2 × 13 1/2
12 × 16	11 1/2 × 15 1/2
12 × 18	11 1/2 × 17 1/2
12 × 20	11 1/2 × 19 1/2
12 × 22	11 1/2 × 21 1/2
12 × 24	11 1/2 × 23 1/2
14 × 14	13 1/2 × 13 1/2
14 × 16	13 1/2 × 15 1/2
14 × 18	13 1/2 × 17 1/2
14 × 20	13 1/2 × 19 1/2
14 × 22	13 1/2 × 21 1/2
14 × 24	13 1/2 × 23 1/2
16 × 16	15 1/2 × 15 1/2
16 × 18	15 1/2 × 17 1/2
16 × 20	15 1/2 × 19 1/2
16 × 22	15 1/2 × 21 1/2
16 × 24	15 1/2 × 23 1/2
18 × 18	17 1/2 × 17 1/2
18 × 20	17 1/2 × 19 1/2
18 × 22	17 1/2 × 21 1/2
18 × 24	17 1/2 × 23 1/2
20 × 20	19 1/2 × 19 1/2
20 × 22	19 1/2 × 21 1/2
20 × 22	19 1/2 × 23 1/2
22 × 22	21 1/2 × 21 1/2
22 × 24	21 1/2 × 23 1/2
24 × 24	23 1/2 × 23 1/2

TABLE 2-2
Principal Grades of U.S. Commercial Lumber—Visual Grading

Select structural
No. 1
No. 2
No. 3
Appearance
Stud
Construction
Standard
Utility

tions. Other available types of plywood include High-Density Overlay (HDO) and Medium-Density Overlay (MDO). These plywoods have an abrasion-resistant resin fiber overlay on one or both faces. Where appearance is a major consideration, plywood is available with rough-sawn, grooved, and other special surfaces.

The standard size of U.S. plywood sheets is 4 ft. × 8 ft. (1.2 m × 2.4 m), with thicknesses of 3/8 in. (1.0 cm) to 1 1/8 in. (2.9 cm). Plywood is also available with tongue-and-groove edges for use in floor and roof construction. Plywood design specifications are provided in Reference 8.

Wood Product Panels

Particleboard is a form of composition board consisting of wood chips bonded together by a resin. It is produced in sheets, usually 4 ft. × 8 ft. (1.2 m × 2.4 m), and in thicknesses of 1/4 in. to 1 1/2 in. (6 mm to 38 mm). Particleboard is primarily used as sheathing and as an underlayment for flooring.

Oriented strand board is a multilayer board manufactured from wood strands bonded together by a resin. The wood strands are placed perpendicular to each other in adjacent layers for increased strength and stiffness.

Waferboard is similar to particleboard except that waferboard is made from larger wood chips than those used in particleboard.

Glued Laminated Timber

Glued laminated timber, often called *glulam*, is fabricated of layers of wood 2 in. (5 cm) or less in thickness glued together to form a solid structural member. Glulam has several advantages over sawn timber. The manufacturing process permits the

TABLE 2-3
Adjustments to Maximum Allowable Lumber Stresses

Loading conditions	*Adjustment factor*
Long duration (fully stressed for more than 10 yrs)	0.90
Load of 2 months duration	1.15
Load of 7 days duration	1.25
Wind or earthquake load	1.33
Impact load	2.00
Wet conditions	0.67–0.97*

*Depends on type of stress

fabrication of either curved or straight members of great size. Because the individual wood pieces used for lamination are thin, they can easily be dried to the desired moisture content prior to fabrication of a member. This produces a dimensionally stable member of high strength. Since high-strength wood can be selectively placed in areas of high stress during manufacture, high-strength beams can be produced at relatively low cost. The product of the glulam manufacturing process is thus a precisely dimensioned, strong, low-cost structural member.

Although glued laminated timber is primarily used in the construction of large buildings such as churches, auditoriums, shopping centers, industrial plants, and sports arenas, it has increasing application in the construction of residential and small commercial buildings. Reference 1 provides information on standard sizes, allowable stresses, and other characteristics of glued laminated timber.

2-2 CONCRETE AND MASONRY

Concrete is one of the world's most versatile and widely used construction materials. Its major weakness is its low strength in tension. To overcome this weakness, reinforcing steel is added to concrete to produce *reinforced concrete.* Almost all concrete employed in today's construction is reinforced concrete.

Components of Concrete

Three components are required to produce concrete: portland cement, aggregate, and water. A fourth component, an additive or admixture, is often added to improve some concrete property.

Cement. Cement is the agent that bonds the aggregates together to produce concrete. The chemical reaction that causes hardening of concrete is called *hydration*, and the heat produced by this reaction is called the *heat of hydration.* The American Society for Testing and Materials (ASTM) has classified a number of types of cement. The principal ones are Types I, II, III, IV, and V. Type I (normal) is a general purpose cement and is the most commonly used. Type II (modified) is more resistant to sulfates (water-soluble SO_4) and produces less heat of hydration than does normal cement. Type III (high early strength) gains strength faster than does normal cement but also produces more heat of hydration. Type IV (low heat) produces less heat of hydration than normal but also gains strength slower than does normal cement. Type V (sulfate-resistant) provides increased resistance to sulfates in soil or water but gains strength slower than does normal cement.

Cements containing an air-entraining agent are designated by the suffice "A." Thus, Types IA, IIA, and IIIA correspond to Types I, II, and III with air-entraining. The effects of air-entraining are discussed in the section on additives. Cements made from a mixture of portland cement and blast-furnace slag are designated by the suffix "S," such as Types IS and IS-A. Cements containing pozzolans are designated by the letter "P." Examples of these are Types IP, IP-A, P, and P-A. *Pozzolans* are finely divided materials such as fly ash, diatomaceous earth, volcanic ash, and calcined shale. Pozzolans reduce the heat of hydration, increase concrete workability, and reduce mix segregation.

Aggregates. Aggregates are added to cement to produce a more economical concrete mix and reduce shrinkage of the concrete as it hardens. Since aggregates make up some 70 percent of concrete volume, their properties have an important influence on the performance of the hardened concrete. To produce quality concrete, aggregates must be clean, strong, chemically stable, and resistant to freezing and thawing damage. To reduce cost by minimizing the void spaces between aggregates (which must be filled by cement paste), at least two sizes of aggregate, commonly sand and gravel, are used. The use of rounded or cubical aggregates also reduces the amount of cement paste required to fill voids.

Water. Water is an essential component of a concrete mix; it must be present to permit hydration to occur. In addition, water serves to make the mix plastic so that it may be placed and finished. However, it has been found that increasing the amount of water in a mix reduces the concrete's strength, watertightness, durability, and wear resistance. The lower the water/cement ratio, the higher the concrete strength and durability. Thus, the amount of water in a mix should be the minimum required for hydration and workability.

The amount of water in a mix is usually expressed as a ratio of water to cement by weight (water/cement ratio) and normally ranges between 0.40 and 0.70. The use of aggregates that are not in a saturated, surface-dry (SSD) condition will add or subtract water from the mix. Methods for adjusting the amount of mix water to compensate for aggregate moisture are presented in Reference 7. The low water/cement ratios necessary to obtain high concrete strength may require the use of one or more of the additives described below to produce a workable mix.

Additives. Various types of additives may be added to concrete mixes to impart special properties to the mix. Some common additives include air-entraining agents, accelerators, retarders, workability agents, water-reducing agents, and pozzolans. *Air-entraining agents* are used to produce concrete that has increased resistance to damage from freezing and thawing as well as scaling caused by the use of deicing chemicals. *Accelerators* speed up the hydration process, thereby increasing early strength and the heat of hydration. *Retarders* produce the opposite effect, and are often used to delay the setting of concrete that is pumped over long distances. They are also often used to produce exposed-aggregate concrete finishes. *Workability agents*, also called plasticizers, are used to increase mix workability. *Water-reducing agents* allow the amount of mix water to be reduced without changing the concrete's consistency. However, some water-reducing agents also act as retarders. The effect of pozzolans has already been described.

Production of Concrete

The concrete required for the construction of small buildings is most often produced and delivered to the construction site by a *ready-mixed concrete* producer. However, it is sometimes necessary to mix concrete on-site for small projects or at remote locations. The following suggestions are provided to help ensure that concrete construction meets required quality standards.

When ordering ready-mixed concrete, the builder must specify the minimum concrete strength and/or maximum water/cement ratio permitted. Recommendations for maximum permissible water/cement ratios for different types of structures and degrees of exposure as well as for slump (workability) are contained in Reference 7. Requirements for additives must also be established based on site conditions and construction techniques to be employed. Concrete delivered by truck mixers should be discharged within 1 1/2 hours after the start of mixing and before the drum has revolved 300 times.

When concrete is to be mixed on site, the concrete components must be carefully proportioned (*batched*). Batching by weight is the most accurate method and is therefore recommended, although batching by volume is sometimes used. Cement for batching small mixers is usually measured by the sack (94 lbs. or 42.6 kg). The following batching and mixing procedure is recommended to help keep the mixer drum clean and provide uniform mixing.

1. Add 10 percent of the mix water before charging the drum.
2. Add 80 percent of the mix water while charging the drum with cement, aggregates, and additives.
3. Add the remaining 10 percent of the mix water after charging is complete.

A minimum mixing time of 1 min., plus 1/4 min. for each cubic yard (0.76 m) of mix over 1 cu. yd., is recommended. Timing of the mixing cycle should not begin until all solid material has been placed into the drum. Aggregates must be clean, and the quantity of mix water adjusted for aggregate moisture. Almost any water suitable for drinking is satisfactory for mix water.

Brick Masonry

Brick is manufactured from fired clay and is produced in thousands of combinations of colors, textures, and shapes. Brick color, which is determined by the clay used and by additives, ranges from near white to almost black. Surface texture ranges from glazed smooth to rough. Among its many sizes and shapes, the most common sizes are the standard nonmodular brick and the standard modular brick illustrated in Figure 2-1. The *standard nonmodular brick* measures 3 3/4 × 2 1/2 × 8 in. (9.5 × 5.7 × 20.3 cm). A *standard modular brick* measures 3 5/8 × 2 1/4 × 7 5/8 in. (9.2 × 5.7 × 19.4 cm). The compressive strength of individual U.S. brick ranges from about 2500 psi (17.2 MPa) to over 22,000 psi (151.7 MPa). However, the overall compressive strength of brick assemblies depends on both the compressive strength of individual brick and the mortar used. Procedures for establishing the design strength of brick structural assemblies involve either testing masonry assem-

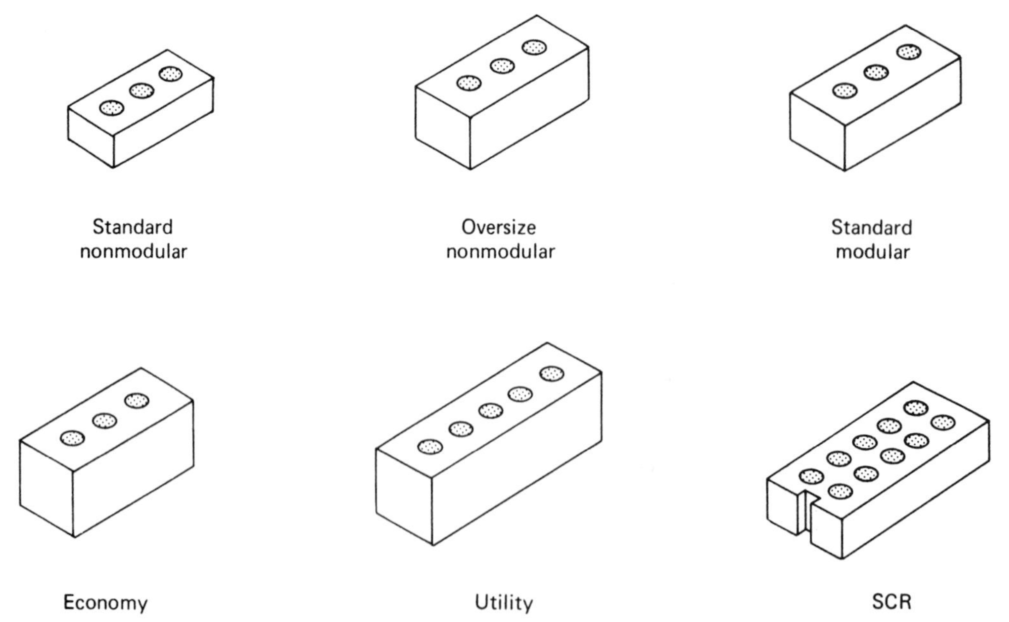

Figure 2-1 Typical clay brick (S. W. Nunnally, *Construction Methods and Management*. 2nd ed. © 1987, p. 334. Reproduced by permission of Prentice-Hall, Inc., Englewood Cliffs, NJ).

blies or assuming strength values based on brick strength and mortar type. The assumed compressive strength values for brick masonry range from 530 psi (3.7 MPa) for 2000 psi (13.8 MPa) brick with Type N mortar to 4600 psi (31.7 MPa) for 14,000 psi (96.5 MPa) brick with Type M mortar and construction inspection by an architect/engineer.

Mortars for brick masonry fall under ASTM Standard C270 (Standard Specifications for Mortar for Unit Masonry) and ASTM C476 (Standard Specifications for Mortar and Grout for Reinforced Masonry). Types M, S, N, and O are the principal mortar types. Their specifications are given in Table 2-4. Type M mortar is a high-strength mortar that should be used whenever high compressive strength and durability are required. Type S is a medium high-strength mortar for general-purpose use. Type N mortar is a medium-strength mortar that should not be used below grade in contact with the earth. Type O mortar is a medium low-strength mortar used primarily for fireproofing and non-load-bearing partitions.

Concrete Masonry

Concrete brick, concrete tile, solid load-bearing concrete block, hollow load-bearing concrete block, and hollow non-load-bearing concrete block are all forms of concrete masonry units. Glazed concrete block is also available. Glazed units combine attractive appearance with ease of cleaning and low cost. Concrete must fill at least 75 percent of a block's cross section for the block to be classified as *solid concrete block*. Block with greater void space is classified as *hollow concrete block*. A typical hollow concrete block has 45 to 50 percent of its cross section filled with concrete. Typical sizes and shapes of concrete masonry units are illustrated in Figure 2-2. The nominal size of standard concrete block is 8 × 8 × 16 in. (20 × 20 × 40 cm). Since the nominal size includes a 3/8 in. (9.5 mm) mortar joint, actual block size is 7 5/8 × 7 5/8 × 15 5/8 in. (19.4 × 19.4 × 39.7 cm). The type of concrete aggregate employed in block manufacture determines whether the block is classified as heavyweight or lightweight concrete block. A typical heavyweight load-bearing concrete block weighs 40 to 50 lbs. (18.1 to 22.7 kg). A similar lightweight concrete block might weigh 25 to 35 lbs. (11.3 to 15.9 kg).

The mortars used for concrete block are the same as those described earlier for brick masonry. However, concrete block may also be laid without mortar. In this case, standard, ground, or interlocking block is stacked without mortar, and a surface bonding agent applied to the outside surfaces. The bonding coat provides waterproofing as well as structural strength. As a matter of fact, the flexural and

TABLE 2-4
Mortar Specifications by Proportion and Strength

	By proportion				*By strength*
Mortar type	*Portland cement*	*Masonry cement*	*Hydrated lime or lime putty*	*Aggregate (damp, loose condition)*	*Ave. 28-day compressive strength psi (MPa)*
M	1	None	1/4	2 1/4 to 3 times the sum of the volumes of cement and lime used	2500 (17.2)
	1	1	None		
S	1	None	1/4 to 1/2		1800 (12.4)
	1/2	1	None		
N	1	None	1/2 to 1		750 (5.2)
	None	1	None		
O	1	None	1 to 2		350 (2.4)

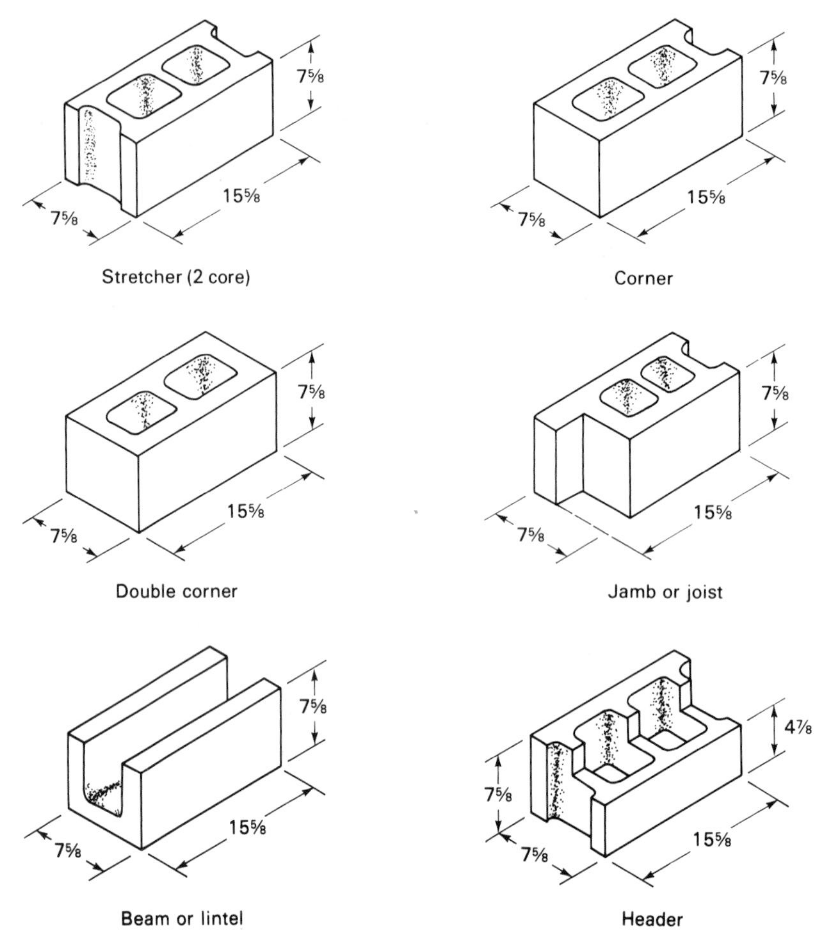

Figure 2-2 Typical concrete block units.

compressive strength of a surface-bonded wall may be greater than that of a conventional block and mortar wall.

2-3 METALS AND PLASTICS

Structural Steel

Structural steel members are frequently employed in the construction of large buildings because of steel's high strength, ductility, and ease of fabrication. Although structural steel finds only limited application in residential and light building construction, it is sometimes used for beams, girders, and columns. Therefore, it is helpful to understand the nomenclature and characteristics of structural steel.

A structural steel member is described by a shape designation, size, weight per unit length, and type of steel. The most common rolled-steel sections are illustrated in Figure 2-3. Standard rolled-steel shapes and their American Institute of Steel Construction (AISC) designations are listed in Table 2-5. Notice that the usual designation code consists of a letter symbol (which identifies the section shape) followed by two numbers. For beams, channels, and piles, the first number indicates the section depth in inches and the second number indicates the weight in pounds per foot of length. Thus, W 18 × 85 indicates an 18 in. deep W shape beam weighing

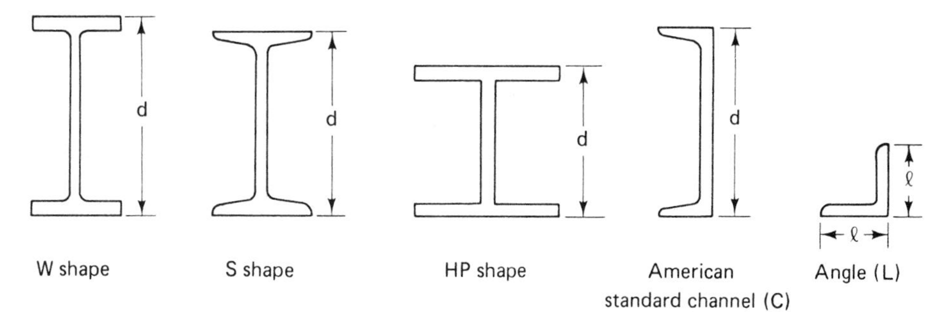

Figure 2-3 Hot-rolled steel section shapes (S. W. Nunnally, *Construction Methods and Management.* 2nd ed. © 1987, p. 317. Reproduced by permission of Prentice-Hall, Inc., Englewood Cliffs, NJ).

85 lbs. per linear foot. Designations for angles, bars, pipes, tubes, and plates are slightly different. Here the designation numbers indicate principal section dimensions in inches. Bar dimensions represent diameter or width and thickness; pipe dimensions represent nominal diameter and a weight code (standard, extra strength, or double-extra strength); tubing dimensions represent outside dimensions and wall thickness; and plate dimensions represent thickness and width. Thus, PL 3/8 × 12 indicates a plate 3/8 in. thick by 12 in. wide.

The type of steel used in a structural steel member is designated by an A followed by the American Society for Testing and Materials (ASTM) designation number. The most common type of structural steel is Type A36, carbon structural steel. High strength steels include Types A572 and A588. Steel strength is designated by the symbol F_y, which indicates the minimum yield stress of the steel. Yield stress is usually expressed in thousands of pounds per square inch (ksi). Type A36 steel has a yield strength of 36 ksi (248.2 MPa). Types A572 and A588 high strength steels are available in yield strengths of 42 ksi (289.6 MPa) to 65 ksi (448.2 MPa).

TABLE 2-5
Hot-Rolled Steel Shape Designations

Type of shape	*Example designation*
W shape	W18×85
S shape	S20×75
M shape	M10×9
American Standard Channel	C10×30
Miscellaneous Channel	MC8×20
Bearing pile	HP12×74
Equal leg angle	L5×5×1/2
Unequal leg angle	L8×6×1
Structural tee cut from:	
W shape	WT12×80
S shape	ST9×35
M shape	MT5×4.5
Flat bar	Bar 2×3/4
Round bar	Bar 2 Ø
Square bar	Bar 2 ⏁
Pipe	Pipe 3 std.
Structural tubing:	
Circular	TS 3 ODx.375
Rectangular	TS 8×2×.375
Square	TS 7×7×.500

Light Metal Structural Members

The *open-web steel joist* illustrated in Figure 2-4 is a form of lightweight built-up steel truss. Because of its light weight, strength, and low cost, it is often used for supporting roofs and floors of small commercial and industrial buildings. When the diagonal members (or web) consist of steel bars, these trusses are often referred to as *bar joists*. Standard joist classifications established by the Steel Joist Institute include types K, LH, and DLH. Series K open web steel joists are available in spans up to 60 ft. (18.3 m) with a maximum depth of 30 in. (76 cm). LH series joists, or longspan steel joists, are available in spans up to 96 ft. (29.3 m) with a maximum depth of 48 in. (122 cm). DLH series joists, or deep longspan steel joists, are available in spans up to 144 ft. (43.9 m) with depths up to 72 in. (183 cm). To offset the deflection of joists due to their own weight, longspan and deep longspan joists are usually cambered (crowned or bowed upward). Steel joists are available with square ends, underslung ends, or extended ends, as shown in Figure 2-5. *Joist girders* are open web steel trusses used as primary framing members. They are available in spans up to 60 ft. (18.3 m) with a maximum depth of 72 in. (183 cm).

Light-gauge cold-formed steel members are available as studs, beams, channels, and joists. Some examples are shown in Figure 2-6. Light metal framing systems fabricated of aluminum are also available.

Plastics

Plastics are finding increasing use in building construction. Probably their major application at present is in piping systems. Plastics commonly used in water piping systems include ABS (acrylonitrile-butadiene-styrene), CPVC (chlorinated polyvinyl chloride), PB (polybutylene), PE (polyethylene), and PVC (polyvinyl chloride). The principal plastics used in drain, waste, and vent piping systems are ABS and PVC.

Some other applications of plastics in building construction include awnings, cabinet and countertop surfacing, door and window frame sheathing, exterior veneer, insulation, partitions, wall panels, and window glazing.

Sheet Metal and Flashing

Sheet metal used in building construction consists of thin metal employed for ductwork, flashing, and gutters and downspouts. Materials commonly used include aluminum, copper, galvanized metal, stainless steel, and terne plate (steel coated with an alloy of lead and tin). *Flashing* is a material installed to prevent the passage of water between joints of a building. A typical chimney flashing is illustrated in Figure 2-7 (see also Chapter 9).

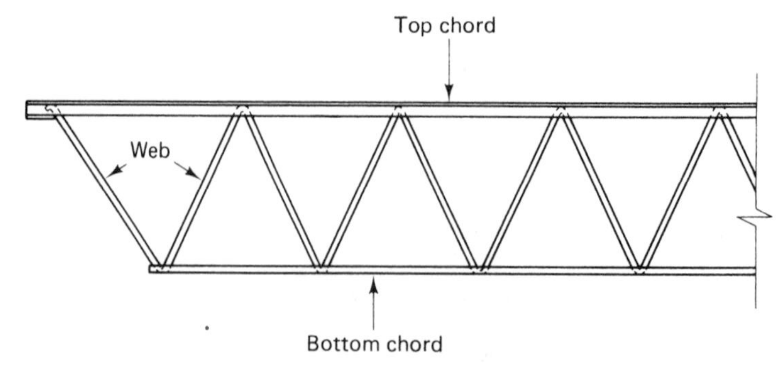

Figure 2-4 Open-web steel joist (S. W. Nunnally, *Construction Methods and Management.* 2nd ed. © 1987, p. 317. Reproduced by permission of Prentice-Hall, Inc., Englewood Cliffs, NJ).

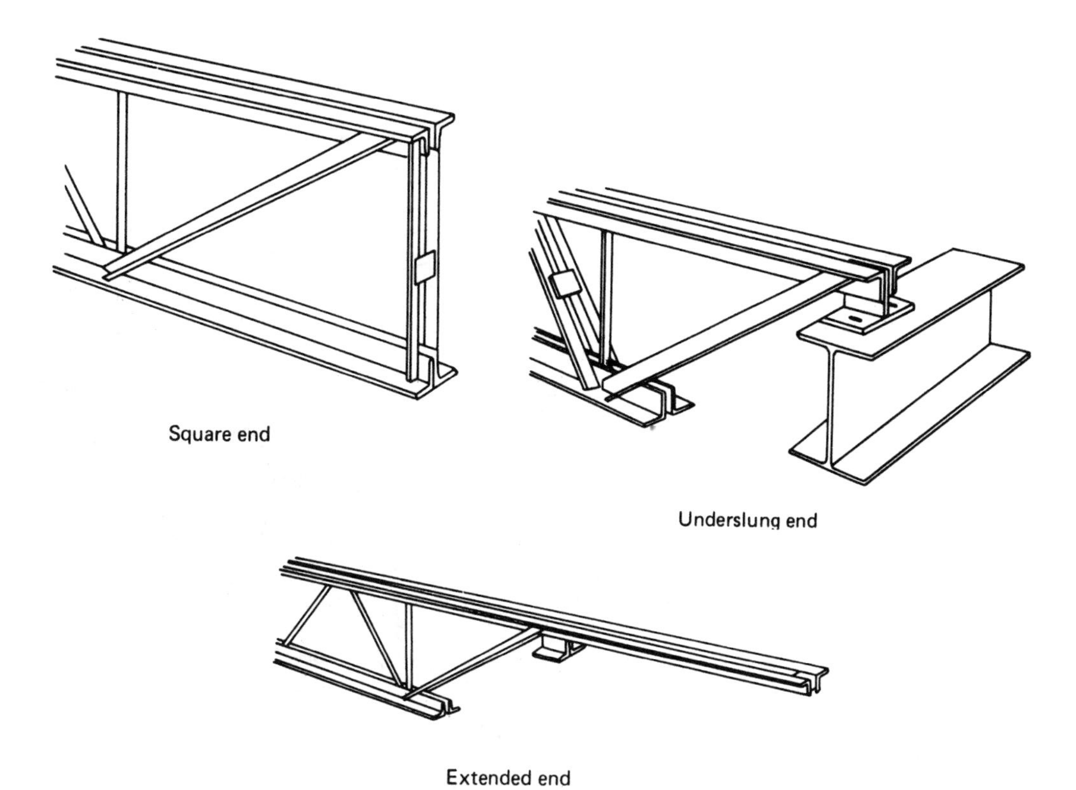

Figure 2-5 Types of steel joist ends (S. W. Nunnally, *Construction Methods and Management.* 2d ed. © 1987, p. 318. Reproduced by permission of Prentice-Hall, Inc., Englewood Cliffs, NJ).

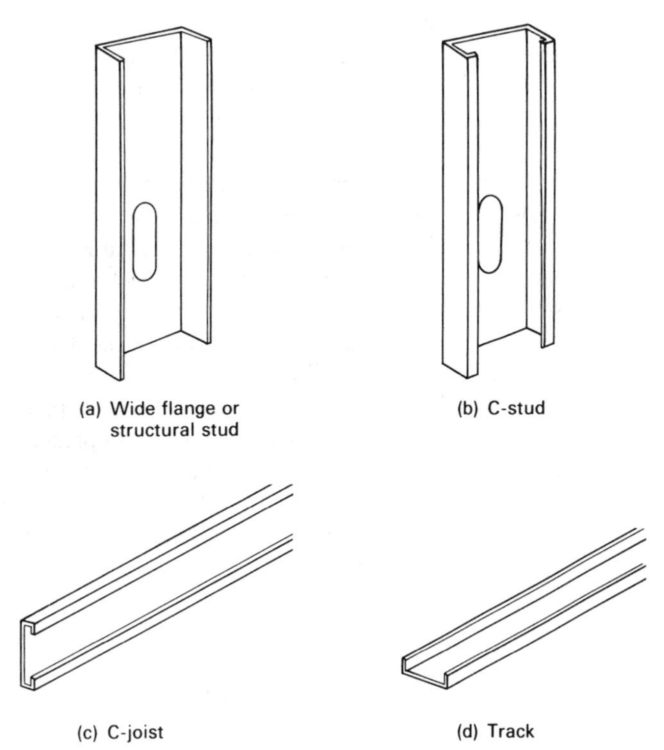

Figure 2-6 Light gauge cold-formed steel.

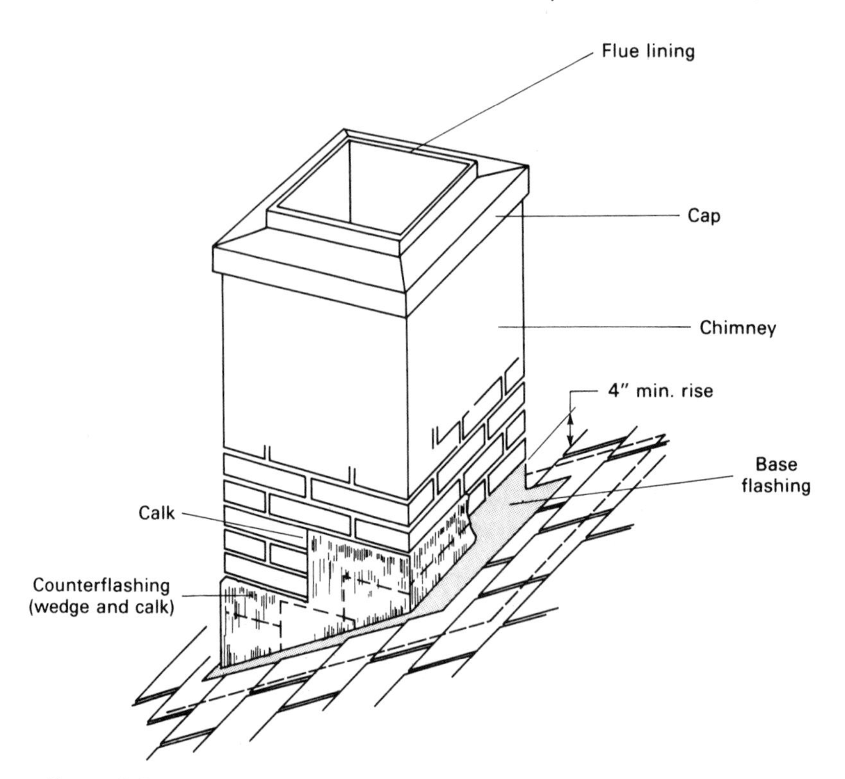

Figure 2-7 Typical chimney flashing (adapted from U. S. Department of Agriculture).

2-4 PROTECTION AGAINST DECAY AND INSECTS

A 1978 study of wood damage in residential buildings in Louisiana showed that approximately one-third of the homes surveyed had suffered damage from decay and/or insects. The major cause of damage was decay, followed by termites and other insects. The most commonly damaged building components were joists, sills, and floors. The principal factor leading to damage was rainwater splash.

Wood decay is caused by a fungus, which requires warmth and moisture for growth. The principal insect causing wood damage is the termite. The two principal varieties of termite are the subterranean termite and the dry-wood termite. Subterranean termites cannot live without obtaining moisture from the soil. However, when there is no direct contact between wood and soil, subterranean termites may construct shelter tubes across foundations and other objects to enable them to travel between the ground and their wood target. Dry-wood termites fly directly to wood and do not require soil contact for growth.

Today the principal methods used to provide protection against decay and insect damage to wood structures is wood preservation and chemical soil treatment. Pressure-treated wood in which preservatives are forced into the wood cells provides the most common form of wood preservation. The use of the techniques described below, in conjunction with wood preservation when required, minimizes wood damage from decay and insects.

Decay

The best protection against decay is obtained by placing dry wood into a dry environment. If wet or green lumber must be used, it is important that the wood be thoroughly dried before it is sealed or covered. Building codes generally require a

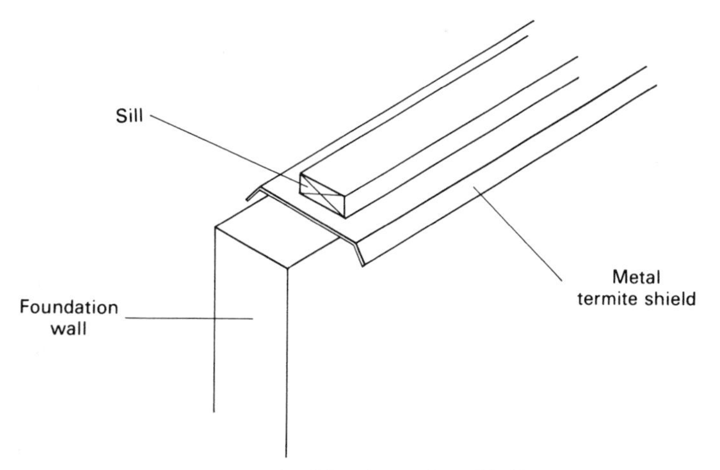

Figure 2-8 Metal termite shield.

minimum clearance between soil and untreated wood. To prevent moisture buildup in crawl spaces, adequate ventilation must be provided. A vapor barrier may also be needed on top of the soil underneath the crawl space. The soil adjacent to foundation walls should be kept dry by the use of roof overhangs, gutters and downspouts, and by sloping the surrounding soil away from the foundation walls.

Insects

In many parts of the United States, it is common practice to chemically treat the soil under building foundations and slabs prior to building construction in order to prevent the growth of subterranean termites. Building codes generally require the following minimum clearance between untreated wood and soil to reduce the danger of insect attack: joists 18 in.; girders 12 in.; sills resting on concrete or masonry walls 8 in.; wood siding, sheathing, and exterior wall framing 6 in. Metal termite shields (see Figure 2-8) may also be placed between foundations and sills or girders to prevent subterranean termites from constructing shelter tubes between the soil and the wood structure. To reduce the chances of dry-wood termite attack, ventilation openings should be covered by a fine wire screen, and the exterior should be carefully sealed and painted.

2-5 PROTECTION OF MATERIAL AT THE JOB SITE

Security

In recent years, theft of material, tools, and equipment at construction sites has become a major problem for builders. Such losses can add significantly to job costs and increase overhead by raising insurance rates. Some measures that may be taken to minimize the loss from theft include:

- Prompt recording of serial numbers of power tools, construction equipment, and installed equipment.
- Lighting and fencing of the construction area.
- Use of security guards.
- Coordination of security measures with local police.

Material Storage

All types of building materials must be protected from physical damage during storage. Framing lumber and plywood sheathing delivered to the job site should be placed in covered storage if possible. It outside storage is necessary, the material should be placed in a well-drained area, raised about 6 in. off the ground using scrap lumber, separated enough to allow air circulation, and protected by a waterproof cover. The delivery of finished woodwork such as cabinets, door and window frames, flooring, and interior trim should be scheduled after the building's walls and roof have been erected and sheathed. In cold, damp weather it may also be necessary to heat the storage area to prevent finished woodwork from absorbing moisture and swelling before installation.

Cement, grout, and plaster must be kept dry during storage. Aggregates may be stored uncovered. However, the moisture content of aggregates must be checked before use and the mix-water adjusted to compensate for aggregate moisture. Concrete masonry units and most clay brick should be stored and laid in a dry condition. An exception, however, is clay brick having a high absorption rate (greater than 20 g of water per minute). Such brick must be laid in a saturated, surface-dry condition so it may be stored uncovered until shortly before use.

Steel structural members should be kept free of dirt, oil, and loose debris during storage because these materials pose a safety hazard during erection. All ice must be removed from steel prior to erection for the same reason. Damaged paint should be touched up promptly to prevent rusting of steel.

PROBLEMS

1. Explain the difference between green lumber and dry lumber in terms of moisture content and strength.
2. Give the actual dimensions of the following lumber (S4S):
   a. 1 × 10
   b. 2 × 43
   c. 2 × 12
3. What is glulam? What are the principal characteristics of this material?
4. What purpose does aggregate serve in a concrete mix?
5. Explain the meaning of the term "water/cement ratio" for concrete and its effect on concrete strength.
6. How is the design strength of brick structural assemblies established?
7. Identify the following structural steel sections.
   a. W 27 × 114
   b. C 12 × 30
   c. L 6 × 6 × $\frac{1}{2}$
8. Briefly describe the extent of damage to U.S. residential buildings caused by decay and insects.
9. What provides the best protection against decay of wood structures?
10. How should finished woodwork be stored at the construction site?

REFERENCES

1. American Institute of Timber Construction. *Timber Construction Manual.* 3rd ed. New York: Wiley, 1985.

2. American Plywood Association. *APA Design/Construction Guide: Concrete Forming.* Tacoma, WA, 1984.
3. Portland Cement Association. *Concrete Masonry Handbook.* Skokie, IL, 1985.
4. Portland Cement Association. *Design and Control of Concrete Mixes.* 13th ed. Skokie, IL, 1988.
5. Gross, James G., and Harry C. Plummer. *Principles of Clay Masonry Construction.* Brick Institute of America, McLean, VA, 1973.
6. National Forest Products Association. *National Design Specification for Wood Construction.* Washington, DC, 1986.
7. Nunnally, S. W. *Construction Methods and Management.* 2nd ed. Englewood Cliffs, NJ: Prentice-Hall, 1987.
8. American Plywood Association. *Plywood Design Specifications.* Tacoma, WA, 1985.
9. Steel Joist Institute. *Standard Specifications, Load Tables & Weight Tables for Steel Joists & Joist Girders.* Myrtle Beach, SC, 1988.

chapter 3

Site Investigation and Preparation

3-1 SITE CONDITIONS

Importance of Field Investigation

Many contractors have learned the hard way that failure to perform an adequate site investigation prior to submitting a construction bid can have costly consequences. A careful site inspection provides valuable information for planning the construction methods and equipment to be employed on the project in addition to disclosing any unusual job conditions. Contractor claims for additional compensation based on changed or unanticipated job conditions are almost invariably denied by the courts when it is determined that such conditions would have been apparent to a competent building professional after a careful site inspection. Therefore, a thorough site investigation should always be performed before bidding on or planning and scheduling a construction project.

Site Inspection Procedures

The following are some of the major items to be investigated during the site inspection:

1. *Site topography*, or nature of the terrain, controls the amount of excavation or fill needed and the difficulty of utilizing construction equipment, and it determines site drainage and flooding potential.
2. *Soil conditions* (see Section 3-2) and the depth of the water table influence excavation procedures, foundation design and construction, site drainage, and the difficulty of movement by vehicles and equipment.
3. *Drainage*, combined with soil conditions, strongly influences vehicular access and construction operation, excavation and compaction procedures, and foundation construction.

4. *Utility services* include availability and cost of providing electrical, water, sewerage, gas, and telephone services.
5. *Site access* includes availability of the necessary right-of-way, availability and condition of existing roads, and the cost of constructing or improving roads up to the construction location.
6. *Building location* in relation to the site boundaries influences access requirements, location of supporting construction facilities, the difficulty of excavation and construction, and environmental effects on surrounding properties.
7. *Obstacles* include trees, streams, lakes, rocks, hills, gullies, utility lines, and other physical features that may interfere with access or construction activities.
8. *Working conditions* include weather (temperature, humidity, rainfall), soil conditions, obstacles, environmental restrictions, and type of construction to be undertaken.
9. *Environmental controls* include restrictions on noise, dust, and hours of work, as well as the need to prevent or control drainage runoff or water pollution.
10. *Construction space* includes the area available for materials storage, construction plant, worker parking and sanitation facilities, and construction equipment storage and maintenance facilities, as well as the area available for actual construction operations.

Site Preparation Costs

The initial, or startup, costs incurred before any major building construction activities can take place are known as *mobilization* costs. These may include the cost of transporting vehicles, construction equipment, and facilities to the job site; constructing or improving access roads or bridges; installing temporary utility services; erecting temporary facilities; and improving drainage. Corresponding costs associated with project completion and cleanup are *demobilization* costs. Mobilization and demobilization costs are commonly treated as part of job overhead costs in the project cost estimate. However, cost accounting and control is much easier if these costs are carried as separate line items in the project cost estimate.

Other site activities, such as clearing and grading, which must be performed before job layout and foundation excavation can begin are commonly treated as separate work items in the project cost estimate.

3-2 SOIL IDENTIFICATION AND CHARACTERISTICS

Soil and Rock

The natural materials on which structures rest are soil and rock. *Rock* is the solid material making up the earth's crust. Builders are primarily concerned with the use of rock for foundation support, as well as with its excavation and removal. Due to its strength and hardness, specialized equipment and procedures are required to excavate and manipulate rock. *Soil*, the material formed from the physical or chemical decomposition of rock, is of more concern to the builder. Soil type, moisture content, and soil density strongly influence excavation procedures, foundation design and construction, site drainage, and the difficulty of movement by vehicles and equipment. Therefore, a basic knowledge of soil identification and classification can be very valuable to the builder.

Soil is usually considered to be composed of the following five fundamental soil types:

1. *Gravel* consists of individual soil particles smaller than 3 in. (76 mm) but larger than 1/4 in. (6 mm) in diameter.
2. *Sand* consists of particles smaller than 1/4 in. (6 mm) in diameter but larger than the No. 200 sieve opening (0.08 mm).
3. *Silt* particles pass the No. 200 sieve but are larger than 0.002 mm in diameter.
4. *Clay* is the smallest fundamental soil type, measuring less than 0.002 mm in diameter.
5. *Organic* soils contain a significant amount of partially decomposed vegetable (organic) matter. Highly organic soil containing visible particles of vegetation is classified as *peat*.

Soil particles that pass the No. 200 sieve are called *fines,* which include both silt and clay as well as organic soils.

Soil Classification

The two principal soil classification systems used in the United States are the Unified System and the AASHTO System. Of these, the Unified System is most useful to the builder; the AASHTO System is used primarily by highway construction agencies.

The Unified System assigns a two-letter symbol to each soil type as shown in Table 3-1. The first letter indicates the predominant soil component. The second letter indicates the soil's gradation or plasticity. When 50 percent or less by weight pass the No. 200 sieve, the soil is called a *coarse-grained soil* and is either a gravel (G) or sand (S). When more than 50 percent by weight of a soil pass the No. 200 sieve, the soil is a *fine-grained soil* and is either a silt (M), clay (C), or organic (O) soil.

Under the AASHTO classification system, soils are classified according to their value as subgrade material. The classification symbols used range from types A-1 to A-7 as shown in Table 3-2.

Field Identification Procedures

The builder performing a site investigation in preparation for bidding on or planning a construction project usually does not have adequate time or facilities to perform a complete soil classification. However, the use of the procedures described below and in Table 3-1 will permit a rapid, reasonably accurate Unified System soil classification to be made.

Basic Procedures. First, all rock particles over 3 in. (76 mm) in diameter are removed and discarded. The remaining soil particles are then separated into two piles, those larger than the No. 200 sieve and those smaller than the No. 200 sieve. (The No. 200 sieve size roughly corresponds to the smallest particle that can be seen by the naked eye.) If more than 50 percent by weight of the sample is larger than the No. 200 sieve, the soil is classified as coarse-grained. If 50 percent or more of the sample is smaller than the No. 200 sieve, the sample is fine-grained. The soil is further classified as described below and in Table 3-1.

To further classify a coarse-grained soil, divide the coarse fraction into two piles separated at the 1/4 in. (6 mm) size. If 50 percent or more of the coarse fraction is larger than 1/4 in. (6 mm), the basic soil type is gravel (G). If more than 50

TABLE 3-1
Unifed System of Soil Classification—Field Identification*

Coarse-grained soils (less than 50% pass No. 200 sieve) Symbol	Name	*Percent of coarse fraction less than 1/4 in.*	*Percent of sample smaller than No. 200 sieve*	*Comments*
GW	Well-graded gravel	50 max.	< 10	Wide range of grain sizes with all intermediate sizes
GP	Poorly graded gravel	50 max.	< 10	Predominantly one size or some sizes missing
SW	Well-graded sand	51 min.	< 10	Wide range of grain sizes with all intermediate sizes
SP	Poorly graded sand	51 min.	< 10	Predominantly one size or some sizes missing
GM	Silty gravel	50 max.	≥ 10	Low-plasticity fines (see ML below)
GC	Clayey gravel	50 max.	≥ 10	Plastic fines (see CL below)
SM	Silty sand	51 min.	≥ 10	Low-plasticity fines (see ML below)
SC	Clayey sand	51 min.	≥ 10	Plastic fines (see CL below)

*Tests on Fraction Passing No. 40 Sieve (Approx. 1/64 in. or 0.4 mm)***

Fine-grained soils (50% or more pass No. 200 sieve) Symbol	Name	*Dry strength*	*Shaking*	*Other*
ML	Low-plasticity silt	Low	Medium to quick	
CL	Low-plasticity clay	Low to medium	None to slow	
OL	Low-plasticity organic	Low to medium	Slow	Color and odor
MH	High-plasticity silt	Medium to high	None to slow	
CH	High-plasticity clay	High	None	
OH	High-plasticity organic	Medium to high	None to slow	Color and odor
Pt	Peat	Identified by dull brown to black color, odor, spongy feel, and fibrous texture		

*S. W. Nunnally, *Construction Methods and Management.* 2nd ed., © 1987, p. 23. Reproduced by permission of Prentice-Hall, Inc., Englewood Cliffs, NJ.

**Laboratory classification based on liquid limit and plasticity index values.

percent by weight of the coarse fraction is smaller than 1/4 in. (6 mm), the basic soil type is sand (S). Next, determine the percentage by weight of fines in the total sample. If less than 10 percent, the second letter symbol is either W (well-graded) or P (poorly graded) according to the soil's gradation. Figure 3-1 illustrates the gradation characteristics of well-graded and poorly graded soils. If 10 percent or more of the total sample is composed of fines, the second letter symbol is either M (silt), C (clay), or O (organic). These are identified as shown in Table 3-1 using the results of the shaking test and the dry strength test described below.

Fine-grained soils are classified by performing the shaking test and dry strength test, then using Table 3-1.

Shaking Test. Add a small amount of water to enough fine material to form a ball about 3/4 in. (19 mm) in diameter. Continue adding water until the sample can be molded in the hand without sticking to the fingers. Placing the sample in the palm of one hand, shake vigorously and observe the speed with which water comes to the surface as indicated by a shiny surface. Only a low-plasticity silt (ML) produces a rapid reaction.

TABLE 3-2
AASHTO System of Soil Classification*

	Group number										
	A-1		*A-2*								
	A-1-a	*A-1-b*	*A-2-4*	*A-2-5*	*A-2-6*	*A-2-7*	*A-3*	*A-4*	*A-5*	*A-6*	*A-7*
Percent passing											
No. 10 sieve	50 max.										
No. 40 sieve	30 max.	50 max.					51 min.				
No. 200 sieve	15 max.	25 max.	35 max.	35 max.	35 max.	35 max.	10 max.	36 min.	36 min.	36 min.	36 min.
Fraction passing No. 40											
Liquid limit	6 max.	6 max.	40 max.	41 min.	40 max.	41 min.		40 max.	41 min.	40 max.	41 min.
Plasticity index			10 max.	10 max.	11 min.	11 min.		10 max.	10 max.	11 min.	11 min.
Typical material	Gravel and sand		Silty or clayey sand or gravel				Fine sand	Silt	Silt	Clay	Clay

*S. W. Nunnally, *Construction Methods and Management*. 2nd ed. © 1987, p. 25. Reproduced by permission of Prentice-Hall, Inc., Englewood Cliffs, N.J.

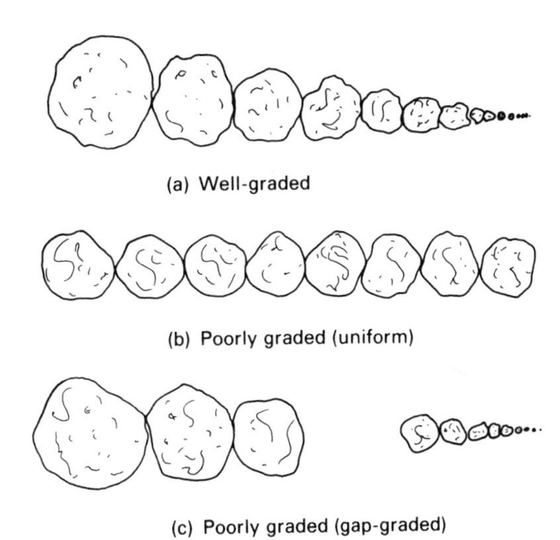

Figure 3-1 Examples of soil gradation.

Dry Strength Test. Add water as needed to form a ball of fine-grained material about 1 1/2 in. (38 mm) in diameter. After the sample has dried thoroughly, attempt to break it using the thumb and forefinger of both hands. A highly plastic soil will not break. If the sample breaks, rub it between the thumb and forefinger of one hand to see if it will erode into powder. A low-plasticity soil will break and powder easily.

Construction Characteristics of Soils

Some important construction characteristics of soils include the ease with which saturated soils can be drained, their value for foundation support, and their susceptibility to frost action (heaving). These characteristics for soils classified under the Unified System are given in Table 3-3.

TABLE 3-3
Construction Characteristics of Soils (Unified System)

Soil type	*Symbol*	*Drainage*	*Value for foundation support (no frost action)*	*Danger of frost action*
Well-graded gravel	GW	Excellent	Excellent	None to slight
Poorly graded gravel	GP	Excellent	Good	None to slight
Silty gravel	GM	Fair to poor	Good to fair	Slight to medium
Clayey gravel	GC	Poor	Fair	Slight to medium
Well-graded sand	SW	Excellent	Good to fair	None to slight
Poorly graded sand	SP	Excellent	Good to fair	None to slight
Silty sand	SM	Fair to poor	Good to poor	Slight to high
Clayey sand	SC	Poor	Poor	Slight to high
Low-plasticity silt	ML	Fair to poor	Fair to poor	Medium to high
Low-plasticity clay	CL	Poor	Fair to poor	Medium to high
Low-plasticity organic	OL	Poor	Poor	Medium to high
High-plasticity silt	MH	Fair to poor	Poor	Medium to high
High-plasticity clay	CH	Very poor	Fair to poor	Medium
High-plasticity organic	OH	Very poor	Very poor	Medium
Peat	Pt	Fair to poor	Not suitable	Slight

3-3 SITE PREPARATION

The first step in the actual construction of a building is site preparation. This involves clearing the area to be excavated, stripping topsoil from the area to be regraded, and laying out and marking the building location.

Clearing

Clearing, sometimes called "clearing and grubbing," consists of removing obstacles such as trees, shrubs, tree stumps, roots, and rocks from the construction area. The equipment commonly used for clearing consists of the front-end loader or dozer, but hand labor is also often needed. For environmental reasons, it is often desirable to retain selected trees and shrubs. If this is to be done, these plants must be clearly marked and carefully protected from construction damage.

After the excavation area is cleared, topsoil is removed or *stripped* from the area to be excavated or regraded. When possible, this topsoil should be stockpiled at the site for reuse in final site grading. Rough grading of the site to bring it to the approximate final contours may also be performed at this time prior to building layout.

During site preparation, adequate drainage must be provided to prevent flooding or ponding of water in the construction area and its access roads. Care must also be taken to control any water runoff to prevent flooding of adjacent property or pollution of lakes and streams.

Building Layout

After the building location has been cleared, the corners of the new buildings are located and marked. This step is referred to as *laying out* or *staking* the building. First, the lot boundaries must be located and marked. This work must usually be done by a registered land surveyor. The layout and staking of the new building may be done by either a surveyor or the builder, but it is most commonly done by the builder.

After the property boundaries have been marked, the building corners are marked by stakes in accordance with the site plan provided by the owner or architect. Corner stakes are often marked with the amount of cut (excavation) required to bring the excavation to the elevation required for the bottom of the foundation. Corner stakes must be checked to ensure that the layout is *square*; that is, all corners are at 90° angles. This is most easily done by ensuring that the length of the diagonal lines between opposite corners of the building are equal, as shown in Figure 3-2. Check also to ensure that the building location is beyond the *setback* lines (minimum required distance from the property boundaries) established by local ordinance or deed restriction. A typical site plan for a house is shown in Figure 3-3.

After the building's corners have been staked, *batter boards* are erected to provide reference lines and elevations that are not likely to be disturbed during construction. Batter boards are commonly constructed of 2 × 4 stakes with 1 × 6 or 1 × 8 boards nailed horizontally. The upper surface of the horizontal boards should be at the same elevation at all locations. Batter boards should be located well beyond the limits of excavation; a minimum distance of 4 ft. (1.2 m) outside the foundation is usually recommended. When the building is not a simple rectangle, additional batter boards are necessary to locate all building lines, as illustrated in Figure 3-4. After the batter boards are set, stretch a taut string between opposite batter boards and adjust their position so that the strings intersect exactly above the building's corner stakes. As shown in Figure 3-5, a plumb bob is used for performing this

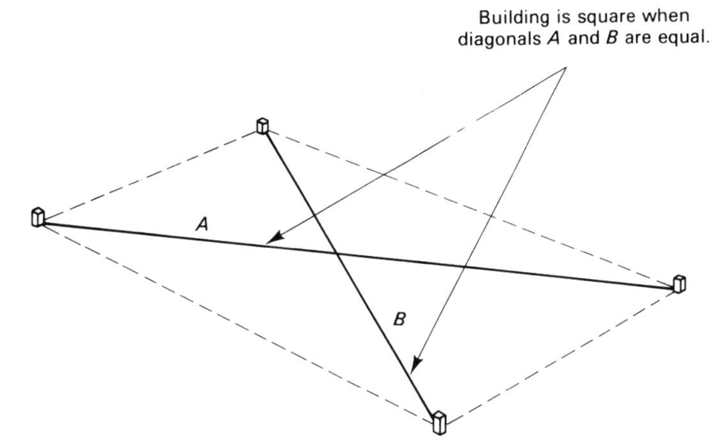

Figure 3-2 Squaring building layout.

alignment. The final position of the string line on the batter board should be marked by a shallow saw cut or a small nail so that the string can be quickly and accurately replaced if damaged or moved during excavation.

Establishing Grade (Elevation)

To establish the elevation of batter boards, foundations, floors, etc., it is necessary to measure the vertical distance from a point of known elevation. The instrument primarily used for this purpose is a *level* (or *builder's level* or *dumpy level*). The level is equipped with a vial or bubble tube so that it may be adjusted to turn in a precise horizontal plane. The level can also be used to measure horizontal angles.

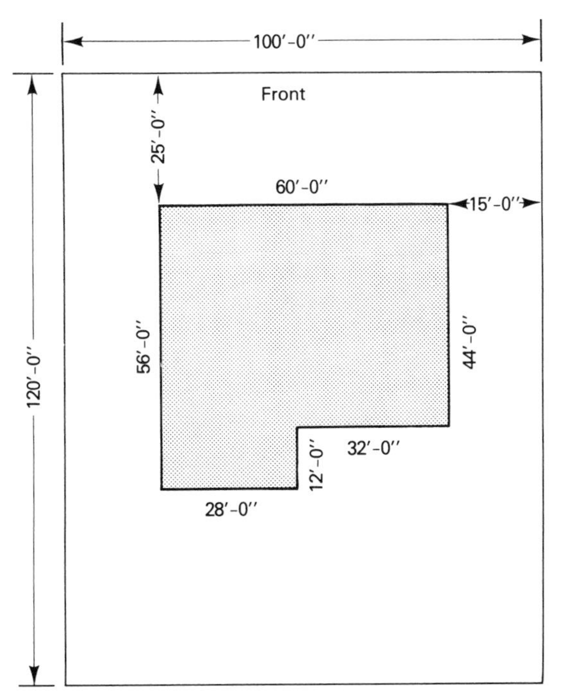

Figure 3-3 Typical plot plan for a house.

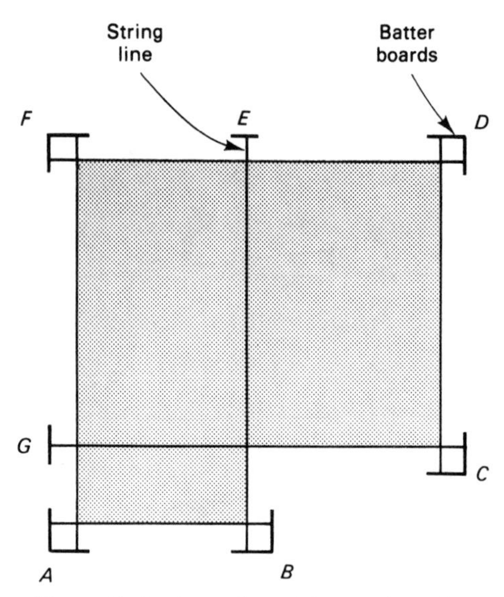

Figure 3-4 Location of batter boards.

A somewhat similar instrument that can also measure vertical angles is the *transit*. A newer form of level, the *laser level*, uses a laser beam rather than a line of sight to establish a precise horizontal reference.

After the level has been properly adjusted, the elevation of the instrument center (*height of instrument*, or H.I.) is computed. The graduated rod used for measuring vertical distance with a level is known as a *leveling rod*, *builder's rod*, or *Philadelphia rod*. The procedure for determining the elevation of an unknown point is as follows (see Example 3-1).

1. Take a rod reading (*backsight*) on the known point, or *benchmark*.
2. Add the rod reading to the elevation of the known point to determine the height of the instrument (H.I.).
3. Subtract the rod reading (*foresight*) at the unknown point to determine its elevation.

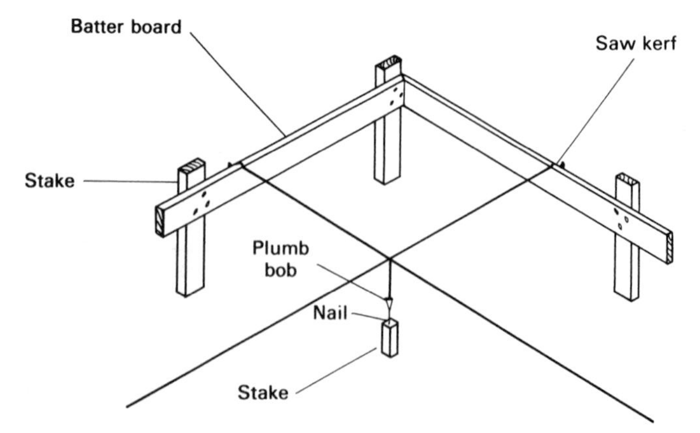

Figure 3-5 Aligning stakes and string lines (adapted from U. S. Department of Agriculture).

EXAMPLE 3-1

Problem: Using the information in the figure below, find the elevation of point B.

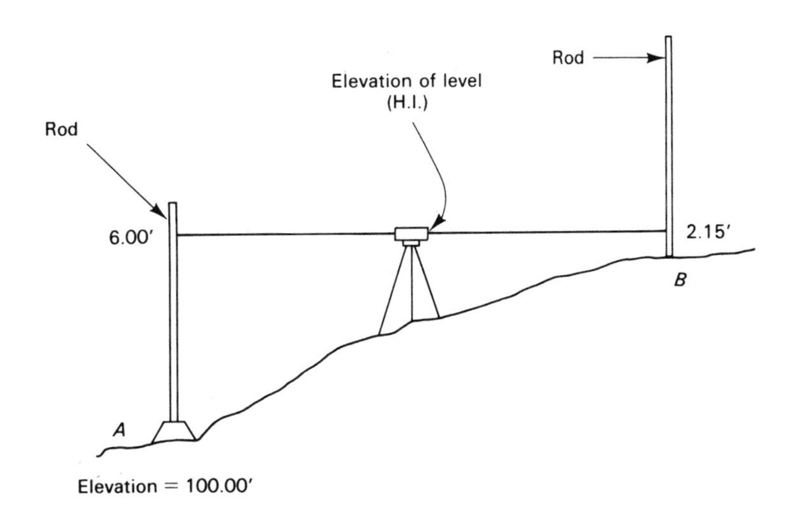

Solution:

$$\text{H.I.} = 100.00' + 6.00' = 106.00'$$
$$\text{Elevation of } B = 106.00' - 2.15' = 105.85'$$

A similar procedure used when it is necessary to set a stake at a precise elevation is illustrated in Example 3-2.

EXAMPLE 3-2

Problem: Given the information in the figure below, find the rod reading required on the top of a stake at point B in order to establish the desired floor elevation of 105.00 ft.

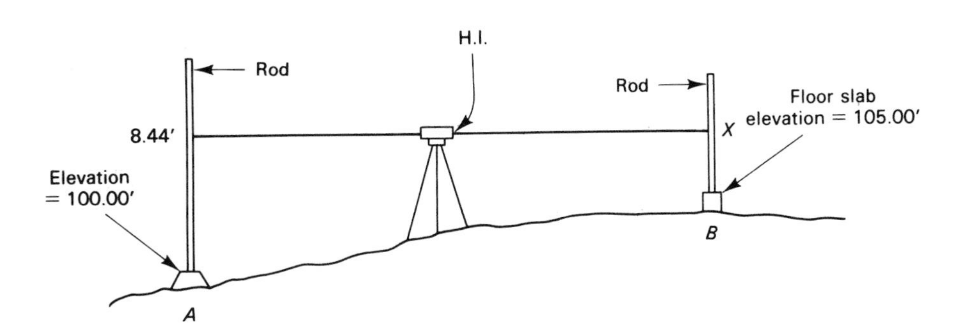

Solution:

$$\text{Required floor elevation} = 105.00'$$
$$\text{H.I.} = 100.00' + 8.44' = 108.44'$$
$$\text{Reading at } B = 108.44' - 105.00' = 3.44'$$

3-4 ESTIMATING CUT AND FILL

When planning or estimating a building construction project, it is often necessary to estimate the volume of natural material (soil or rock) that must be excavated or placed as fill. The procedures described in this section facilitate such computations.

Soil Volume Change

When estimating cut and fill, it is important to realize that the volume occupied by a given weight of soil varies according to its condition or state. The three principal soil conditions or states are:

Bank Soil in its natural state. A unit volume is commonly identified as a *bank cubic yard* (foot, meter, etc.). The usual abbreviation for bank cubic yard is BCY.

Loose Soil that has been excavated but not compacted. A unit volume is commonly identified as a *loose cubic yard* (foot, meter, etc.). The usual abbreviation for loose cubic yard is LCY.

Compacted Soil that has been compacted. A unit volume is commonly identified as a *compacted cubic yard* (foot, meter, etc.). The usual abbreviation for compacted cubic yard is CCY.

When a soil is excavated, it increases in volume because the soil grains are loosened by excavation and air fills the additional void spaces. This phenomenon is called *swell*. Swell for some typical soils is given in Table 3-4.

TABLE 3-4
Typical Soil Volume Change

Soil	*Swell (%)*	*Shrinkage (%)*
Clay	30	20
Common earth	25	10
Gravel and sand	12	12

Swell for a specific soil may be calculated by Equation 3-1. Since we are only concerned with the ratio of unit weights, any convenient unit of volume may be used in this equation; i.e., cubic foot, cubic yard, cubic meter, etc.

$$\text{Swell (\%)} = \left(\frac{\text{weight/bank volume}}{\text{weight/loose volume}} - 1\right) \times 100 \qquad (3\text{-}1)$$

EXAMPLE 3-3

Problem: Find the swell of a soil that weighs 98 lbs./cu. ft. in its natural state and 70 lbs./cu. ft. after excavation.

Solution:

$$\text{Swell} = \left(\frac{98}{70} - 1\right) \times 100 = 40\%$$

That is, one bank cubic foot of in-place soil expands to 1.4 loose cubic feet after excavation. Therefore, to find the loose volume resulting from excavation, multiply the in-place volume of material to be excavated by the factor "1 + swell," where swell is expressed as a decimal. The procedure is expressed by Equation 3-2 and illustrated by Example 3-4.

$$\text{Loose volume} = \text{bank volume} \times (1 + \text{swell}) \tag{3-2}$$

EXAMPLE 3-4

Problem: Find the loose volume of soil resulting from a basement excavation that contains 160 bank cubic yards. The soil's swell is 25 percent.

Solution:

$$\text{Loose volume} = 160 \times (1 + 0.25) = 200 \text{ BCY}$$

The reverse phenomenon, *shrinkage*, occurs when a soil is compacted; that is, the compacted soil occupies a smaller volume than it did in the bank condition. Typical soil shrinkage is shown in Table 3-4. Shrinkage expressed as a percentage may be calculated by the use of Equation 3-3. Note, however, that rock always occupies a larger volume after excavation and compaction than it does in its natural state. Therefore, the shrinkage of rock is always negative.

$$\text{Shrinkage (\%)} = \left(1 - \frac{\text{weight/bank volume}}{\text{weight/compacted volume}}\right) \times 100 \tag{3-3}$$

EXAMPLE 3-5

Problem: Find the shrinkage of a soil that weighs 104 lbs./cu. ft. in its natural state and 130 lbs./cu. ft. after compaction.

Solution:

$$\text{Shrinkage} = \left(1 - \frac{104}{130}\right) \times 100 = 20\%$$

To find the compacted volume resulting from the compaction of a given volume of a soil, simply multiply the bank volume of the soil by the factor "1 − shrinkage." This relationship is represented by Equation 3-4, where shrinkage is expressed as a decimal.

$$\text{Compacted volume} = \text{bank volume} \times (1 - \text{shrinkage}) \tag{3-4}$$

Note that both swell and shrinkage are measured from the in-place, or bank, condition. Thus, it is necessary to apply both swell and shrinkage factors to directly move between loose and compacted states. However, a simpler method is available when unit weights are known. This relationship is expressed by Equation 3-5, and its use is illustrated in Example 3-6.

$$\text{Compacted volume} = \frac{\text{weight/loose volume} \times \text{loose volume}}{\text{weight/compacted volume}} \tag{3-5}$$

EXAMPLE 3-6

Problem: Find the compacted volume of 100 LCY when the soil's unit weights are: 73 lbs./cu. ft. loose, 104 lbs./cu. ft. bank, and 130 lbs./cu. ft. compacted.

Solution:

$$\text{Compacted volume} = \frac{73 \times 100}{130} = 56.15 \text{ CCY}$$

Estimating Volume of Excavation

A simple, commonly used estimating procedure to find the volume of an excavation is to multiply the horizontal area of the excavation by the average depth of excavation. This procedure is expressed as Equation 3-6 below. In using this equation, the average depth of excavation is calculated as the average of the cut at each corner of the excavation. The procedure is illustrated in Example 3-7.

$$\text{Excavation volume} = \text{horizontal area} \times \text{average depth} \tag{3-6}$$

EXAMPLE 3-7

Problem: Estimate the volume of excavation, in bank cubic yards, for the basement shown below. The figure at each corner indicates the depth of cut at that location.

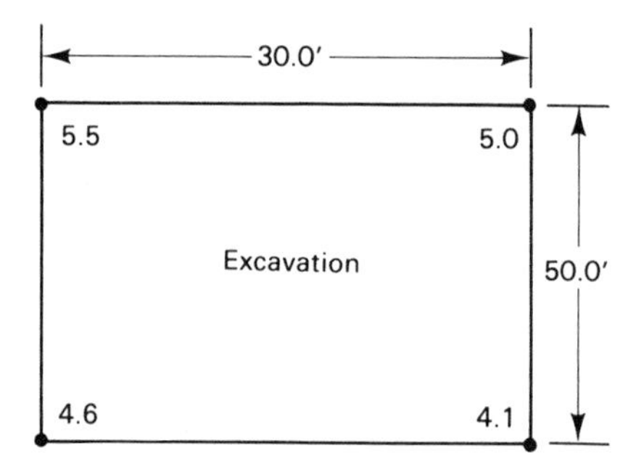

Solution:

$$\text{Volume} = (30 \times 20) \times \left(\frac{5.5 + 5.0 + 4.1 + 4.6}{4}\right) = 2880 \text{ B cu. ft.}$$

$$\frac{2880}{27} = 106.7 \text{ BCY}$$

When using this procedure, the area to be excavated is divided into a convenient set of rectangles, triangles, or circular segments for the purpose of calculation. The volume of each portion is then calculated and the results added to find the total volume of excavation.

PROBLEMS

1. Explain the importance of making a site inspection prior to bidding on a construction project.
2. What does the term *mobilization* mean in construction? What costs are involved?
3. Identify the five fundamental soil types.
4. A soil is classified as ML under the Unified System. Identify the soil type. What would you expect this soil's drainage, foundation value, and susceptibility to frost action to be?
5. What is clearing and grubbing?
6. How are batter boards used for building layout?
7. Given the following construction survey data (in feet), calculate the elevation of point *B*.

Location	*Backsight*	*Foresight*	*Elevation*
A	6.00		100.00
B		4.00	

8. Using the following construction survey data (in feet), calculate the elevation of the floor slab at point *C*.

Location	*Backsight*	*Foresight*	*Elevation*
A	11.48		210.75
B	3.19	7.44	
C		4.81	

9. A soil weighs 80 lbs./cu. ft. loose, 100 lbs./cu. ft. in-place, and 115 lbs./cu. ft. compacted. Find the swell and shrinkage of the soil.
10. Calculate the volume of excavation (cubic yards bank measure) required for the basement shown below. Excavation depths are given in feet.

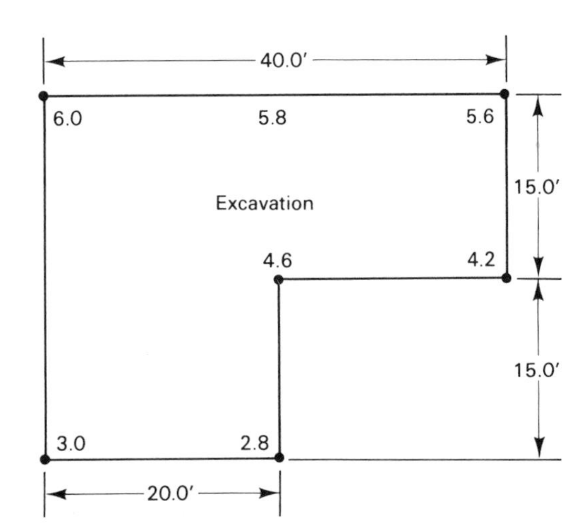

REFERENCES

1. Ahlvin, Robert G., and Vernon A. Smoots. *Construction Guide for Soils and Foundations*. 2nd ed. New York: Wiley, 1988.

2. Barry, B. Austin. *Construction Measurements*. 2nd ed. New York: Wiley, 1988.
3. Calter, Paul. *Practical Math Handbook for the Building Trades*. Englewood Cliffs, NJ: Prentice-Hall, 1983.
4. Hornung, William J. *Estimating Building Construction*. 2nd ed. Englewood Cliffs, NJ: Prentice-Hall, 1986.
5. National Association of Home Builders. *Land Development Manual*. Washington, DC, 1969.
6. Nunnally, S.W. *Construction Methods and Management*. 2nd ed. Englewood Cliffs, NJ: Prentice-Hall, 1987.

Chapter 4

Foundations and Excavations

4-1 FOUNDATION FUNDAMENTALS

Purpose

The purpose of a building *foundation* is to transfer the weight of the building and its loads to the underlying soil or rock. Thus, the foundation is a part of the building's *substructure,* that portion of the building below ground. You should be aware, however, that when a *foundation failure* occurs it is usually caused by the movement or settlement of the soil or rock supporting the foundation rather than by a failure of the foundation structure itself. Such a failure is usually caused by improper foundation design or construction resulting from an inadequate soil investigation and analysis.

Types of Foundations

The principal types of building foundations include spread footings, piles, piers, and mat (or raft) foundations, as illustrated in Figure 4-1. The method by which the first floor of a building is supported also defines the type of foundation structure. When the first floor rests directly on the ground as illustrated in Figure 4-2(a), it is *slab-on-grade.* When the first floor is suspended a short distance (less than a full floor height) above the ground, it is *crawl space* construction [Figure 4-2(b)]. The crawl space simplifies the installation of below-the-floor utilities and provides convenient access to utility lines. When one or more full stories are provided below ground level, the levels are called *basements* [Figure 4-2(c)]. A basement provides additional living or storage space at relatively low cost. However, unless carefully constructed, they are often troubled by dampness or water leakage.

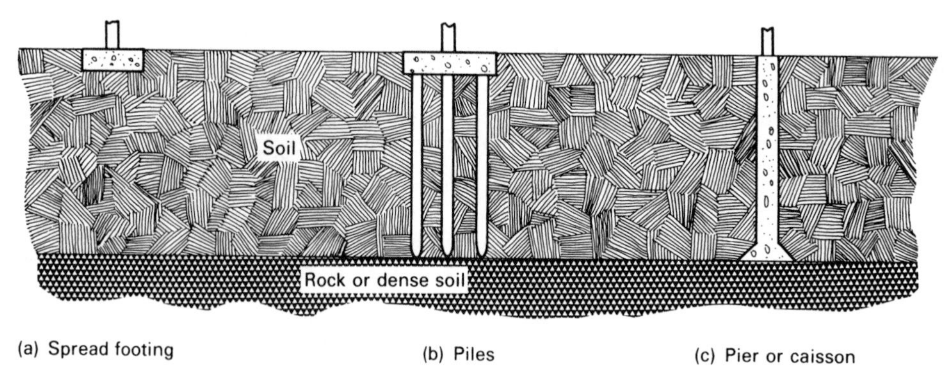

Figure 4-1 Foundation types.

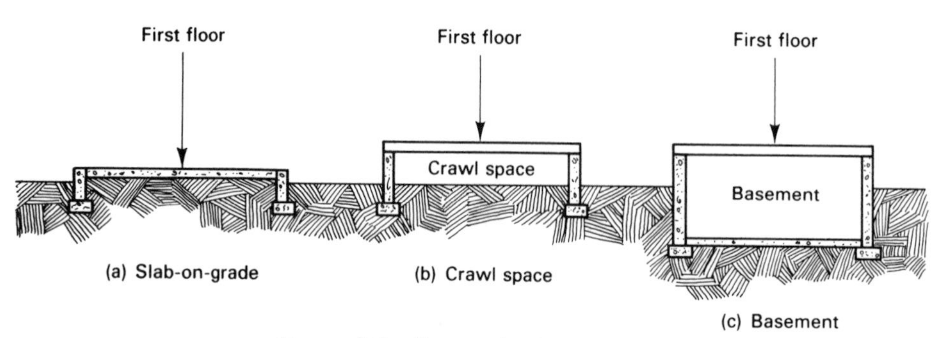

Figure 4-2 Types of substructure.

Foundation Settlement

Unless a building foundation rests directly on sound rock, some *settlement* (sinking) of the building occurs over time. The amount of settlement the building can tolerate without damage depends on the design of the building. If the settlement is uniform under all parts of the foundation, there is little probability of significant structural damage. However, *differential settlement* occurs when some portions of the foundation settle more than others. Differential settlement often results in cracking or leaking of the structure and may even result in building collapse.

Building codes provide minimum footing size and foundation wall thickness for small structures on common stable soils. For unstable soils, areas of high seismic (earthquake) danger, and large structures, foundation structures must be designed by a qualified engineer.

4-2 FOOTINGS AND FOUNDATION WALLS

Spread Footings

Spread footings, often just called *footings,* are the simplest and most widely used type of foundation for small buildings. In principle they simply distribute foundation loads over an area of soil large enough so that foundation pressure does not exceed the soil's allowable bearing pressure. Although usually constructed of reinforced concrete, footings are sometimes made of pressure-treated wood or other materials.

There are several types of spread footings, as illustrated in Figure 4-3. *Individual footings* include wall footings [Figure 4-3(a)] and single, or isolated, footings [Figure 4-3(b)]. When a single footing supports several columns or a wall and column, it is known as a *combined footing* [Figure 4-3(c)]. *Stepped footings* [Figure 4-3(d)] are used when the surface under the footing is sloped. When the footing extends over the entire area under a building, the foundation is called a *mat* or *raft foundation* [Figure 4-3(e)]. Such foundations consist of a heavily reinforced concrete slab and generally involve deep excavation and a large volume of concrete.

For slab-on-grade construction, the footing and floor slab may be combined, as shown in Figure 4-4(a). Such construction is known as a *thickened-edge slab* or a *monolithic slab*. This is often an inexpensive method for constructing foundations for houses on clean granular soils. When the edge of the floor slab rests on the foundation wall, it is called an *edge-supported slab* [Figure 4-4(b)]. When the floor slab terminates inside the foundation wall, it is a *floating slab* or *isolated slab* [Figure 4-4(c)].

Isolated footings like those shown in Figure 4-3(b) are used to support single columns, posts, or piers. They are commonly used to provide intermediate support to girders or beams supporting first-floor joists. The construction requirements for such footings are similar to those listed below for wall footings. Minimum size and depth requirements are normally specified by local building codes.

Foundation Walls

All exterior footings except the thickened-edge slab use a foundation wall on top of the footing to support the exterior bearing walls of the structure. A typical concrete footing and foundation wall are illustrated in Figure 4-5. When the footing and foundation wall are constructed separately, a key, or recess, is provided in the top of the footing to anchor the foundation wall into the footing. No key is necessary when the footing and foundation wall are constructed in one concrete pour. Foundation walls may also be constructed of masonry or wood (Figure 4-7). Foundation walls must extend at least 8 in. (20 cm) above finished grade as illustrated in Figure

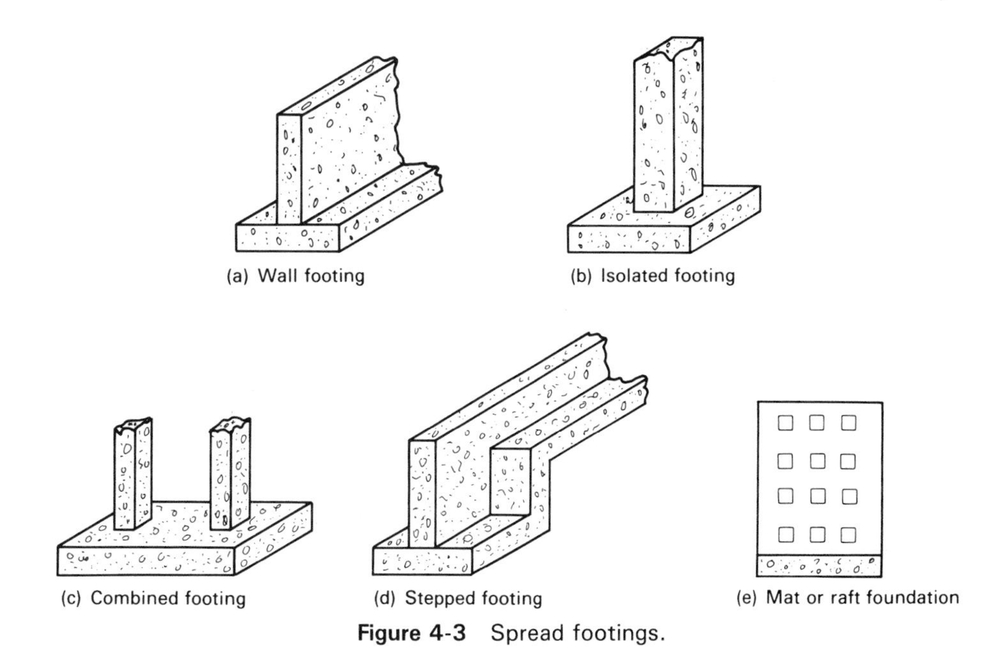

Figure 4-3 Spread footings.

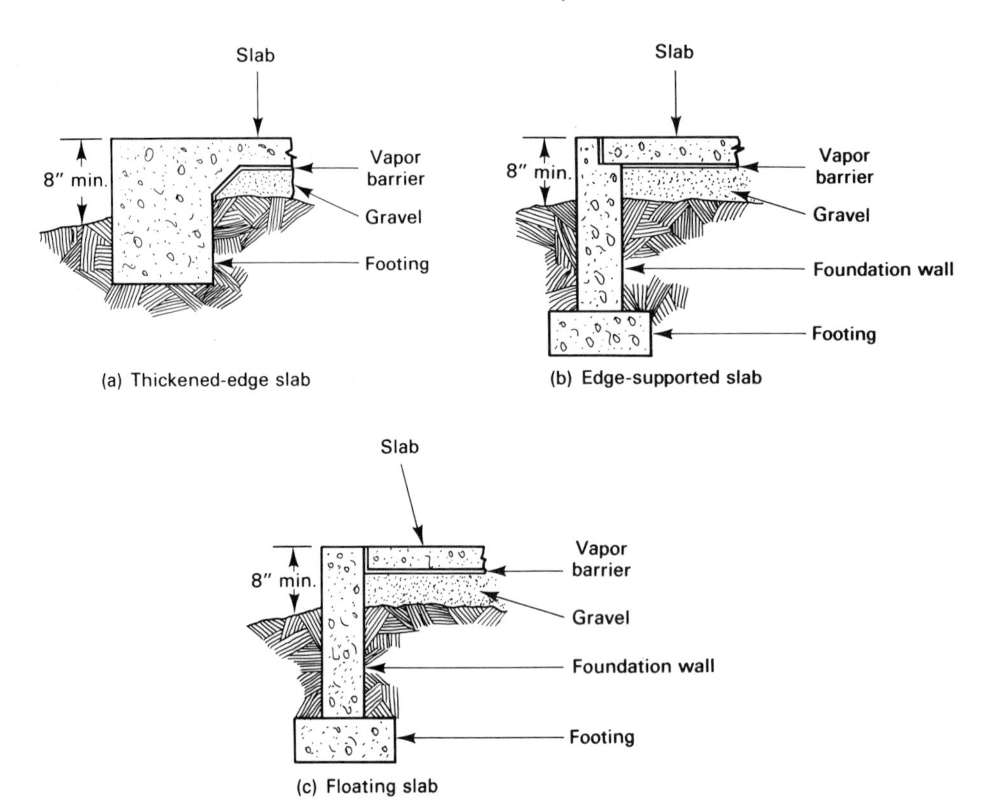

Figure 4-4 Slab on grade construction.

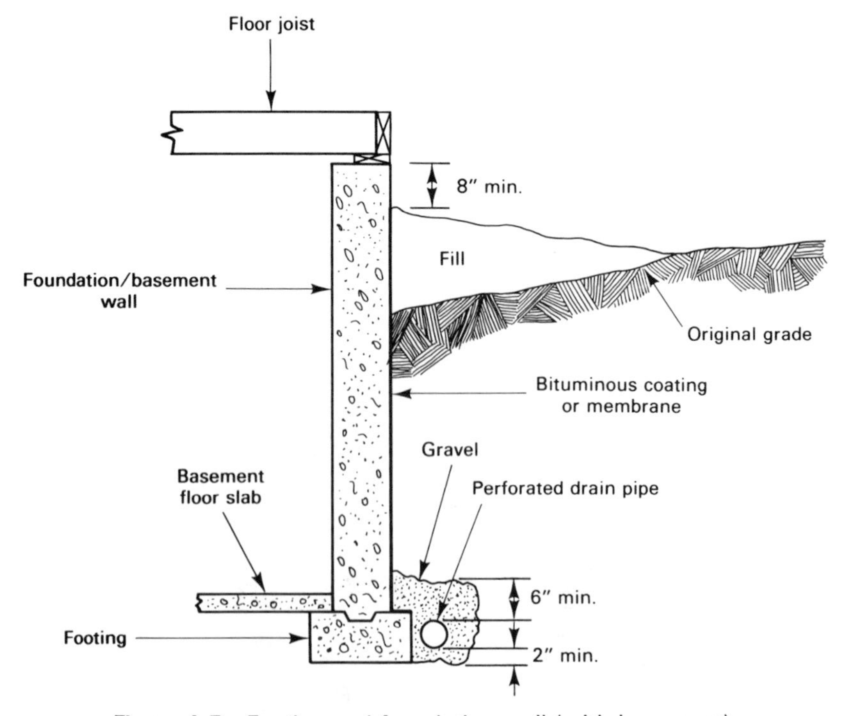

Figure 4-5 Footing and foundation wall (with basement).

4-5. Notice the wood sill plate anchored to the top of the foundation wall. In typical wood frame construction, the floor joists rest directly on the sill plate.

When foundation walls enclose a basement, a drainage system must be installed to ensure a dry basement. A typical drainage system is illustrated in Figure 4-5. This system consists of perforated drain pipe placed on a base of washed gravel or crushed stone at least 2 in. (5 cm) thick and covered with at least 6 in. (15 cm) of gravel. The gravel used should be one sieve size larger than the pipe perforations. The drain pipe must drain by gravity or by pump to an approved drainage system.

To dampproof the foundation walls, the exterior of concrete and masonry foundation walls are covered with a bituminous coating approved for this purpose from the footings to the finished grade. The exterior of masonry foundation walls should be covered ("parged") with at least 3/8 in. (1 cm) of portland cement mortar before dampproofing. Foundation walls of concrete or masonry enclosing habitable walls should be waterproofed by covering their exterior with a waterproof membrane extending from the footings to the finished grade. Waterproofing for wood foundation walls requires sealing the plywood panel joints with waterproof caulking followed by covering the wall with a waterproof membrane.

A *grade beam* (Figure 4-6) serves the same purpose as a foundation wall. However, the grade beam is supported by piles or piers rather than a footing.

Footing Construction

Some of the common requirements of building codes as well as other recommendations for the construction of concrete footings are described below and illustrated in Figure 4-5.

1. The bottom of footings must extend below the frost line and at least 12 in. (30 cm) below grade.
2. The minimum footing thickness should be 6 in. (15 cm) but not less than the foundation wall thickness.
3. The top surface of the footing should be horizontal or have a maximum slope of 1 in 10. Use stepped footings for steeper slopes.
4. The vertical portion of stepped footings should be at least 6 in. (15 cm) thick, and the difference in elevation (step) between adjacent footing levels should not exceed 2 ft. (61 cm).
5. The footing should extend approximately one-half the foundation wall thickness beyond the wall on each side, but not more than 6 in. (15 cm).

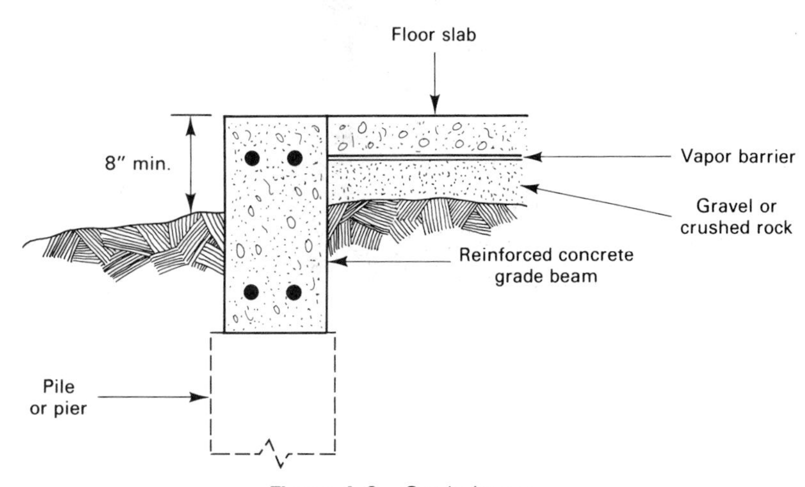

Figure 4-6 Grade beam.

6. Use reinforcing steel in the footing as required by the building code or specified by the architect/engineer.

In cohesive soils, concrete footings may be formed in place using the sides of the footing excavation as forming (see Figure 4-12). When necessary, wood or metal forms the height of the footing are used. Formwork for concrete footings and foundation walls is described in Chapter 5.

Wood Foundations

A method of wood foundation construction, designated the Permanent Wood Foundation System by the National Forest Products Association, has been approved by the major building codes. This system is illustrated in Figure 4-7. Major elements of the system design include the use of pressure-treated wood and plywood, careful waterproofing of foundation walls, and adequate drainage of the material under

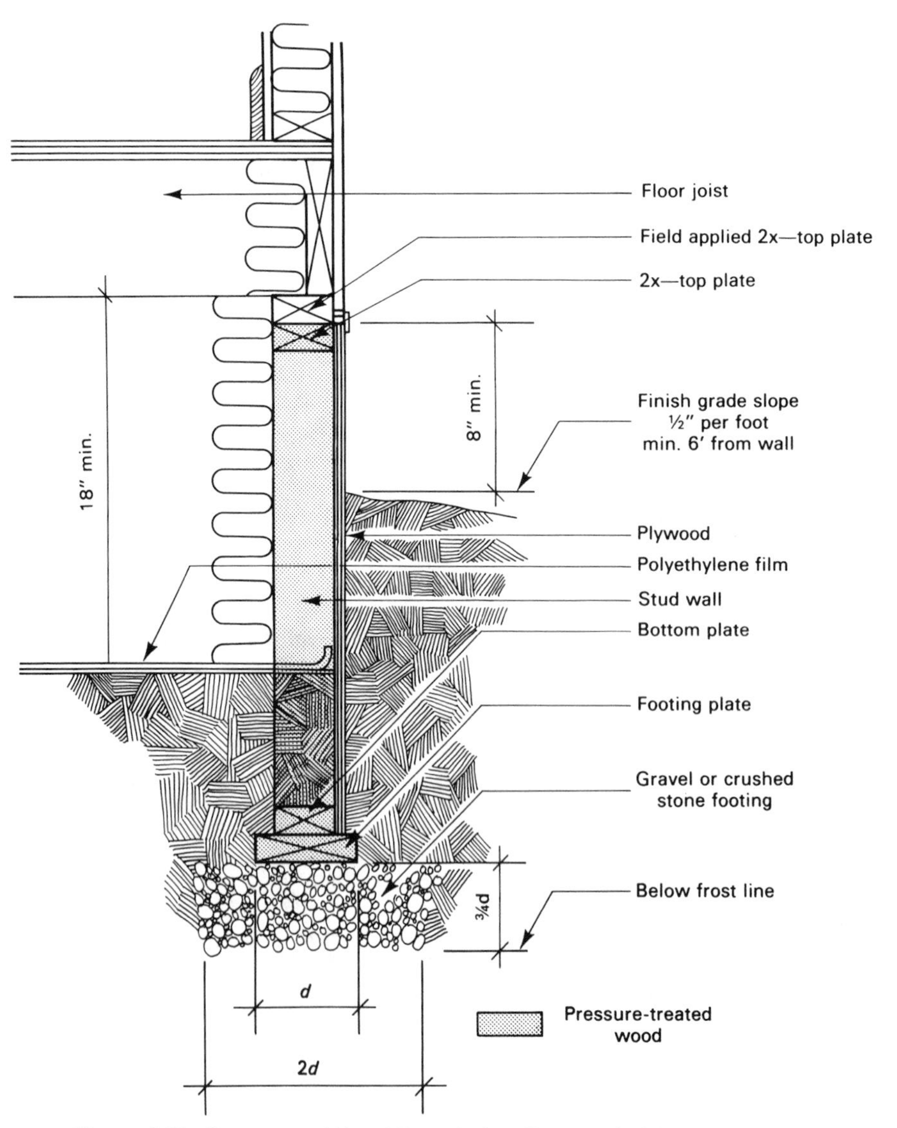

Figure 4-7 Permanent Wood Foundation System (with crawl space) (courtesy National Forest Products Association, Washington, DC).

and around the footings and floor slabs. Design and construction requirements are detailed in Reference 6.

Some of the construction advantages claimed for this foundation system include:

- Lower cost than conventional foundations.
- Fast installation.
- Construction that can be carried out in bad weather.
- A warmer, drier, and more attractive basement than with conventional construction.

Increasing Soil Bearing Strength

By increasing the allowable soil bearing pressure it is sometimes possible to employ spread footings for relatively high foundation loads. With clean granular soil this can be done by densifying the soil under the footings. One such technique for soil densification is known as *Vibroflotation.* This is a process of soil densification employing a vibratory probe or vibroflot. The probe is inserted into the soil by jetting (spraying water at high velocity into the soil below the probe) and then allowed to penetrate to the required depth. The probe's vibrator is then switched on as the probe is slowly raised to the surface. As the soil around the probe densifies and settles, clean granular material is added from the surface. The process results in a column of densified soil, as illustrated in Figure 4-8. The process is then repeated at each footing location. The use of soil densification can be quite effective on clean granular soils and may permit the use of allowable bearing pressures of 5 tons/sq. ft. (479 kPa) or more. Some other soil densification methods include vibratory compaction, dynamic consolidation, and the Terraprobe method.

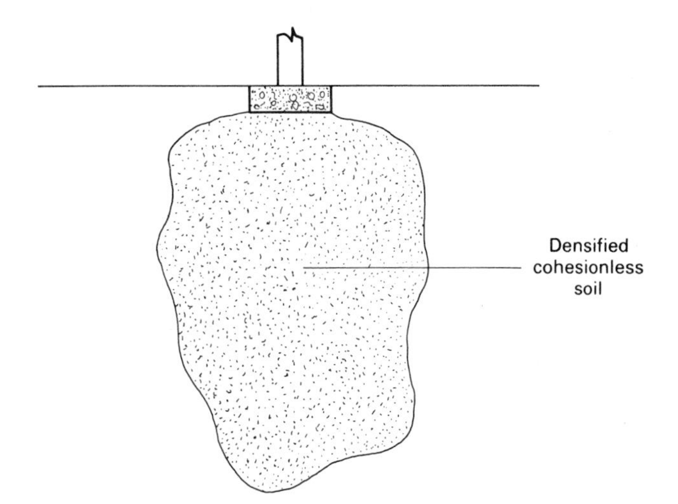

Figure 4-8 Soil densification under footing.

4-3 PILES, PIERS, AND POLE STRUCTURES

Piles

When the ground surface is unstable or unable to support foundation loads, pile foundations are often used. A pile (Figure 4-9) is simply a column driven into the ground to transfer foundation loads to a deeper and stronger layer of soil or rock.

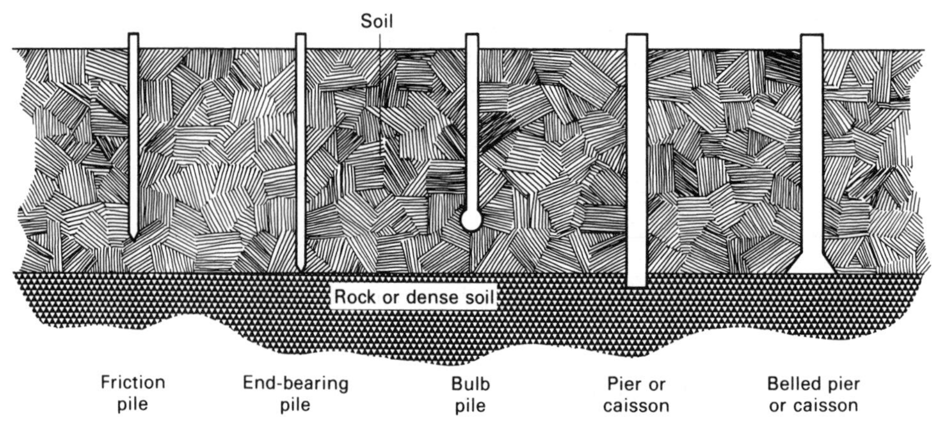

Figure 4-9 Piles and piers.

Piles may be designed as either end-bearing or friction piles. An *end-bearing pile* transfers its load directly to the layer of soil or rock at the bottom end of the pile. A *friction pile* distributes its load to the soil along the surface of the pile by skin friction. In actual practice, virtually all piles support their loads by a combination of end-bearing and skin friction.

Piles are classified according to the materials of which they are made and the method by which they are installed. Principal pile types include timber, precast concrete, cast-in-place concrete, steel, composite, and bulb piles. *Timber piles* are easy to handle and drive, inexpensive, and easy to cut and splice. However, they are subject to insect damage and decay, are limited in available length, and are subject to splintering during hard driving. *Precast concrete piles* may be cast in almost any desired shape and size, carry high loads, and resist decay. However, they are difficult to cut and splice and require careful handling because of their brittleness. *Cast-in-place piles,* also called "shell piles," are constructed by driving a metal shell into the ground and filling it with concrete. A core or mandrel is usually placed inside the shell during driving to reduce shell damage during driving. Shells are normally left in place, but some types may be pulled as the hole is filled with concrete. *Steel piles* can carry high loads, resist driving damage, and are relatively easy to cut and splice. Their principal disadvantage is their high cost. *Composite piles* are made of two or more different materials. For example, the bottom section might consist of a timber pile, while the upper section might consist of a shell pile. Such a pile might be economical when the environment at the bottom section is not conducive to decay but the upper portion is subject to decay. *Bulb piles* (also known as pressure-injected footings, compacted concrete piles, and Franki piles) are forms of cast-in-place concrete piles with an enlarged base. The enlarged base increases its load-carrying ability as a friction pile.

Pile driving is a specialized construction task and will not be described in detail (see Reference 3). Basically, the pile or shell is driven into the ground using one of the various types of available pile drivers. The drop hammer mounted on a crane is often used for small pile-driving projects. Powered hammers, which include single-acting hammers, double-acting hammers, differential hammers, diesel hammers, and vibratory hammers, are more commonly used for larger projects. Powered hammers usually provide faster and less expensive results than do drop hammers.

The safe load-carrying capacity of a pile may be estimated by one of several techniques. Most pile-driving equations are based on hammer energy and pile penetration per blow. Building codes usually specify a maximum safe pile capacity based on soil and pile type or specify a pile capacity equation to be used during driving. Safe pile capacity may also be determined by performing pile load tests. Although expensive, load testing often justifies the use of higher pile loads and therefore reduces pile costs for a project.

Piers and Caissons

Piers (Figure 4-9) are nothing more than columns constructed in the ground to transfer foundation loads to a deeper and stronger layer of soil or rock. Thus, they function like end-bearing piles, which are constructed in an excavation rather than being driven. Piers may be constructed in an open excavation, a lined excavation, or a drilled excavation. A common method of construction uses a drilled hole stabilized by a slurry of clay and water (such as bentonite slurry). Concrete is then placed in the hole using a tremie (a tube and hopper used for depositing concrete underwater), displacing the slurry. A *caisson* is a structure designed to exclude water or soil from an excavation. Since most piers are installed in lined excavations, the foundation terms *pier, caisson,* and *drilled pier* are often used interchangeably.

In cohesive soils, piers are often widened, or belled, at the bottom (Figure 4-9) to increase the area bearing on the supporting rock or soil. This type of construction is a *belled pier* or *belled caisson.* While belled piers often increase allowable pier load, belled excavations are more difficult to construct, inspect, and fill with concrete than are straight piers.

Pole Structures

A *pole building* or *pole house* (Figure 4-10) is a structure supported by poles (columns) that serve as both a foundation and part of the building superstructure. Although poles may be placed by pile-driving methods, they are usually set into drilled holes and rest on concrete punching pads at the bottom of the holes. Poles are commonly made of pressure-treated wood, although steel and other materials may be used. Poles usually extend to the top of the building's upper floor and form part of the building frame. However, platform construction may be used. In this case, only the lowest floor (platform) is a part of the pole frame.

Pole buildings are especially well suited to construction on steep sites or where it is necessary to minimize disturbance to existing site vegetation. Pole construction also provides a relatively inexpensive method of building construction.

4-4 *EXCAVATING PRACTICES AND SAFETY*

Excavating Procedures

Some excavation is required when constructing all types of foundations except pile foundations. After topsoil has been removed from the building area, excavation for slab-on-grade and crawl space construction is usually performed by hand shovel or by a small backhoe-loader. For basement construction, the construction equipment commonly employed for excavation includes the front-end loader (wheel or track), backhoe-loader, and hydraulic excavator (see Chapter 16).

When selecting construction equipment for foundation excavation, always consider equipment capabilities and productivity. An efficient plan of excavation must be developed based on the soil, site conditions, and equipment to be used. Wheel and track loaders are capable of excavating and hauling material but may require a ramp to exit the excavation. Track loaders are able to operate on steeper grades and in softer soils than wheel loaders but travel slower than wheel loaders. Backhoes mounted on loaders make versatile excavators but have limited depth and reach capabilities. Crawler-mounted hydraulic excavators have higher productivity,

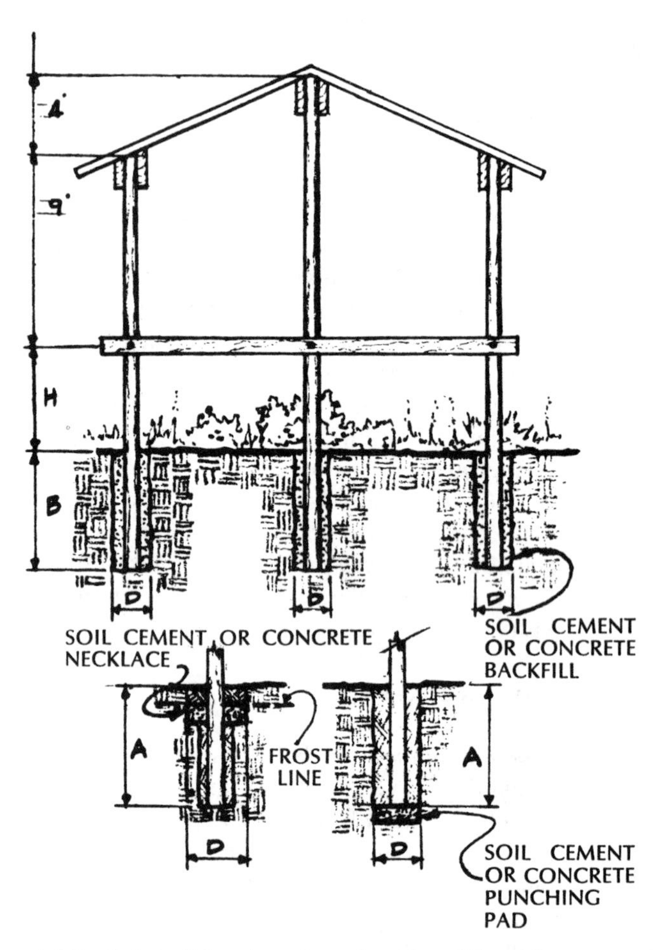

Figure 4-10 Pole building (courtesy American Wood Preservers Institute).

dig deeper, and have a greater reach than backhoe-loaders but must be transported to the site on equipment trailers.

The area of excavation must be wide enough to provide space outside the foundation line for drainage, footing and wall forms, and construction workers. After excavating to the level of the top of the footings, the lines of footing excavation are located and marked, as shown in Figure 4-11.

Slope Stability

An understanding of the basic principles of slope stability is a prerequisite to safe excavating practice. As discussed in Chapter 3, soils may be classified as either cohesive or cohesionless. The soil grains of a cohesive soil tend to stick together, and the soil's shear strength depends on this cohesion. A highly cohesive soil, such as a high plasticity clay, typically fails, as shown in Figure 4-12(a). Notice that a large mass of soil has moved along a surface called a *slip plane.* The natural shape of this failure surface approximates an ellipse.

Cohesionless soils derive their shear strength from friction between the soil grains. Since an open slope has no confining pressure, the stable *angle of repose* that the slope forms with the horizontal equals the soil's angle of internal friction. An embankment of a cohesionless soil typically fails as shown in Figure 4-12(b).

Soils actually encountered in construction usually fall somewhere between the

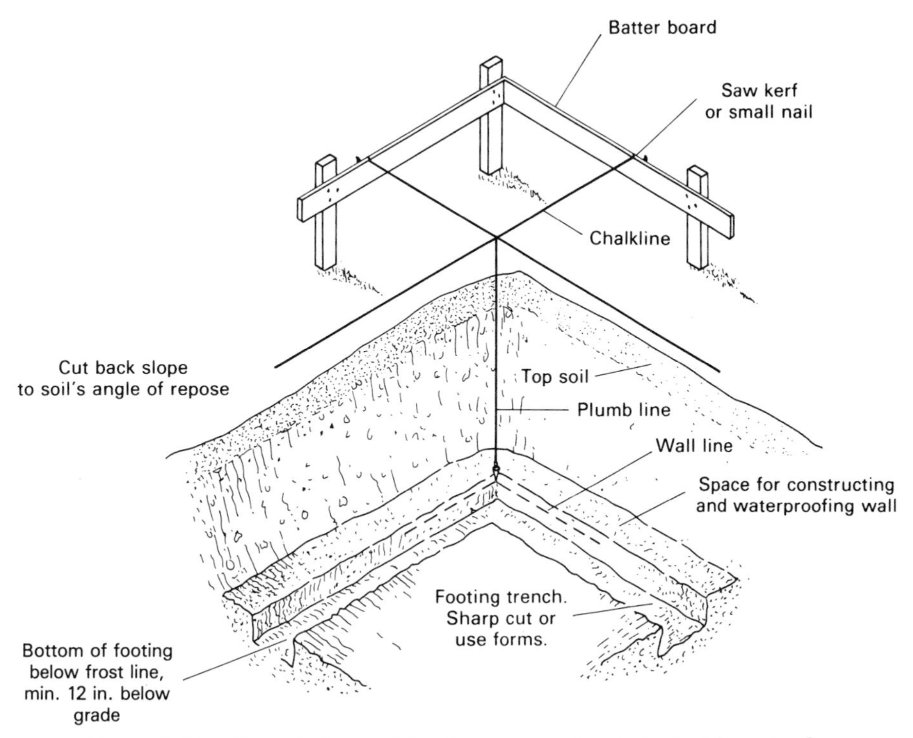

Figure 4-11 Foundation and footing excavation (adapted from U. S. Department of Agriculture).

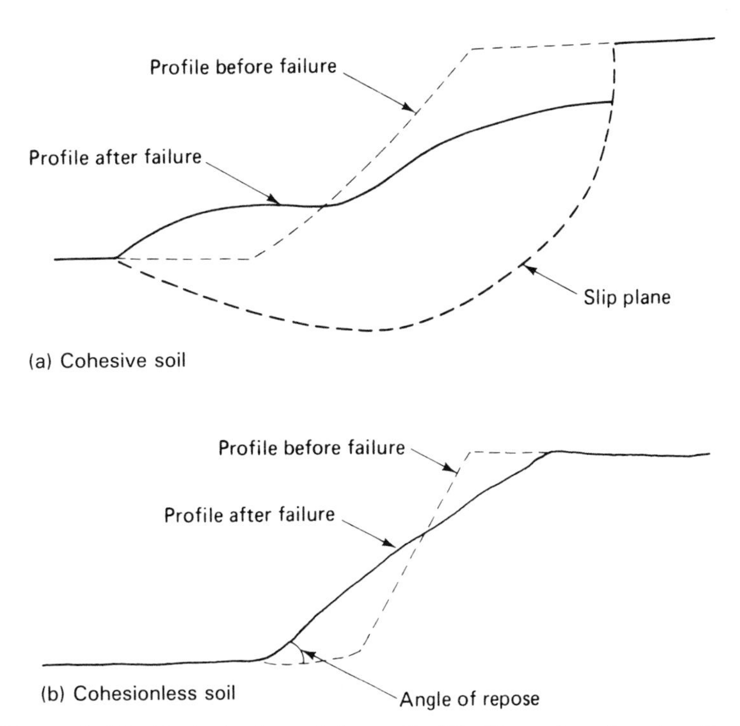

Figure 4-12 Typical slope failure (S. W. Nunnally, *Construction Methods and Management.* 2nd ed. © 1987, pp. 208–209. Reproduced by permission of Prentice-Hall, Inc., Englewood Cliffs, NJ).

two extremes described above. However, the behavior of a highly plastic clay is close to that of a perfectly cohesive soil, whereas the behavior of a clean sand approximates that of a perfectly cohesionless soil.

Embankment Failure

In theory, a slope of cohesionless soil cannot stand at an angle greater than the soil's angle of internal friction. The angle of internal friction for sands ranges from about 28° for loose sands to as high as 46° degrees for dense sands. In practice, almost all sands contain some cohesive material. Moist sand also exhibits a certain amount

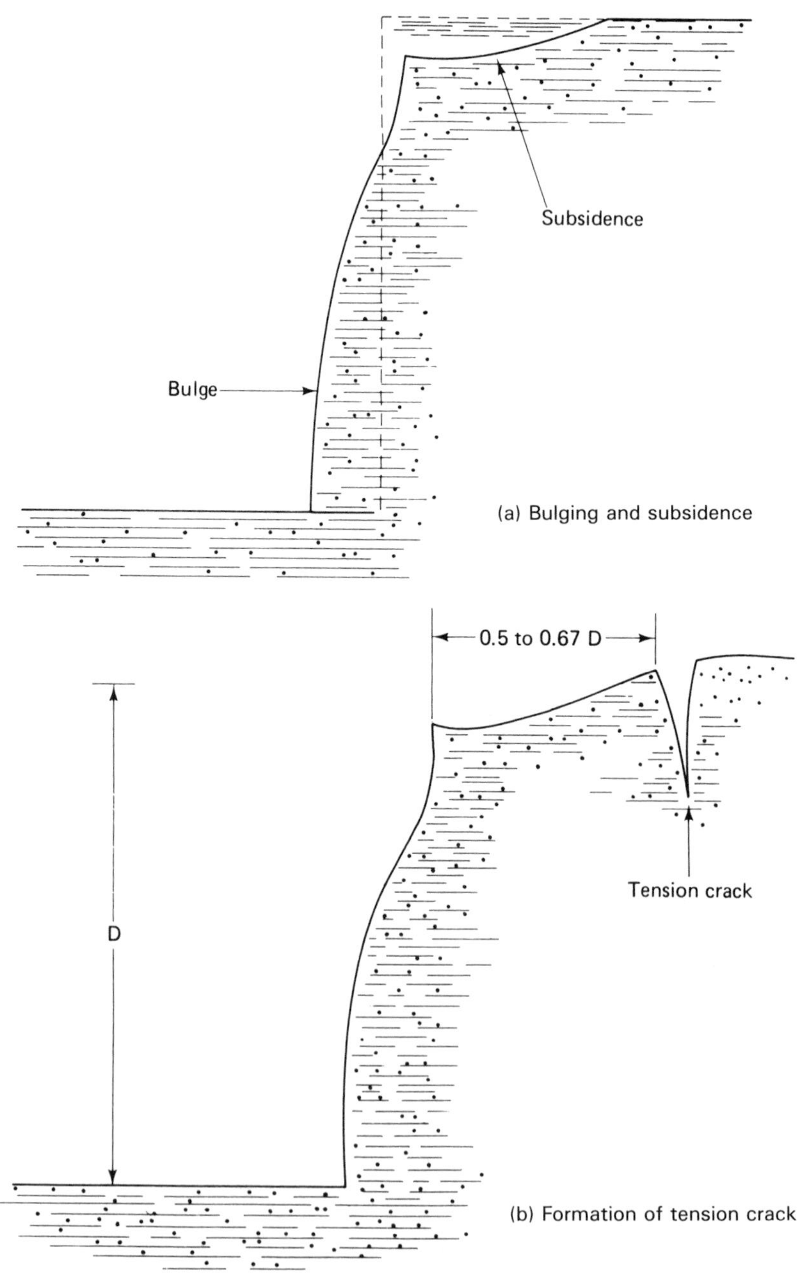

Figure 4-13 Excavation in cohesive soil. (S. W. Nunnally, *Construction Methods and Management.* 2nd ed. © 1987, p. 210. Reproduced by permission of Prentice-Hall, Inc., Englewood Cliffs, NJ).

of cohesion, referred to as "apparent" cohesion. When excavated, sands often stand at an angle somewhat greater than their angle of internal friction for a limited period until they are disturbed or dry out.

A vertical excavation in cohesive soil can theoretically stand to a depth that depends on the soil's cohesive strength and angle of internal friction. This depth ranges from under 5 ft. (1.5 m) for a soft clay to about 18 ft. (5.5 m) for a medium clay. The safe depth for a stiff clay is actually less than for a medium clay due to the cracks and fissures found in stiff clay. As clay is excavated, however, the sides and top of the excavation begin to bulge and subside as shown in Figure 4-13(a). With time, cracking will occur, as shown in Figure 4-13(b), followed by slope failure.

Avoiding Slope Failure

To fully prevent failure of the sides of an excavation, the sides must be properly shored or cut back to the angle of repose of the soil. In fact, OSHA regulations require that banks over 5 ft. (1.5 m) high be shored, cut back to a stable slope, or otherwise protected. In addition, when workers are required to work in a trench over 4 ft. (1.2 m) deep, an adequate means of emergency exit must be located within 25 ft. (7.6 m) of each worker. Unfortunately, accidents and fatalities resulting from slope failure continue to occur all too often, particularly in trenching work.

Some factors that increase the danger of embankment collapse include:

- Operation of construction equipment or vehicles near the upper edge of the cut.
- Storage of excavated material (spoil) or construction supplies near the edge of the cut.
- Wet conditions, including a high water table.
- Vibration from construction equipment operations or from off-site sources.

The presence of any of these factors indicate that extra attention must be paid to preventing slope failure.

Shoring

When it is not feasible to cut the sides of an excavation back to the soil's angle of repose, excavation walls should be physically restrained by shoring, or the workers should be protected by an enclosure such as a trench shield. *Shoring* provides lateral support for the sides of an excavation. The most common types of shoring include sheeting, lagging, and sheet piling. CAUTION: Unless the shoring is constructed to allow water to drain freely from the embankment, the shoring must be designed to withstand the full hydrostatic pressure that can occur behind the shoring.

Sheeting (Figure 4-14) consists of vertical members (usually timber) placed against the sides of an excavation and supported by horizontal beams (wales or stringers). For trenches or other narrow excavations, wales are held in place by horizontal braces called struts, cross braces, trench jacks, or trench braces. For wide excavations, wales are restrained by inclined braces or by anchors embedded in the sides of the excavation.

Lagging consists of horizontal members placed against the sides of an excavation. Vertical members (soldier beams or soldier piles) support the lagging. Soldier beams are supported by horizontal wales and braces or anchors. Soldier piles are driven into the ground and anchored or braced at the top when necessary.

Sheet piling consists of steel, timber, or concrete sheets driven into the ground by a pile driver. Unless the piling is self-supporting by forming a circular cofferdam, horizontal support is required at the upper end of the sheet piling.

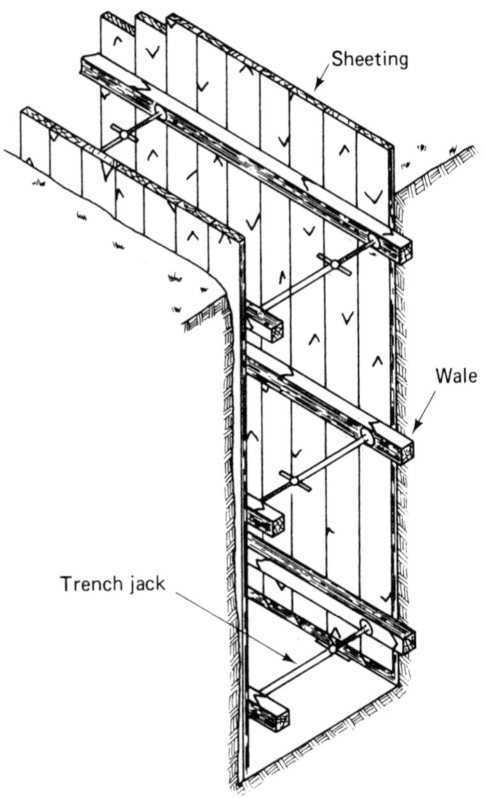

Figure 4-14 Trench shoring with sheeting (S. W. Nunnally, *Construction Methods and Management.* 2nd ed. © 1987, p. 214. Reproduced by permission of Prentice-Hall, Inc., Englewood Cliffs, NJ).

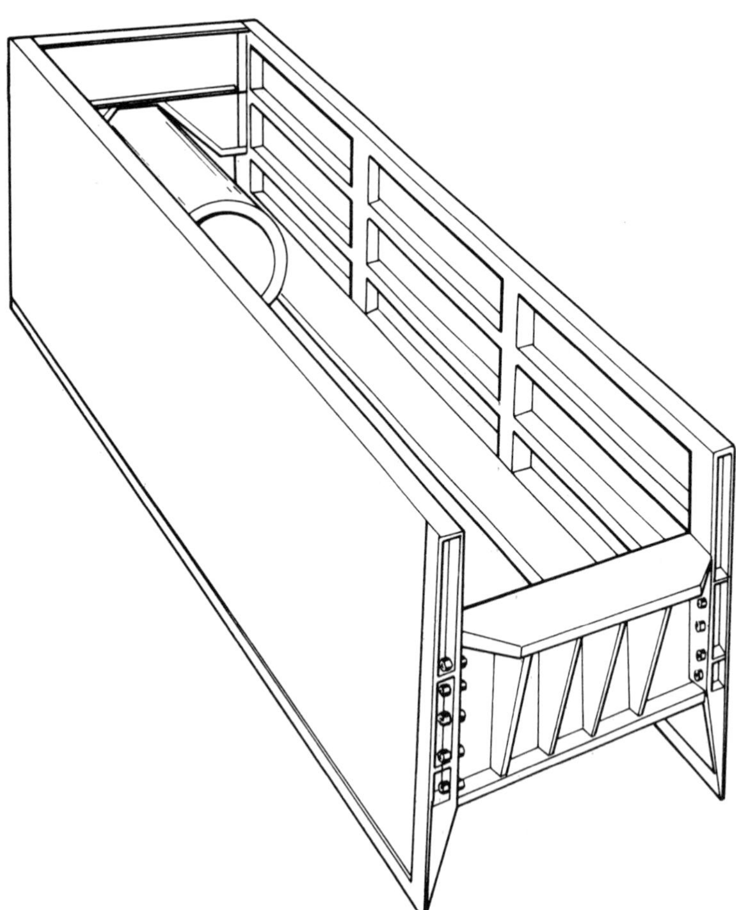

Figure 4-15 Trench shield. (S. W. Nunnally, *Construction Methods and Management.* 2nd ed. © 1987, p. 215. Reproduced by permission of Prentice-Hall, Inc., Englewood Cliffs, NJ).

In trenches and other excavations requiring workers to be in the excavation for only a short period, a *trench shield* (Figure 4-15) or trench box may be used to protect the workers. Use of a trench shield can eliminate the requirement for shoring to provide worker safety.

PROBLEMS

1. Explain the purpose of a foundation and describe the major types of building foundations.
2. What is the meaning of the term *foundation failure?*
3. What are stepped footings and when are they used?
4. Describe the three types of slab-on-grade construction.
5. Sketch and describe the major features of a Permanent Wood Foundation System for a house with basement.
6. Explain how Vibroflotation is carried out for a cohesionless soil. What type of foundation would be used in conjunction with this process?
7. How do piles transfer foundation loads to subsurface soil or rock?
8. What is a pole structure? What are the advantages of this type of construction?
9. In addition to the basic soil characteristics, there are a number of site and construction conditions that increase the danger of embankment collapse at an excavation. Name at least three such conditions.
10. Describe the major OSHA safety requirements for carrying out a trench excavation 6 ft. (1.8 m) deep.

REFERENCES

1. Ahlvin, Robert G., and Vernon A. Smoots. *Construction Guide for Soils and Foundations.* 2nd ed. New York: Wiley, 1988.
2. American Wood Preservers Institute. *FHA Pole House Construction.* 2nd ed. Vienna, VA, 1975.
3. Nunnally, S.W. *Construction Methods and Management.* 2nd ed. Englewood Cliffs, NJ: Prentice-Hall, 1987.
4. U.S. Department of Labor. *OSHA Safety and Health Standards: Construction Industry* (OSHA 2207). Washington, DC, 1983.
5. Patterson, D. *Pole Building Design.* American Wood Preservers Institute, Vienna, VA, 1969.
6. National Forest Products Association. *The All-Weather Wood Foundation System: Basic Requirements.* Technical Report No. 7. Washington, DC, 1982.

Chapter 5

Concrete Construction

5-1 CONCRETE FORMWORK

The Nature of Formwork

Plastic concrete must be restrained until it has hardened sufficiently to support its own weight and any additional loads that may be imposed. *Formwork* is used for this purpose. In some instances, soil or rock may be excavated and shaped so that it will replace formwork. This method was used for placing the concrete footings illustrated in Figure 4-11. In all other cases, formwork of wood, metal, plastic, or other material must be constructed.

Formwork must satisfy three principal requirements:

1. Be strong enough to safely support the weight of plastic concrete, construction loads, and external forces.
2. Produce the desired shape and surface texture in the hardened concrete.
3. Be as economical as possible.

The design of concrete formwork is not usually included in the project plans provided by the owner. As a result, the contractor must either design the formwork, utilize the design services of a specialty formwork supplier, or engage an engineer for this purpose. The design loads for concrete formwork must be carefully established to include the weight of the formwork itself, construction workers, equipment and materials, and external loads. While minimum design loads are often specified by building codes and standards, these may not be sufficient in all cases to ensure construction safety. A number of construction failures resulting in fatalities have involved concrete formwork. As a result, some localities now require that all but the simplest formwork be designed by professional engineers. Additional information on the design and construction of concrete formwork is provided in the end-of-chapter references.

The requirements for concrete member shape and surface texture are normally contained in the construction plans and specifications for the project. While plywood is most commonly used for sheathing concrete forms, metal and plastic forms produce a smoother concrete surface and have a longer life. Plastic and fiberglass forms can be molded to provide almost any desired concrete shape.

Formwork Economy

The distribution of concrete construction costs for a typical multistory reinforced concrete building is shown in Figure 5-1. Notice that the cost of formwork accounts for almost one-half of the total concrete construction costs. In some cases, the cost of formwork may be as great as 60 percent of the total cost of concrete construction. Therefore, every effort must be made to minimize the cost of formwork. When selecting a formwork system for a major project, a cost analysis should be made of all feasible forming systems and methods of construction. The objective is, of course, to select the construction plan that safely produces concrete meeting all quality requirements at the minimum cost. A concrete construction cost analysis should include:

- The cost of labor, equipment, and materials for constructing and erecting formwork.
- The cost of reinforcing steel and its placement.
- The cost of concrete materials, and its placing, curing, and finishing.

The repetitive use of concrete forms also reduces unit formwork cost. Either standard commercial forms or custom-made forms may be used in most instances. When fabricating custom-made forms, construct the forms at ground level using assembly line techniques whenever possible. For maximum reuse of forms:

- Use care in form removal.
- Clean and oil the forms after each use.
- Promptly move the forms to their next position or to a protected storage area.

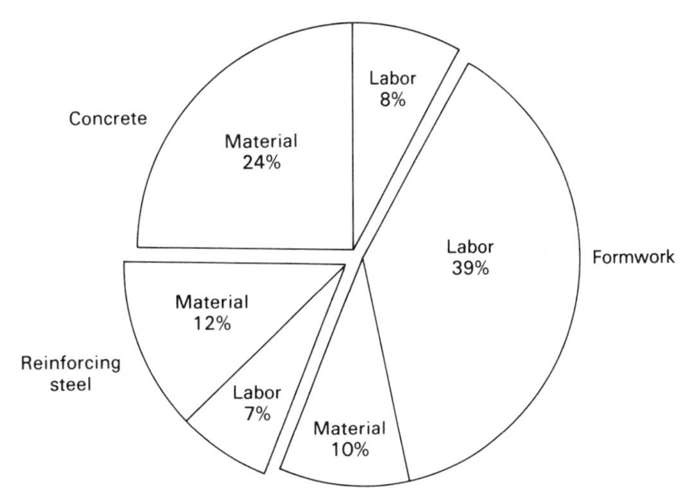

Figure 5-1 Distribution of concrete construction costs.

Wall and Column Forms

Almost any type of concrete wall can be cast in-place, but foundation walls are the most common application of vertical forms in residential and light construction. A typical form for a foundation wall is shown in Figure 5-2. The sheathing for the form is usually plywood, but board sheathing may be used. The pressure of the plastic concrete is resisted by the form ties acting against the wales, studs, and sheathing. Bracing is provided only to prevent form movement and to resist wind load and other external loads. Double wales as illustrated are commonly used so that ties may be inserted without drilling the wales.

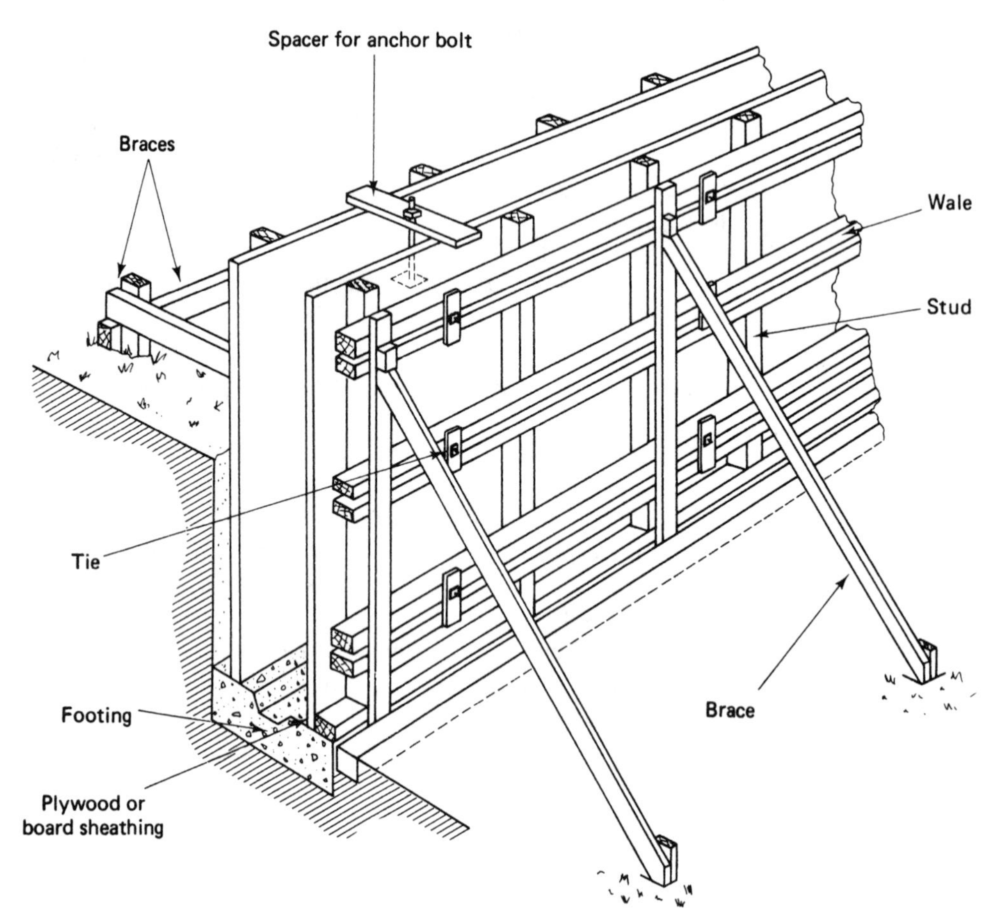

Figure 5-2 Typical foundation wall form (S. W. Nunnally, *Construction Methods and Management.* 2nd ed. © 1987, p. 240. Reproduced by permission of Prentice-Hall, Inc., Englewood Cliffs, NJ).

Some typical form ties are illustrated in Figure 5-3. Notice that the snap tie and the coil tie incorporate a spreader device to provide the desired spacing between form walls while concrete is placed and cured. If ties do not incorporate spreaders, external spreaders must be used. The two principal categories of form ties are continuous single-member ties and internally disconnecting ties. A *continuous single-member tie* is pulled out after the concrete has hardened or is broken off at a weak point just below the surface after forms are removed. The snap tie shown in Figure

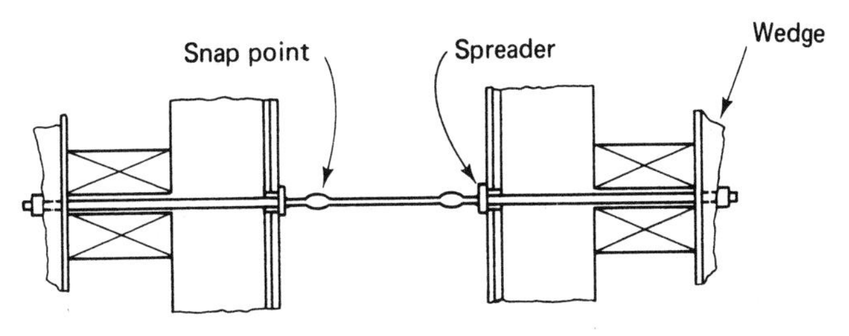

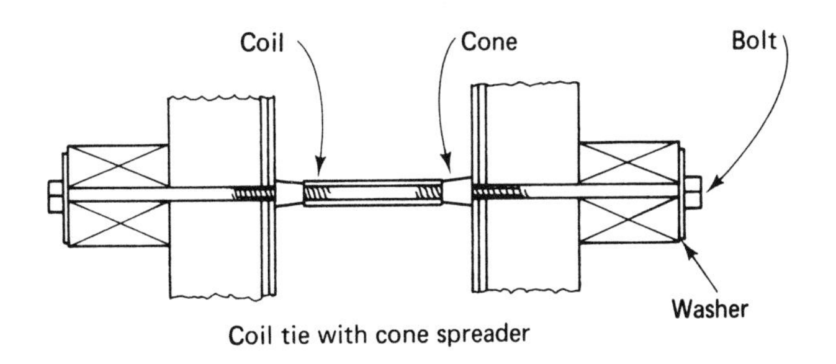

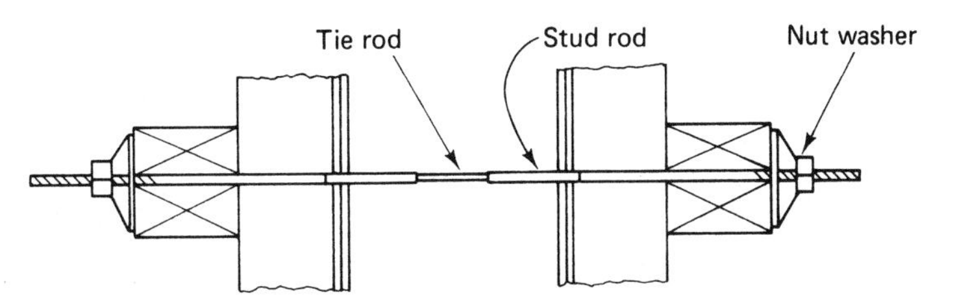

Stud rod (she-bolt) tie

Figure 5-3 Typical form ties (S. W. Nunnally, *Construction Methods and Management.* 2nd ed. © 1987, p. 241. Reproduced by permission of Prentice-Hall, Inc., Englewood Cliffs, NJ).

5-3 is a continuous single-member tie. *Internally disconnecting ties* permit the tie ends to be unscrewed to allow form removal. After the forms are removed, the holes in the concrete surface are plugged or grouted. The coil tie and stud rod tie shown in Figure 5-3 are examples of internally disconnecting ties.

A typical column form is illustrated in Figure 5-4. Although similar to a wall form, notice that the pressure of the plastic concrete is now resisted by column clamps rather than by ties. Forming for round columns often consists of fiber tubes or steel reinforced fiberglass tubes. High narrow forms such as column forms often contain openings called windows or doors to facilitate the placing of concrete in the form. A cleanout door should be provided near the bottom of the form as illustrated in Figure 5-4 to permit removal of debris prior to placing concrete in the form. Special fittings may be inserted near the bottom of vertical forms to permit pumping concrete into the form from the bottom. Also, notice the chamfer strips placed in the corners of the form shown in the figure. The use of chamfer strips is recommended to reduce damage to the edges of the concrete column during and after form removal.

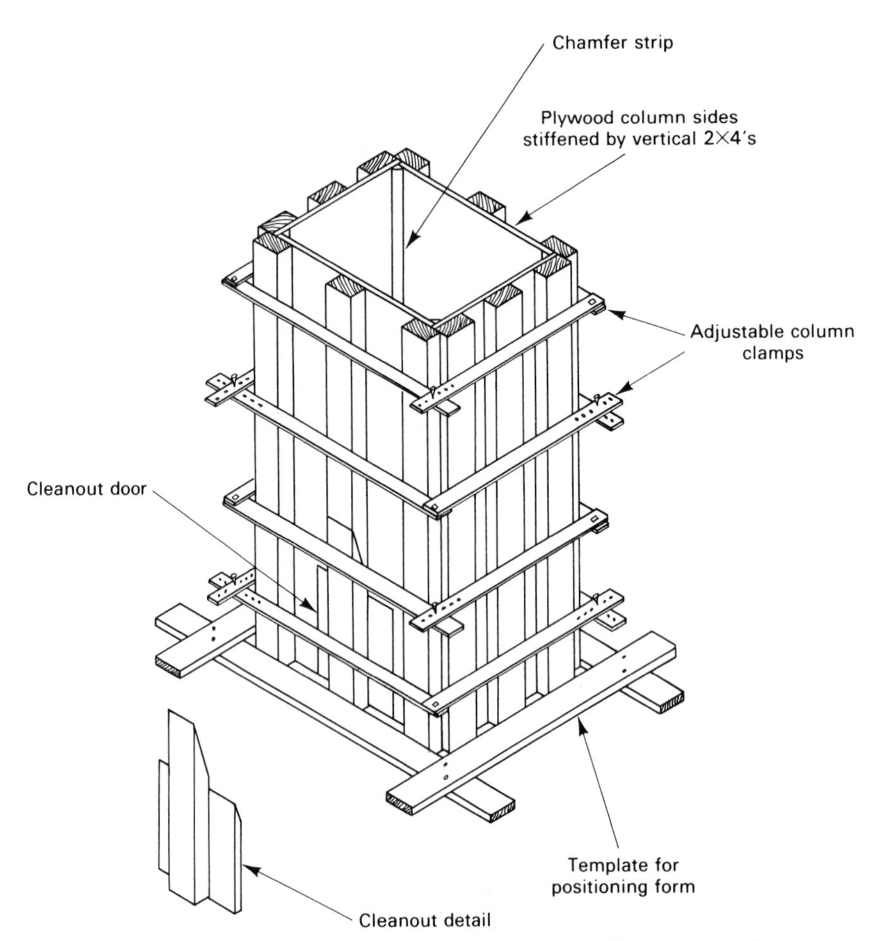

Figure 5-4 Typical column form (courtesy American Concrete Institute).

Floor, Roof, and Other Forms

The forms used for placing horizontal concrete members such as floor slabs and roofs are referred to as slab forms. A typical form for an elevated slab is illustrated in Figure 5-5. Notice that the weight of plastic concrete and vertical construction loads is transferred to the underlying soil or structure by the sheathing, joists, stringers, shores, and footings (mudsills). However, wind loads and other external horizontal loads must be resisted by cross bracing or external bracing. The forming for a slab with an integral beam is illustrated in Figure 5-6.

There are many other concrete building components that may require concrete forms. The formwork suitable for constructing a stairway up to 3 ft. (0.9 m) wide is illustrated in Figure 5-7. There is a growing trend toward precasting such components and lifting them into place during building construction.

5-2 REINFORCING STEEL

Identification

Reinforcing steel and its placement are critical elements in the performance of reinforced concrete construction. The builder must be able to identify the reinforcing steel specified on construction plans and ensure that the steel is properly placed

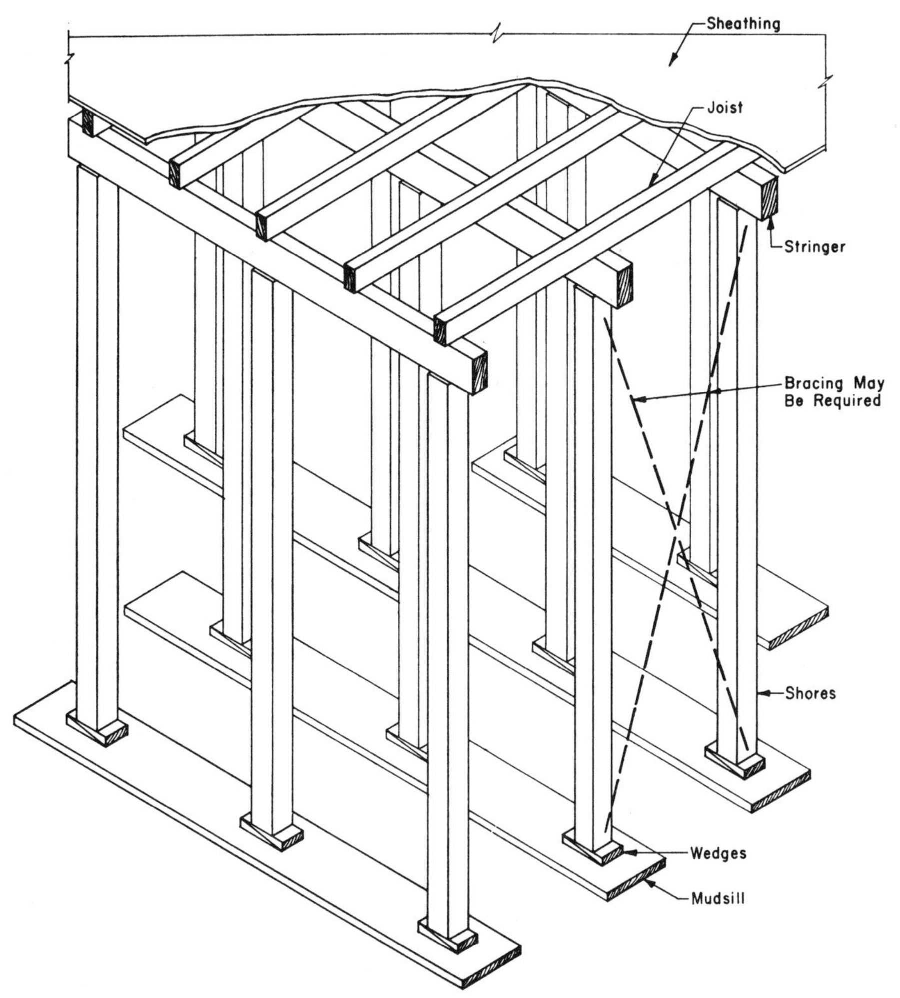

Figure 5-5 Typical form for elevated slab (courtesy American Concrete Institute).

during construction. The major categories of reinforcing steel include reinforcing bars, spirals, and welded wire fabric.

Reinforcing bars are the most commonly used form of reinforcing steel. Smooth bars are available, but deformed bars are most commonly used. Deformed bars are manufactured with ridges that provide increased bonding with the surrounding concrete. The marking code established by the American Society for Testing and Materials (ASTM) for standard deformed bars is illustrated in Figure 5-8. Notice that two marking systems are used, the continuous line system and the number system. Bar grade corresponds to the steel's rated yield stress in thousands of pounds per square inch (ksi). Four types of steel are available as indicated, but new billet steel is most common. Bars are available in the 11 ASTM standard sizes listed in Table 5-1. Notice that the bar size number approximately equals the bar diameter in eighths of an inch. For example, a No. 4 bar has a diameter of 4/8 in. or 0.5 in. (12.7 mm).

The reinforcing steel commonly used in concrete slabs and pavements is *welded wire fabric* (Figure 5-9). Again, it is available with either smooth wire or deformed wire and can be obtained with a galvanized finish. Notice that fabric length is defined as the tip-to-tip length of longitudinal wires, and fabric width is defined as the center-to-center distance between the outside wires. The industry stan-

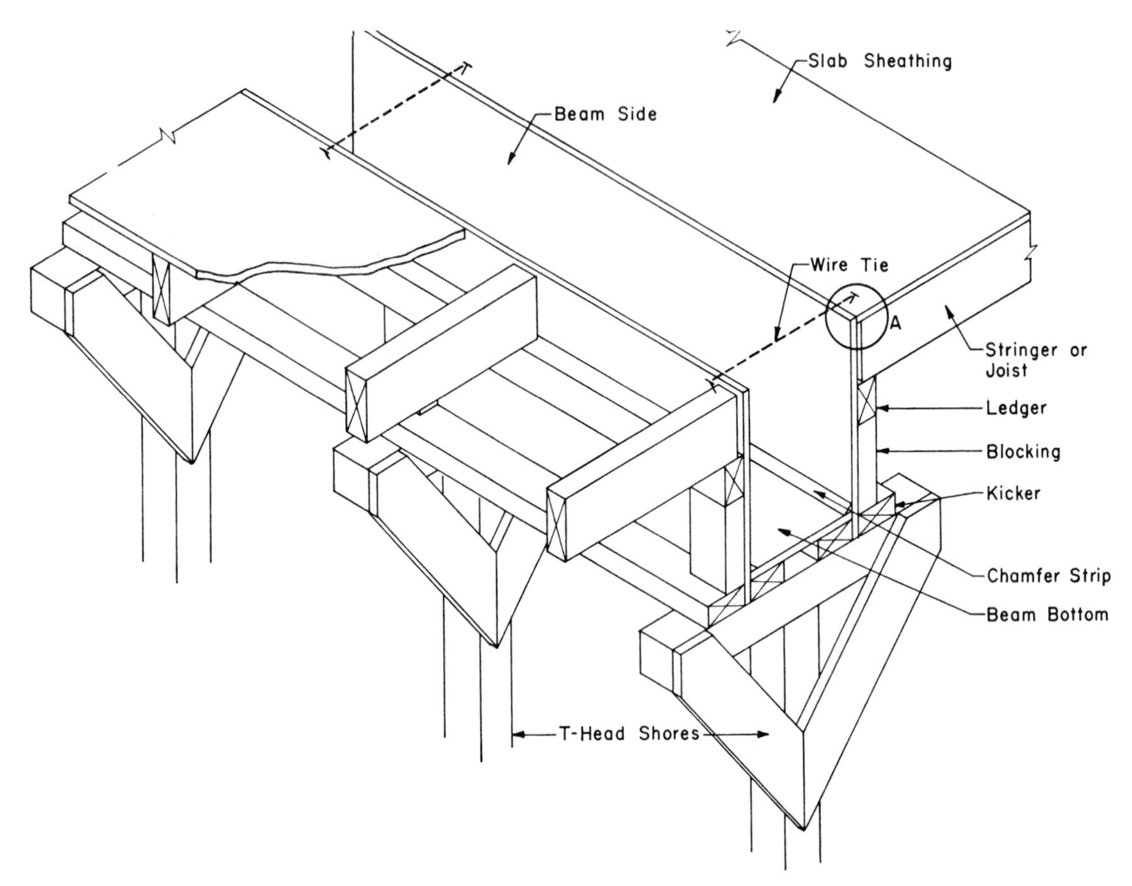

Figure 5-6 Beam and slab form (courtesy American Concrete Institute).

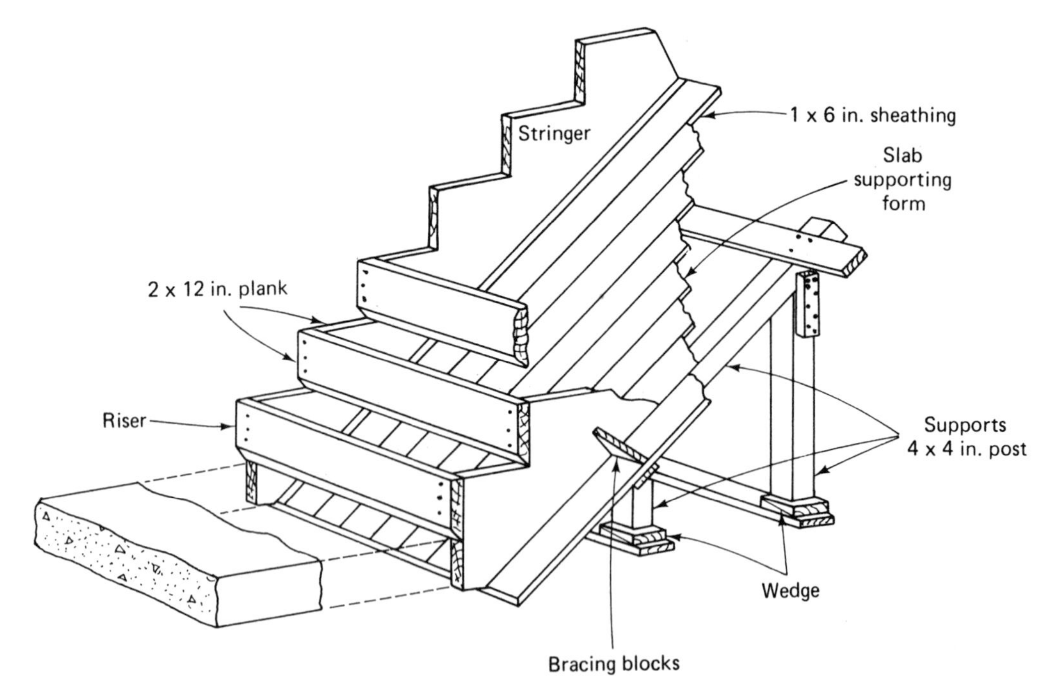

Figure 5-7 Form for stairway (U. S. Department of the Army).

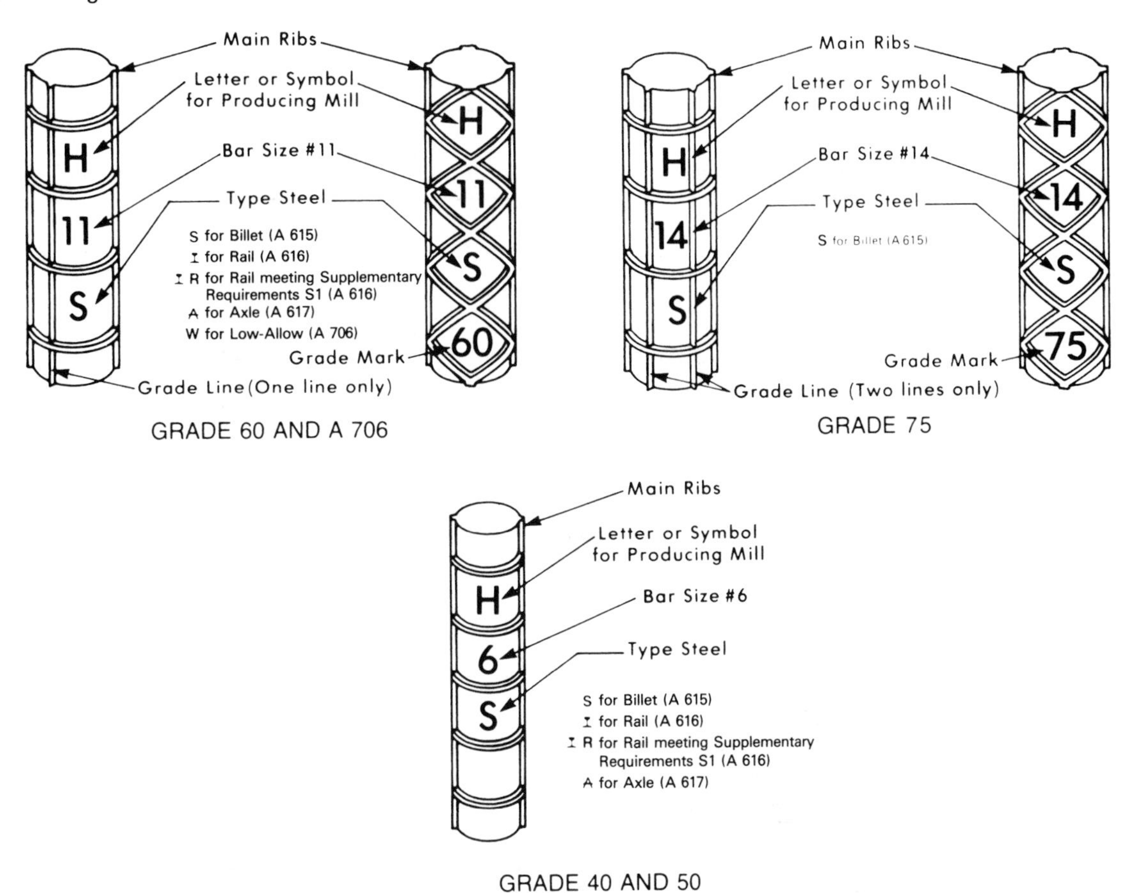

Figure 5-8 Identification of reinforcing bars (courtesy Concrete Reinforcing Steel Institute).

dard method for designating wire style uses the letters WWF followed by the longitudinal wire spacing (in. or mm), the transverse wire spacing (in. or mm), the longitudinal wire size (1/100 sq. in. or mm^2), and transverse wire size (1/100 sq. in. or mm^2). Data on wire size is given in Table 5-2. The letter M preceding wire size indicates metric size while the letter D indicates deformed wire.

Example: WWF 6 × 6-W2.0 × W2.0 or
WWF 152 × 152-MW13 × MW13 (metric)

Longitudinal wire spacing = 6 in. (152 mm)
Longitudinal wire size = 0.02 sq. in. (13 mm^2)
Transverse wire spacing = 6 in. (152 mm)
Transverse wire size = 0.02 sq. in. (13 mm^2)

Steel *spirals* (Figure 5-10) are often used for reinforcing round concrete columns. Three standard bar sizes are available: 3/8 in. (0.95 cm), 1/2 in. (1.27 cm), and 5/8 in. (1.59 cm). The outside diameters of standard steel spirals range from 11 in. (28 cm) to 39 in. (99 cm). Their pitch, or distance between centers of adjacent spirals, ranges from 1 3/4 in. (4.4 cm) to 3 in. (7.6 cm) by 1/4 in. (0.64 cm) increments. They are available in steel grades 40, 60, and 70.

TABLE 5-1

Standard Reinforcing Bar Sizes (ASTM)*

Size number	Weight		Diameter		Section area	
	lbs./ft.	*kg/m*	*in.*	*mm*	*sq. in.*	*mm²*
3	0.376	0.560	0.375	9.52	0.11	71
4	0.668	0.994	0.500	12.70	0.20	129
5	1.043	1.552	0.625	15.88	0.31	200
6	1.502	2.235	0.750	19.05	0.44	284
7	2.044	3.042	0.875	22.22	0.60	387
8	2.670	3.973	1.000	25.40	0.79	510
9	3.400	5.059	1.128	28.65	1.00	645
10	4.303	6.403	1.270	32.26	1.27	819
11	5.313	7.906	1.410	35.81	1.56	1006
14	7.650	11.384	1.693	43.00	2.25	1452
18	13.600	20.238	2.257	57.33	4.00	2581

*S. W. Nunnally, *Construction Methods and Management.* 2nd ed., © 1987, p.248. Reproduced by permission of Prentice-Hall, Inc., Englewood Cliffs, NJ.

TABLE 5-2

Steel Wire Data for Welded Wire Fabric*

Wire size number		Diameter		Area		Weight	
Smooth	*Deformed*	*in.*	*mm*	*sq. in.*	*mm²*	*lbs./ft.*	*kg/m*
W31	D31	0.628	16.0	0.31	200	1.054	1.569
W28	D28	0.597	15.2	0.28	181	0.952	1.417
W26	D26	0.575	14.6	0.26	168	0.934	1.390
W24	D24	0.553	14.1	0.24	155	0.816	1.214
W22	D22	0.529	13.4	0.22	142	0.748	1.113
W20	D20	0.505	12.8	0.20	129	0.680	1.012
W18	D18	0.479	12.2	0.18	116	0.612	0.911
W16	D16	0.451	11.5	0.16	103	0.544	0.810
W14	D14	0.422	10.7	0.14	90	0.476	0.708
W12	D12	0.391	9.9	0.12	77	0.408	0.607
W11	D11	0.374	9.5	0.11	71	0.374	0.557
W10	D10	0.357	9.1	0.10	65	0.340	0.506
W9.5		0.348	8.8	0.095	61	0.323	0.481
W9	D9	0.338	8.6	0.09	58	0.306	0.455
W8.5		0.329	8.4	0.085	55	0.289	0.430
W8	D8	0.319	8.1	0.08	52	0.272	0.475
W7.5		0.309	7.8	0.075	48	0.255	0.379
W7	D7	0.299	7.6	0.07	45	0.238	0.354
W6.5		0.288	7.3	0.065	42	0.221	0.329
W6	D6	0.276	7.0	0.06	39	0.204	0.311
W5.5		0.265	6.7	0.055	35	0.187	0.278
W5	D5	0.252	6.4	0.05	32	0.170	0.253
W4.5		0.239	6.1	0.045	29	0.153	0.228
W4	D4	0.226	5.7	0.04	26	0.136	0.202
W3.5		0.211	5.4	0.035	23	0.119	0.177
W3		0.195	5.0	0.030	20	0.102	0.152
W2.9		0.192	4.9	0.029	19	0.099	0.147
W2.5		0.178	4.5	0.025	16	0.085	0.126
W2		0.160	4.1	0.02	13	0.068	0.101
W1.4		0.134	3.4	0.014	9	0.048	0.071

*S. W. Nunnally, *Construction Methods and Management.* 2nd ed., © 1987, p. 249. Reproduced by permission of Prentice-Hall, Inc., Englewood Cliffs, NJ.

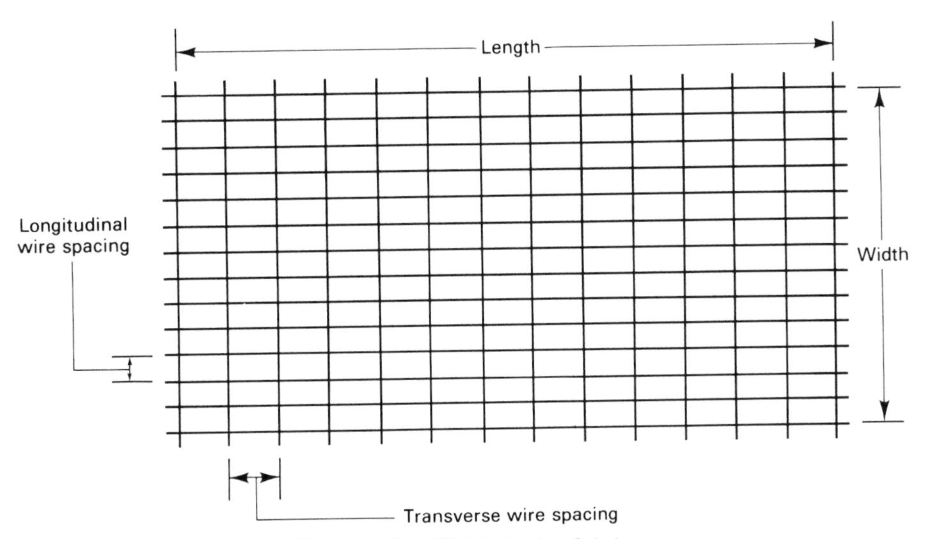

Figure 5-9 Welded wire fabric.

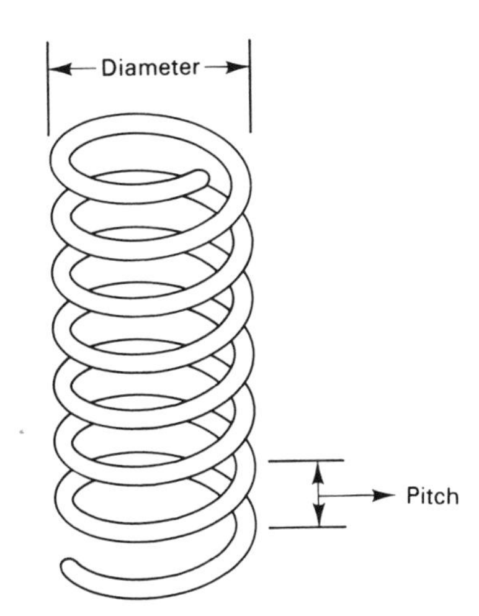

Figure 5-10 Spiral reinforcement.

Placing Reinforcing Steel

Because of concrete's low strength in tension, reinforcing steel is used to prevent concrete from failing or cracking because of tension forces. Some of the ways in which tension occurs in concrete members under load are illustrated in Figure 5-11. Notice how the concrete member deforms under load and the way in which reinforcing steel is placed to resist tension and shear. Reinforcing steel also reduces the cracking produced by the shrinkage of concrete due to temperature changes and hydration. It is important to realize that the *positioning of reinforcing steel in a concrete member is critical to the ability of the member to safely carry its design load.*

To prevent corrosion of reinforcing steel, it must be protected from water and corrosive elements by an adequate thickness of concrete cover. The minimum required cover is often specified by building codes. However, the American Concrete Institute (ACI) recommends the following minimum thickness of concrete cover be provided in any case:

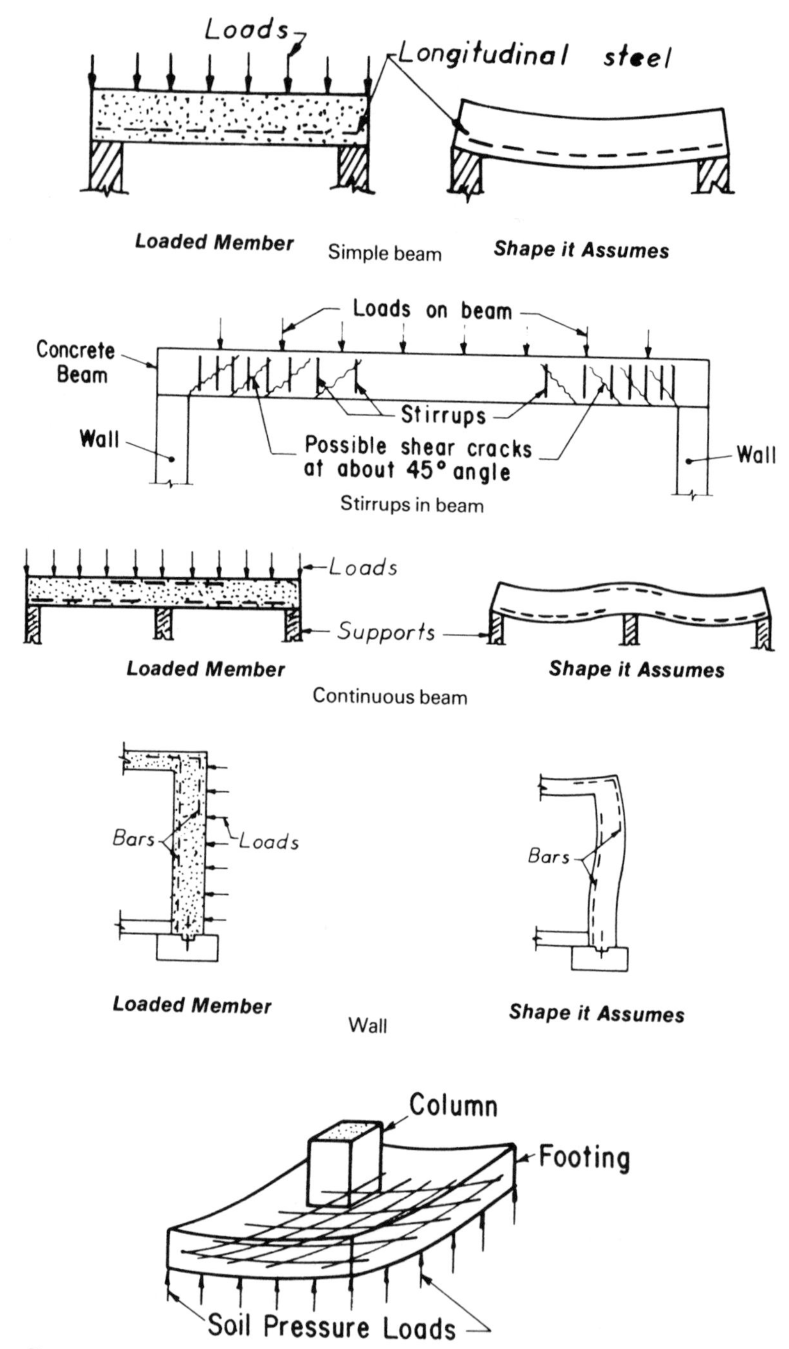

Figure 5-11 Use of reinforcing steel (courtesy Concrete Reinforcing Steel Institute).

- Concrete placed directly on the ground: 3 in. (7.6 cm)
- Concrete placed in forms but exposed to weather or the ground: 1 1/2 in. (3.8 cm)
- Beams, columns, and girders not exposed to weather or the ground: 1 1/2 in. (3.8 cm).
- Walls, slabs, and joists not exposed to weather or the ground: 3/4 in. (1.9 cm).

Sufficient space must be provided between reinforcing steel to allow concrete to completely fill the forms without creating cavities. It is recommended that the larger of the following minimum clear distances be provided between parallel reinforcing bars:

- For all members except columns: 1 in. (2.5 cm), 1 bar diameter, or 1 1/3 times the maximum aggregate size.
- For columns: 1 1/2 in. (3.8 cm), 1 1/2 bar diameters, or 1 1/2 times the maximum aggregate size.

In reality it is extremely difficult to place and maintain reinforcing steel in a precise position during concrete construction operations. Therefore, the Concrete Reinforcing Steel Institute (CRSI) has suggested the use of the following tolerances for steel placement:

- Lengthwise position of bar ends: sheared bars: ± 2 in. (5.1 cm); bars with hooked ends: ± 1/2 in. (1.3 cm).
- Horizontal spacing of bars in slabs and walls: ± 1 in. (2.5 cm).
- Spacing of outside top, bottom, and side bars in beams, joists, and slabs: ± 1/4 in. (0.64 cm).
- Distance between adjacent stirrups: ± 1 in. (2.5 cm).

Reinforcing bars may be spliced end to end by overlapping the bars and tying them together with tie wire, by arc welding, by mechanical butt connectors, or by the use of an exothermic (heat-generating) welding process such as thermite welding. Bars that cross should only be fastened together by wire ties. The minimum length of bar lap required for tie splices depends on a number of variables and must be specified by the designer or the building code.

Several items of reinforcing steel terminology should be understood. *Stirrups* (Figure 5-12) are U-shaped reinforcing rods used in beams, joists, and slabs. *Chairs* or *boosters* are wire bar supports used to maintain a specified clearance between a reinforcing rod and a concrete form or other surface. A *hooked bar* is a reinforcing bar having a 180° bend at the end. A reinforcing *cage* is a preassembled group of reinforcing steel inserted into a concrete form as an assembly.

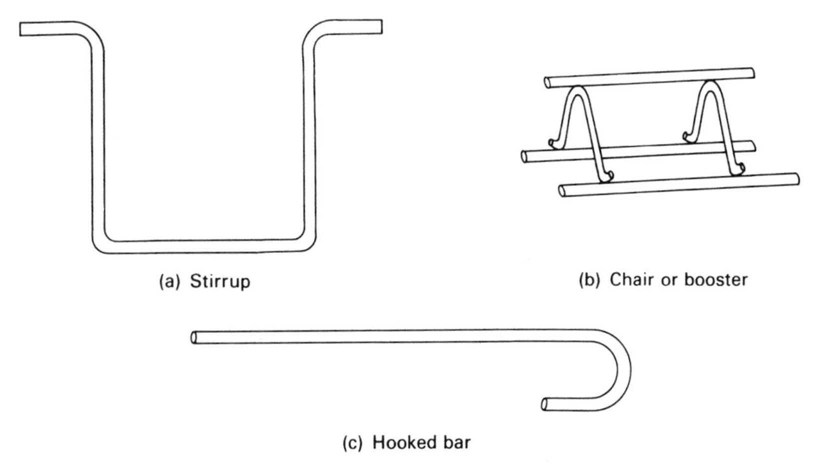

Figure 5-12 Reinforcing steel and accessories.

5-3 CONCRETE CONSTRUCTION PRACTICES

The full concrete construction process involves concrete batching, mixing, transporting, placing, consolidating, finishing, and curing. To obtain a concrete product that meets all design, safety, and appearance requirements, it is essential that each phase of the concrete construction process be carefully executed. Proper concrete batching and mixing procedures were presented in Chapter 2. Some of the basic principles and recommended practices involving the remaining phases of concrete construction are described below.

Transporting and Placing Concrete

Concrete may be transported from the mixer to its point of use in many ways. In urban areas, plastic concrete usually arrives at the construction site in a truck, or transit, mixer. However, for small jobs or remote sites, concrete may be mixed on site. The process of moving concrete into its final position (usually within forms) is called *placing* or *pouring*. In addition to the truck mixer, equipment used for transporting and placing concrete includes wheelbarrows, concrete buggies, chutes, conveyors, concrete buckets attached to cranes, and concrete pumps.

The two major hazards involved in concrete transporting and placing are partial hardening resulting from delays or high temperatures, and segregation resulting from improper handling. Under normal temperature conditions, concrete delivered in truck mixers should be discharged within 1 1/2 hours after the start of mixing and before the drum has revolved 300 times. To avoid segregation, concrete should not be allowed to fall freely more than 5 ft. (1.5 m) without using a downpipe (also called an "elephant trunk") or concrete ladder. A downpipe at least 2 ft. (0.6 m) long should be used at the discharge end of a concrete conveyor. When chuting concrete from a truck mixer, the slope of the chute must be steep enough to keep the chute clean but not steep enough to produce segregation. Vertical forms may also be filled by pumping concrete into the form from the bottom. This technique has the advantage of reducing air voids during concrete placement. As a result, little or no vibration is required. However, the formwork must be designed and constructed to withstand the additional concrete pressure involved in using this method of placement.

The surface onto which concrete is placed and the interior of all concrete forms must be properly prepared before pouring concrete. If concrete is to be placed directly onto a subgrade, moisten the subgrade or seal it with a plastic membrane to prevent the subgrade from absorbing water from the plastic concrete. Before concrete is placed on top of hardened concrete, the surface of the hardened concrete should be roughened and coated with grout or a thin layer of mortar to increase bonding between the old and new concrete.

Some recommended construction practices involving formwork and placing concrete into forms include:

- Carefully align and brace formwork before placing concrete.
- Tighten formwork joints to prevent loss of concrete paste, which will lead to voids in the finished concrete.
- Inspect the interior of forms prior to concrete placement. The forms should be clean, tight, and the interior coated with form oil or parting compound to permit form removal without damage to the concrete.
- Do not exceed the specified rate of pour, so that formwork design loads are not exceeded.

- Do not attempt to vibrate hardened concrete because this can cause form damage or rupture.
- When pumping concrete into vertical forms from the bottom, fill the form as rapidly as possible. The pressure developed in attempting to lift partially hardened concrete can rupture forms.

Consolidating Concrete

The process of removing air voids from concrete as it is placed is *consolidating* concrete. Since concrete consolidation is usually accomplished by vibration, this process is also called *vibrating* concrete. Concrete may be consolidated by hand rodding or spading. However, vibrating by immersion-type electric, hydraulic, or pneumatic vibrators is most commonly employed. While it is possible to vibrate the entire concrete form by attaching a form vibrator to the outside of the form, this procedure requires that the form be designed and constructed to withstand these stresses. As a result, this method of consolidation is rarely used outside of concrete prefabrication plants.

When vibrating concrete with hand-held vibrators, insert the vibrator vertically into the concrete and allow it to penetrate about 1 in. (2.5 cm) into the previously placed layer of concrete. Withdraw the vibrator and move it to another location when cement paste becomes visible around the top of the vibrator. Overvibration and using the vibrator to move concrete horizontally can cause segregation in the plastic concrete. However, vibration of previously placed concrete is not harmful as long as the concrete becomes plastic when vibrated.

Finishing and Curing Concrete

The process of bringing a concrete surface to its final elevation and surface texture is called *finishing*. The steps in this process include screeding, floating, troweling, and brooming or texturing. *Screeding* consists of striking off and bringing the concrete surface to the required elevation. *Floating* smoothes and compacts the concrete surface as well as embeds aggregate particles. Floating is accomplished with a wood or metal float when the concrete has begun to harden. It is usually begun when the pressure of a worker's foot makes only a slight impression in the concrete surface. When a smooth, dense surface is desired, *troweling* with a steel trowel follows floating. Power floats or trowels may be used to speed up the finishing process. Finally, *brooming* or *texturing* of the surface using a broom or texturing tool may be used to produce a skid-resistant surface.

Curing is the process of maintaining the proper moisture and temperature conditions in concrete for a period of time to allow cement hydration to proceed under favorable conditions. A curing method often employed involves spraying the concrete surface with a curing compound promptly after finishing. The curing compound seals the concrete surface to prevent the loss of moisture during curing. Other curing techniques often used include water sprays, ponding of water on the surface, and covering the concrete with wet burlap or straw.

Hot and Cold Weather Concreting

When temperatures exceed the optimum temperature of 50° to 60°F (10° to 15.5°C), the rate of hardening of concrete is greatly accelerated. As a result, the heat of hydration produced by the cement is also increased, further compounding the problem. A temperature of about 90°F (32°C) is considered the practical upper limit for normal concrete operations. Above this temperature, special hot weather procedures must be employed. Some techniques for cooling concrete include cooling the aggregate or mix-water before mixing, using Type IV (low-heat) cement, adding a retarder to the mix, or installing cooling coils in the concrete structure. For hot weather concreting, it is also recommended that the maximum time between mixing and pouring of concrete be reduced from the normal 1 1/2 hours to 1 hour or less. Weather conditions involving wind, low humidity, and high temperature may also lead to cracking of concrete surfaces because of drying shrinkage. The use of retarders, sun shades or covers, and water sprays help prevent or reduce such surface cracking.

Cold weather concreting presents similar but opposite problems. Concrete that is allowed to freeze within 24 hours after placing may suffer permanent damage. A minimum temperature of 50°F (10°C) is often specified for pouring concrete. This minimum temperature should be maintained for at least 3 days after the concrete is placed. The temperature of concrete can be raised by heating the aggregate or mix water, using Type III (high early strength) cement, or adding an accelerator. Constructing a heated enclosure around the concrete will allow a satisfactory temperature to be maintained during placing, consolidating, and curing. However, unvented heaters should not be used during the first 36 hours after placing to avoid producing a concrete surface that dusts after hardening. Concrete must be allowed to cool gradually when heating is stopped, or cracking may result.

5-4 SAFETY AND QUALITY CONTROL

Safety in Concrete Construction

Major failures of concrete construction usually occur during concrete placement, during form removal or reshoring, or after forms and temporary supports have been removed. In view of the many serious construction accidents that have occurred during concrete construction, every precaution must be taken to prevent such failures, some of which include:

- Ensure that the rate and location of concrete placement is controlled so that formwork design loads are not exceeded.
- Construct adequate formwork foundations on stable material. Mudsills, as illustrated in Figure 5-6, should be used under shoring that rests on the ground.
- Check nearby excavations and embankments to ensure that embankment failure does not lead to formwork collapse.
- Carefully brace all formwork. The bracing of shores for elevated slabs (horizontal formwork) is particularly critical. Consider horizontal loads produced by power buggies or other equipment that operate on the formwork.
- Monitor formwork during concrete placement to detect any movement of forms or bracing.
- Do not remove forms or supports until the concrete has reached the required strength. Use special care during cold weather when the concrete gains strength more slowly than usual.
- Carry out *reshoring* (the process of placing temporary shores under slabs as the forms are stripped) exactly as specified by the formwork designer. Strip and reshore only a limited area at one time. Do not allow construction loads on slabs being reshored.
- Promptly pull nails from forms after stripping to protect workers from puncture wounds.

Quality Control of Concrete Construction

Careful attention to quality control is required to obtain a concrete construction product that possesses the required strength, durability, and appearance. Poor quality control procedures and inadequate supervision produces inferior concrete. Some of the common deficiencies in concrete construction that have been observed by the U.S. Army Corps of Engineers in their worldwide operations include:

Concrete slabs on grade

- Saturation and damage to subgrade caused by water standing around foundation walls and/or inadequate storm drainage.
- Poor compaction of subgrade evidenced by slab settlement.
- Uneven surface finish.
- Inadequate curing.

Structural concrete

- Poor alignment of reinforcing steel and exceeding prescribed placement tolerances.
- Poor form alignment and inadequate form bracing. Often indicated by form bulging or spreading or poorly aligned structural members.
- Excessively honeycombed wall areas.
- Inadequate concrete consolidation.
- Obvious cold joints in walls.
- Belated form tie removal, form stripping, and patching.

The phases of proper concrete quality control include:

- Mix design.
- The quality of concrete materials.
- Batching and mixing procedures.
- Transporting, placing, consolidating, finishing, and curing procedures.
- Site testing of plastic and hardened concrete.

Items to be checked in the mix design phase include the type and gradation of aggregate, the type of cement, and the quantity of each component in the design mix. Aggregates should be tested for gradation, fines content, resistance to abrasion, moisture content, and organic impurities. Concrete batching and mixing procedures should be checked for the accuracy of batching and the use of proper mixing procedures. The procedures used for transporting, placing, consolidating, finishing, and curing must comply with all specifications and accepted construction practices.

Concrete delivered to the job site should be tested as prescribed by the specifications and the building code. Common tests on plastic concrete include the slump test for workability and tests for air and cement content. The concrete temperature should also be checked when engaged in hot-weather or cold-weather concreting. Strength testing of samples taken at the job site are usually performed after 7 and 28 days of curing. The cylinders used for samples to be compression-tested usually measure 6 in. (15.2 cm) in diameter by 12 in. (30.5 cm) high. Samples to be used for beam flexure testing are usually 6 in. (15.2 cm) square by 20 in. (50.8 cm) long. The procedure recommended by the American Concrete Institute for evaluating concrete compression test results (ACI 214) is widely used.

PROBLEMS

1. Name and briefly explain the principal requirements that concrete formwork must satisfy.
2. How is the pressure of the plastic concrete within a wall form resisted?
3. Give the approximate diameter of a No. 5 steel reinforcing bar.
4. Describe the welded wire fabric that is identified on the construction drawings as WWF 6x12-W16xW18.
5. Briefly explain how an immersion-type concrete vibrator should be used to consolidate concrete being placed in a wall form 4 ft. (1.2 m) high.
6. What items should be checked when inspecting concrete formwork prior to placing concrete?
7. When are major failures of concrete construction most likely to occur?
8. List at least five safety precautions that should be observed during concrete construction.
9. What tests are often required to be performed on plastic concrete delivered to the job site? Why are these tests needed?
10. Explain why the position (placement) of reinforcing steel within a concrete member is critical to the safe performance of the member.

REFERENCES

1. American Plywood Association. *APA Design/Construction Guide: Concrete Forming.* Tacoma, WA, 1984.
2. Portland Cement Association. *Design and Control of Concrete Mixtures.* 13th ed. Skokie, IL, 1988.
3. National Forest Products Association. *Design of Wood Formwork for Concrete Structures.* Wood Construction Data No. 3. Washington, DC, 1961.
4. Hurd, M. K. *Formwork for Concrete.* 5th ed. American Concrete Institute, Detroit, MI, 1981.
5. Concrete Reinforcing Steel Institute. *Manual of Standard Practice.* 24th ed. Chicago, IL, 1986.
6. Nunnally, S. W. *Construction Methods and Management.* 2nd ed. Englewood Cliffs, NJ: Prentice-Hall, 1987.
7. Concrete Reinforcing Steel Institute. *Placing Reinforcing Bars.* 5th ed. Chicago, IL, 1989.

chapter 6

Masonry Construction

6-1 CONCRETE BLOCK

Uses

Concrete masonry units include concrete block, concrete brick, and concrete tile. Of these, *concrete block* is most frequently used in residential and light construction, for foundation walls, piers, and exterior walls. However, it can also be used to construct interior walls and partitions, chimneys, privacy walls and fences, retaining walls, and similar structures. In this section we focus on its application to the construction of concrete block walls.

Types and Sizes

Concrete block is manufactured from a mixture of portland cement, aggregate, and water. In the manufacturing process, a no-slump concrete mixture is machine-molded into the desired shape and then steam-cured to accelerate the curing process. The principal types of concrete block are solid load-bearing block, hollow load-bearing block, and hollow non-load-bearing block. The specifications of the American Society for Testing and Materials governing hollow load-bearing concrete block (ASTM C90) provide for two grades (N and S), two types (I and II), and three weights (normal weight, medium weight, and lightweight). Grade N units are suitable for general use, and grade S units are limited to use above grade when protected from the weather. Type I (moisture-controlled) units are intended for use where shrinkage of block due to drying must be controlled. Type II units are non-moisture-controlled. By weight, blocks are classified as normal weight (also called ''heavyweight''), medium weight, or lightweight, depending on the type of aggregate employed in their manufacture.

Hollow load-bearing block is most commonly used in construction. Hollow block is defined as block having less than 75 percent of its horizontal cross sectional area made up of concrete. The empty core of typical hollow block makes up 30 to 50 percent of the total horizontal section. The average strength of hollow load-bearing block ranges from about 1000 psi (6.9 MPa) for regular block to over 2500 psi (17.2 MPa) for extra-high-strength block. A typical lightweight block weighs 25 to 35 lbs. (11.3 to 15.9 kg) per unit while a similar normal weight block might weigh 40 to 50 lbs. (18.1 to 22.7 kg) per unit. Medium weight block falls between these extremes. Although not widely used, *solid concrete block* having a core area of 25 percent or less is available. It is sometimes employed for special construction applications such as high bearing loads, exterior facings, and fire protection. *Hollow non-load-bearing block* is utilized for partition walls, curtain walls, and other nonstructural applications. Concrete block is also available with special surface facings (*prefaced* units). Ceramic glazed masonry units are often utilized for walls in hallways, restrooms, swimming pools, and similar areas. Prefaced block is also available with resin and portland cement surfaces.

The most common size of concrete block is the nominal 8 × 8 × 16 in. (20 × 20 × 40 cm) unit illustrated in Figure 2-2. This nominal size is achieved by combining the actual size of 7 5/8 × 7 5/8 × 15 5/8 in. (19 × 19 × 39 cm) with a 3/8 in. (1 cm) joint. This produces a unit that conforms to the basic 4 in. (10 cm) modular dimension employed in U.S. construction practice. Some other common styles and sizes of block are illustrated in Figure 2-2. Half-height and half-width (4 in. or 10 cm) block and half-length (8 in. or 20 cm) block are also available.

Building a Block Wall

Some of the steps in laying up a concrete block wall are illustrated in Figure 6-1. Since a vertical section of masonry one unit thick is called a *wythe* (see Figure 6-6), this wall is a single-wythe wall. The footing or other surface on which the first course (row) of block is to be laid must be clean and level as illustrated in (a). The mason may first lay out the initial course of block without mortar to check block spacing and alignment. A full bed of mortar is then spread and furrowed (divided into two rows) with the trowel to ensure a full bed of mortar under each face of the block.

Construction is started by erecting a building corner (called a "lead"—see Figure 6-9) as a guide for laying up the wall. Lay the initial block by pushing it down into the mortar bed until mortar squeezes out of the bed (bottom) joint. Several blocks on each side of the corner are then laid. Each block is laid by spreading mortar on the head (end) joint, then pushing the block down into the mortar bed and sideways against the previously placed block until mortar squeezes out of both joints. Check the alignment, level, and plumb of the block. Use the handle of the mason's trowel to tap block into proper position.

Construction of the corner continues until the block is four or five courses above the perimeter course. For the second and succeeding courses, place a bed of mortar on top of the underlying course. Place mortar on the head joint of a new block and push it downward and sideward into place (b). Mortar for the head joint may be placed on both the new and abutting block. Check for alignment (c) and level. As the corner is being raised, check the level (d), plumb (e), and block spacing. After the corner has been raised four or five courses, proceed to the next adjacent corner.

When filling in units between corners, a string line may be used to keep blocks in vertical and horizontal alignment. Joints are tooled (smoothed and compacted) for appearance and watertightness as the mortar begins to harden. A round jointing tool, which creates a concave joint, is most commonly used. The horizontal joints should be tooled first followed by the vertical joints.

During construction, the wall must be cleaned of mortar droppings. Wait until

(a) (b) (c) (d) (e)

Figure 6-1 Building a concrete block wall.

the droppings are almost dry, then remove them with a trowel. Any remaining mortar should be removed after the mortar has dried by rubbing it with a small piece of concrete block and brushing it. The residue from tooling should also be removed by brushing.

Bond Patterns

Some commonly used concrete block bond patterns are illustrated in Figure 6-2. The running bond is most often used for construction purposes as shown in Figure 6-1. However, other patterns such as the basket weave bond, the horizontal stack bond, and the square stack bond are frequently employed for architectural effect.

Reinforced Concrete Block Construction

Concrete block construction strengthened by reinforcing steel and grout placed in joints and cavities is *reinforced hollow-unit masonry* or *reinforced concrete masonry*. Reinforcement provides additional structural strength and helps prevent cracking. Building codes usually require that concrete block construction be rein-

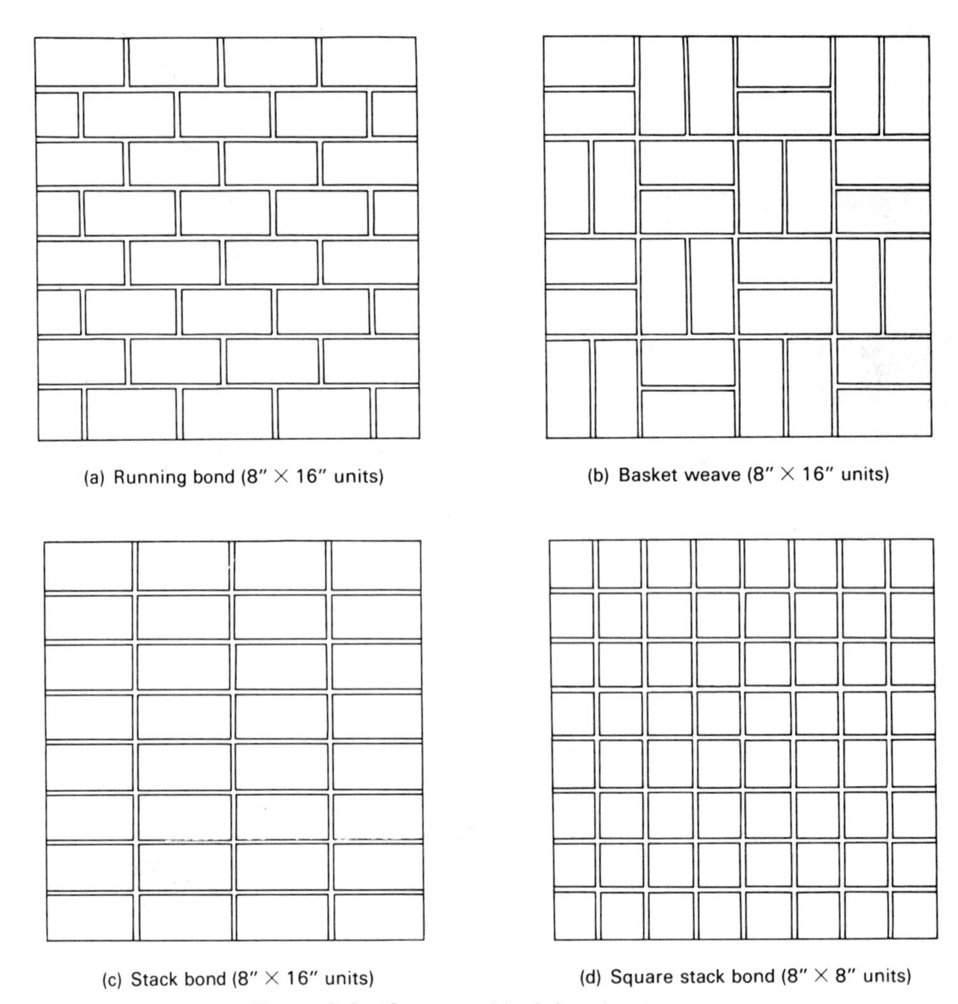

Figure 6-2 Concrete block bond patterns.

forced in locations subject to high winds or seismic activity. Reinforced concrete masonry walls have been used to construct high-rise residential and commercial buildings.

Vertical reinforcement is usually accomplished by placing reinforcing bars in some or all of the block cores and then filling these cores with grout, as illustrated in Figure 6-3. Horizontal reinforcement is provided by steel wire reinforcement placed in the horizontal joint mortar (Figure 6-4) or by horizontal reinforcing bars surrounded by grout (Figure 6-3). Figure 6-5 shows grout being pumped into the cores of a reinforced concrete block wall.

Two masonry wythes separated by an air space at least 2 in. (5 cm) wide and tied together by metal ties make up a masonry *cavity wall* [Figure 6-4(c)]. Advantages of cavity walls include better thermal and acoustical insulation, increased resistance to moisture penetration, and greater fire resistance. When constructing a cavity wall, building codes usually specify a maximum horizontal and vertical spacing of ties. The cavity of a cavity wall can be filled with grout (grouted masonry construction) or reinforcing steel and grout (reinforced grouted masonry construction) for greater structural strength.

Horizontal joint reinforcement (Figure 6-4) can be used to reinforce single wythe walls or to bond the two wythes of cavity and composite walls. Building codes usually require horizontal reinforcement of stack bond walls and walls subject to

Figure 6-3 Reinforced concrete block wall (courtesy Portland Cement Association).

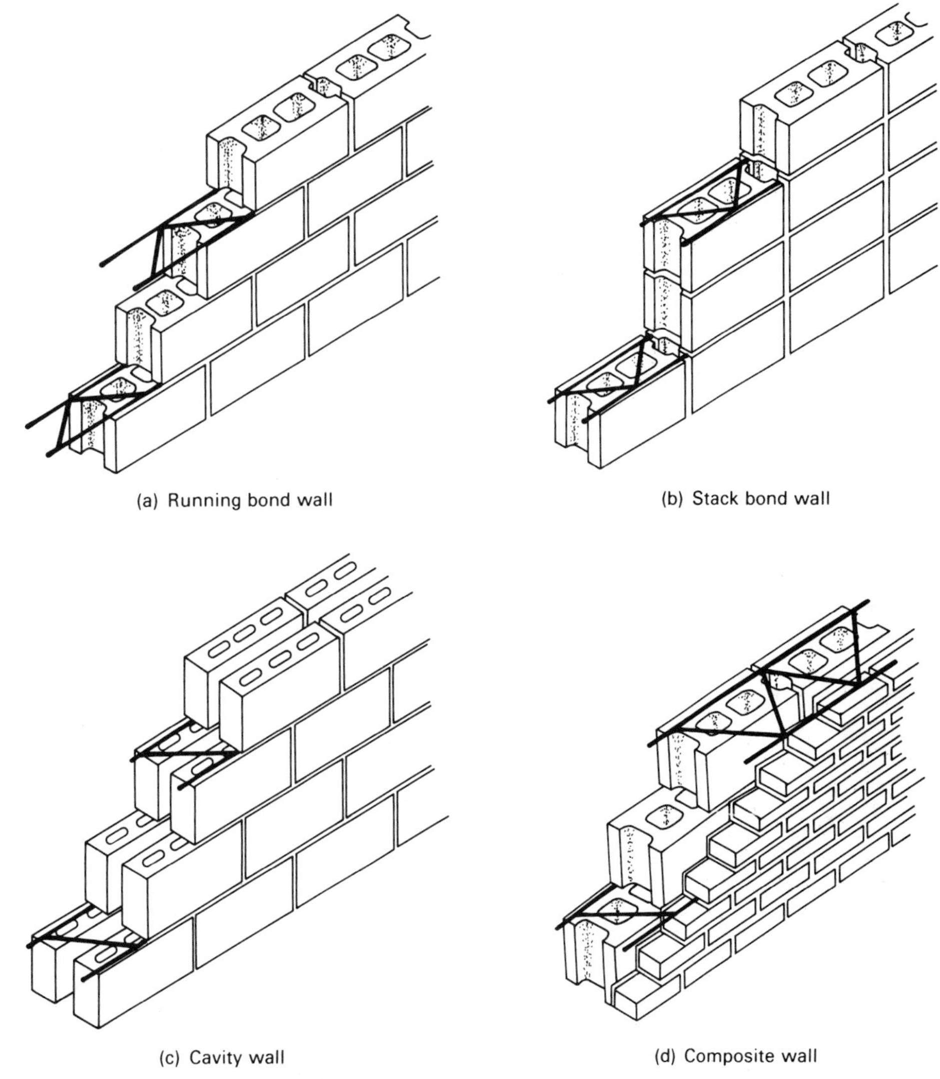

Figure 6-4 Horizontal joint reinforcement.

Figure 6-5 Pumping grout into reinforced concrete block wall [courtesy Portland Cement Association (Reference 5)].

high winds or seismic activity. They also specify a minimum amount of reinforcement to be used.

The minimum joint thickness for reinforced horizontal joints should be twice the diameter of the reinforcing bar or wire. Horizontal joint reinforcement should be protected by at least 5/8 in. (16 mm) of cover from an exposed face. Other reinforcing bars should have a minimum cover of 3/4 in. (19 mm) or one bar diameter when not exposed to soil or weather. When exposed to soil or weather, the minimum cover should be increased to 2 in. (51 mm).

6-2 BRICK

Uses

Clay brick, commonly simply called *brick,* is the principal form of clay masonry used in the United States. In residential construction, it is primarily used as a veneer (see Section 6-3). However, it is also used to construct solid masonry bearing walls, columns, chimneys, privacy walls, and pavements for residential and commercial buildings.

Some of the terms applied to masonry construction are illustrated by the brick wall in Figure 6-6. A masonry *course* is a horizontal row of brick. A *wythe* is a vertical section one brick thick. Horizontal mortar joints are known as *bed joints,* and vertical mortar joints are *head joints.* The mortar joint between two adjacent wythes is a *collar joint.* Individual bricks are also described by the position in which they are placed, as shown in Figure 6-7. Thus, a brick placed perpendicular to the plane of a wall is a *header,* while a *stretcher* lies parallel to the plane of the wall. However, note that a header brick placed vertically is a *rowlock* and a stretcher brick placed vertically is a *shiner.*

Types and Sizes

Brick is made from clay, shale, or a mixture of clay and shale. After being molded to the desired size and shape, the unit is baked in a kiln. A wide range of sizes and shapes is produced in the United States, and some common shapes are illustrated in Figure 6-8. The actual size of the standard nonmodular brick is 3 3/4 × 2 1/4

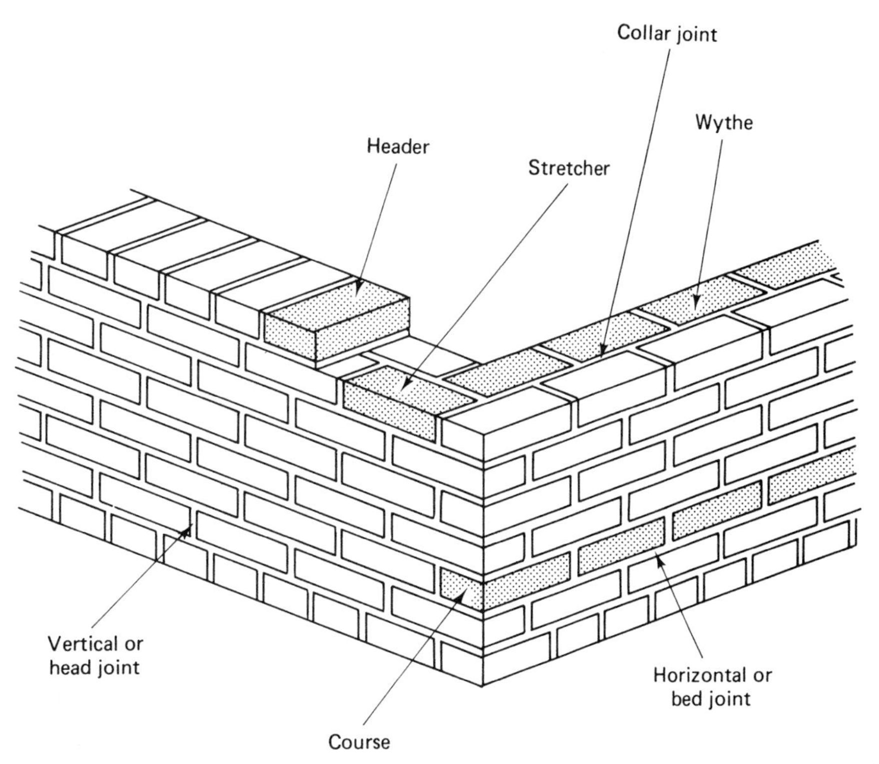

Figure 6-6 Elements of a brick wall (S. W. Nunnally, *Construction Methods and Management.* 2nd ed. © 1987, p. 333. Reproduced by permission of Prentice-Hall, Inc., Englewood Cliffs, NJ).

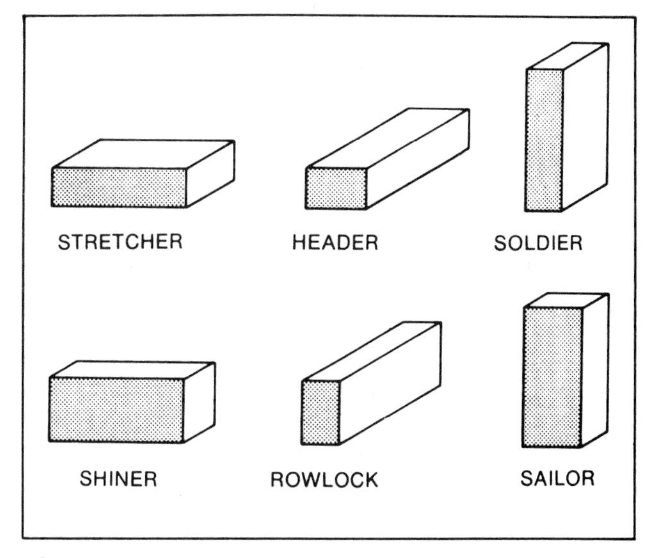

Figure 6-7 Terms applied to brick positions (courtesy Brick Institute of America).

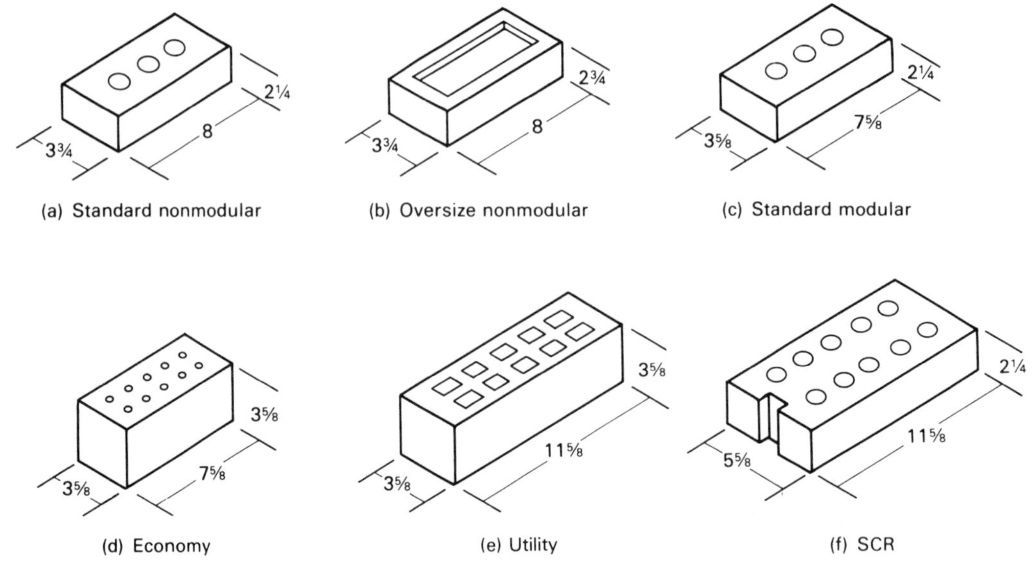

Figure 6-8 Typical brick sizes and shapes.

× 8 in. (95 × 57 × 203 mm) and that of the oversized nonmodular brick is 3 3/4 × 2 3/4 × 8 in. (95 × 70 × 203 mm). Modular brick is sized so that a brick plus its mortar joint produces a modular dimension; that is, a multiple of 4 in. (100 mm). Since a mortar joint of 3/8 in. (10 mm) is commonly used, actual modular brick sizes are as follows. The standard modular brick is 3 5/8 × 2 1/4 × 7 5/8 in. (92 × 57 × 194 mm); the economy brick is 3 5/8 × 3 5/8 × 7 5/8 in. (92 × 92 × 194 mm); the utility brick is 3 5/8 × 3 5/8 × 11 5/8 in. (92 × 92 × 295 mm); and the SCR brick is 5 5/8 × 2 1/4 × 11 5/8 in. (143 × 57 × 295 mm).

Individual bricks produced in the United States range in compressive strength from about 2,000 psi (13.8 MPa) to over 22,000 psi (151.7 MPa). However, the compressive strength of a brick assembly depends on the mortar strength and the quality of construction workmanship as well as individual brick strength. As a result, the allowable compressive strength for unreinforced brick masonry permitted by most building codes ranges from 75 psi (517 kPa) for 1,500 to 2,500 psi (10.3 to 17.2 MPa) brick and Type O mortar to 400 psi (2.8 MPa) for 8,000+ psi (55.2+ MPa) brick and Type M mortar. However, when the engineered brick masonry design and construction procedures of Reference 1 are followed, the allowable brick masonry compressive strength can go as high as 4,600 psi (31.7 MPa) for 14,000 psi (96.5 MPa) brick with Type M mortar and construction inspection by an architect or engineer.

Building a Brick Wall

The steps involved in building a brick wall are similar to those involved in building a concrete block wall, as described in Section 6-1. The wall must be constructed on a firm, level foundation such as a footing. As a first step, the mason usually lays out the initial course of brick without mortar to check brick spacing and alignment. If possible, the width of head joints is adjusted so that an integer number of whole brick may be used. However, it may be necessary to include one cut brick in the course to achieve the desired spacing.

A building corner (called a "lead") is first erected as a guide for laying up the wall. A bed of mortar large enough to support the first course of the lead is placed on the foundation. The initial corner brick is then placed and pushed down into the mortar bed until the bed joint reaches the desired thickness. Each succeeding brick is then mortared on the end and placed into position against the previously placed brick. When the first course of the lead has been completed, a mortar bed is placed for the next course. This process continues until the lead is raised to a height of six or seven courses. A completed lead is illustrated in Figure 6-9. During the construction of the lead, brick surfaces must be checked frequently for alignment, level, and plumb.

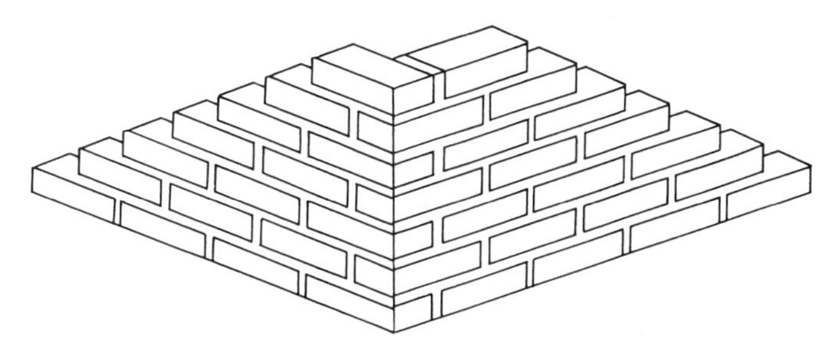

Figure 6-9 Completed brick corner lead.

After the first corner lead is completed, an adjacent corner lead is constructed. The units between two corner leads are then filled in, using a string line to keep brick in vertical and horizontal alignment. When the wall between two leads reaches the height of the leads, the corner leads are raised another six or seven courses, and the process is repeated.

Bond Patterns

Brick assemblies are bonded by mortar, by ties or reinforcing placed in mortar joints, and by the placement of brick to produce an interlocking action. As in concrete block construction, the pattern produced by the manner in which the brick is assembled is known as a *bond pattern* or simply *bond.* Some common bond patterns include running bond, English bond, common bond, Flemish bond, stack bond, and cross bond. These bond patterns are illustrated in Figure 6-10. When used, headers serve to provide structural bonding between wythes. Header courses occur in common bond at regular intervals; typically every fifth, sixth, or seventh course. Since stack bond provides no interlocking action, building codes usually require horizontal reinforcing to be used with this pattern. The cross bond, like English bond, consists of alternating courses of stretchers and headers. However, note that in cross bond, each stretcher course is offset by one-half brick from the stretcher course below it. This results in a series of interlocking crosses, as seen in the shaded area of the figure.

Hollow Walls

As noted in Section 6-1, masonry *cavity walls* consist of two wythes of masonry separated by an air space of at least 2 in. (5 cm) and tied together by metal ties or horizontal reinforcing. A brick cavity wall is illustrated in Figure 6-11. Masonry cavity walls often combine an exterior wythe of brick with an interior wythe of concrete block or other masonry.

A *masonry bonded hollow wall* is similar to a cavity wall except that the two wythes are bonded by masonry headers instead of metal ties. An example of this type of construction, called a *utility wall,* is illustrated in Figure 6-11. Masonry bonded hollow walls do not resist moisture penetration as well as cavity walls but perform satisfactorily when properly constructed. Recommended construction practice for this type of wall includes:

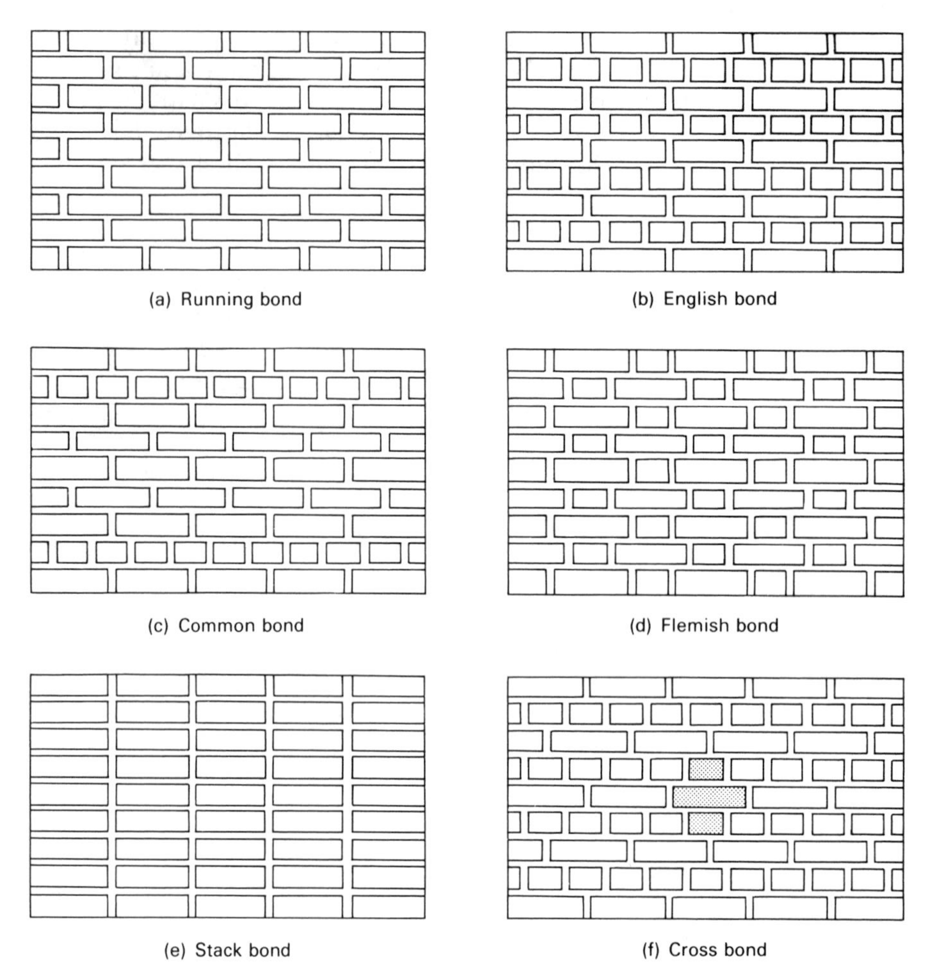

Figure 6-10 Common brick bond patterns.

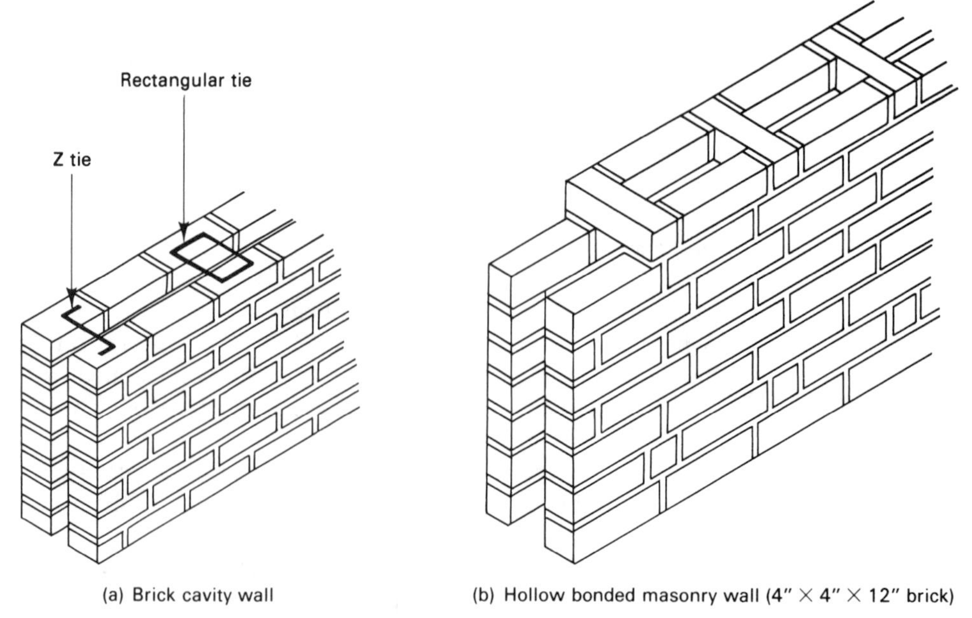

Figure 6-11 Brick cavity and masonry bonded hollow walls.

- Bonding every sixth course with alternating headers and stretchers as illustrated.
- Using type S mortar.
- Tooling exterior joints with a concave joint tool.
- Installing flashing (see Section 6-4) at the bottom of the wall.
- Providing weep (drainage) holes along the bottom exterior course at 24 in. (61 cm) intervals.

Reinforced Walls

When the space between wythes of a cavity wall is filled with grout, the form of construction is *grouted masonry*. The grout increases the moisture resistance, fire resistance, and structural strength of the wall. Type M or S mortar or a grout conforming to American Society for Testing and Materials (ASTM) C476 should be used as a grouting agent. The cavity between masonry bonded hollow walls may be filled with insulation, but building codes often do not allow grouting of this type of construction.

Brick construction in which reinforcing steel has been embedded is *reinforced brick masonry*. The most common method of constructing a reinforced brick masonry wall is by filling the cavity of a cavity wall with reinforcing and grout as illustrated in Figure 6-12. This type of construction is also known as *reinforced grouted masonry*. Requirements for the design and construction of reinforced brick masonry are presented in Reference 1.

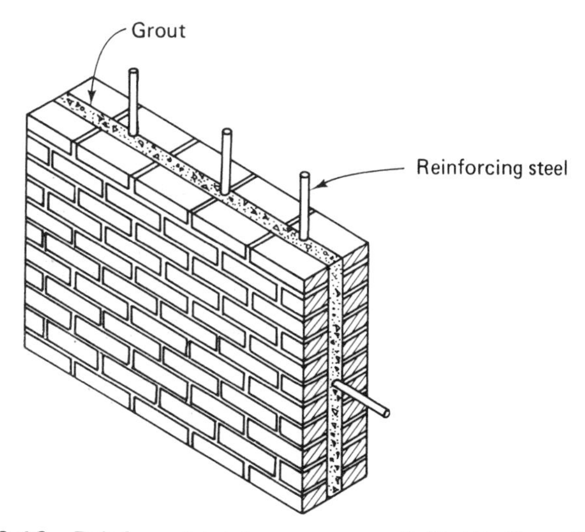

Figure 6-12 Reinforced brick masonry wall (S. W. Nunnally, *Construction Methods and Management*. 2nd ed. © 1987, p. 337. Reproduced by permission of Prentice-Hall, Inc., Englewood Cliffs, NJ).

6-3 VENEERED AND COMPOSITE WALLS

Brick is frequently used as an exterior surface for wood frame or concrete block construction. When an exterior wythe of masonry is provided for architectural effect but does not carry any load except its own weight, it is known as a *veneer*. A brick veneer over wood frame construction is illustrated in Figure 6-13. An air space of at least 1 in. (25 mm) should be provided between the brick and sheathing to

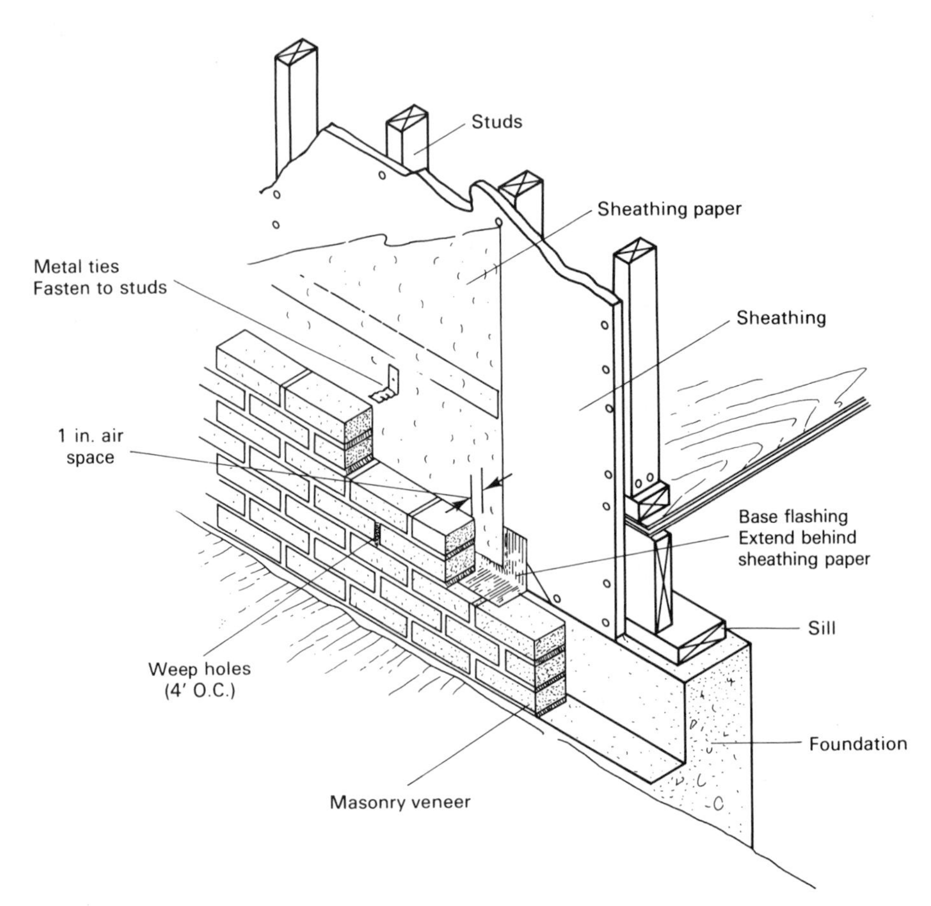

Figure 6-13 Brick veneer over wood frame (adapted from U. S. Department of Agriculture).

facilitate laying of the brick and to provide drainage for any moisture that may penetrate the brick veneer. Note the use of flashing and weep holes to control and drain any moisture that penetrates the veneer. Corrosion-resistant metal ties bond the veneer to the frame structure. Building codes usually specify the maximum spacing of metal ties but commonly require that one tie be provided for each 3.25 sq. ft. (0.3 m^2) of wall area. In areas subject to high seismic or wind loads, each tie should support not more than 2 sq. ft. (0.2 m^2) of veneer. Waterproof building paper or membrane is usually installed over the sheathing as shown. However, some building codes do not require the use of building paper over plywood sheathing when a 1 in. (25 mm) air space is provided (see Figure 9-9). The maximum spacing of weep holes is specified by building code but is usually 4 ft. (1.2 m).

An alternate method of bonding brick veneer to wood frame is permitted by most building codes. In this method, a paperbacked, welded wire fabric is attached directly to the wood frame, and the cavity between the frame and veneer is filled with grout; no weep holes or metal ties are used. This method of construction is also known as *reinforced masonry veneer.*

Brick veneer may be applied to concrete block walls in a manner similar to that described above. More often, however, such a wall is constructed as a composite wall, cavity wall, or reinforced masonry wall with an exterior wythe of brick and an interior wythe of concrete block. In composite wall construction, the two wythes are bonded together by masonry headers, metal ties, or horizontal reinforcement [Figure 6-4(d)].

Stone veneer is sometimes used over wood frame (Figure 6-14) or other con-

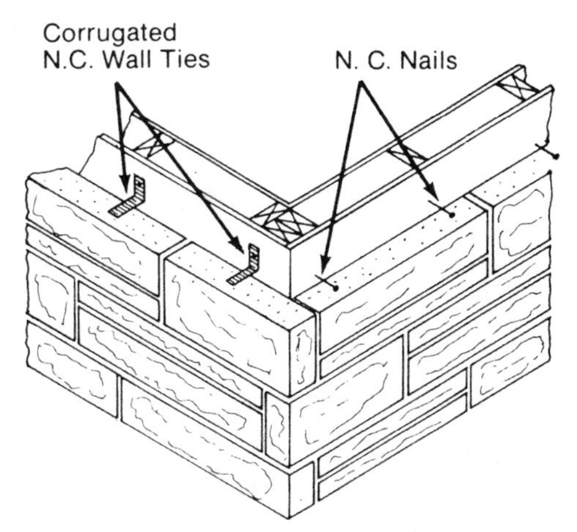

Stone Veneer Anchored to Wood Frame Backing

N.C. = noncorrosive

Figure 6-14 Stone veneer over wood frame (courtesy Indiana Limestone Institute of America, Inc.).

struction. The stone veneer is bonded to the structure's frame by metal ties as in brick veneer construction. Composite wall and cavity wall construction may also be used with an exterior wythe of stone and an interior wythe of concrete block or other masonry.

6-4 CONSTRUCTION PRACTICES

Bracing Masonry During Construction

Many masonry walls under construction have collapsed in high winds because of inadequate bracing. Proper bracing of a concrete block wall under construction is shown in Figure 6-15. The graph of Figure 6-16 shows the maximum safe unsupported height of some common masonry walls under construction as a function of peak wind velocity. Although this graph does not include any specific safety factor, it neglects all bonding provided by the partially set mortar. This should provide an adequate safety factor except in unusual cases. The peak wind pressure used in preparing the graph is based on the recommendations of ANSI A58.1-1972 for a wall of 600 sq. ft. (56 m^2) or greater. The peak wind velocity for a construction site can be obtained from local weather records. However, the velocity assumed should not be less than that prescribed by local building codes. Methods for calculating the maximum safe unsupported height for other masonry walls are presented in Reference 4.

Braces, forms, shores, and other supports must not be removed until the masonry assembly has gained adequate strength to support construction loads. Under normal conditions, no concentrated load should be applied to a masonry wall or column during the first three days after construction.

Structural Features

The masonry above windows, doors, and other openings may be supported by a masonry arch or a lintel, as illustrated in Figure 6-17. A *lintel* is a beam of steel, concrete, reinforced masonry, or other material, which spans a masonry opening. Reinforced masonry or reinforced concrete lintels are most commonly used today because of their appearance and ease of maintenance.

Figure 6-15 Bracing of a concrete block wall under construction. [courtesy Portland Cement Association (Reference 5)].

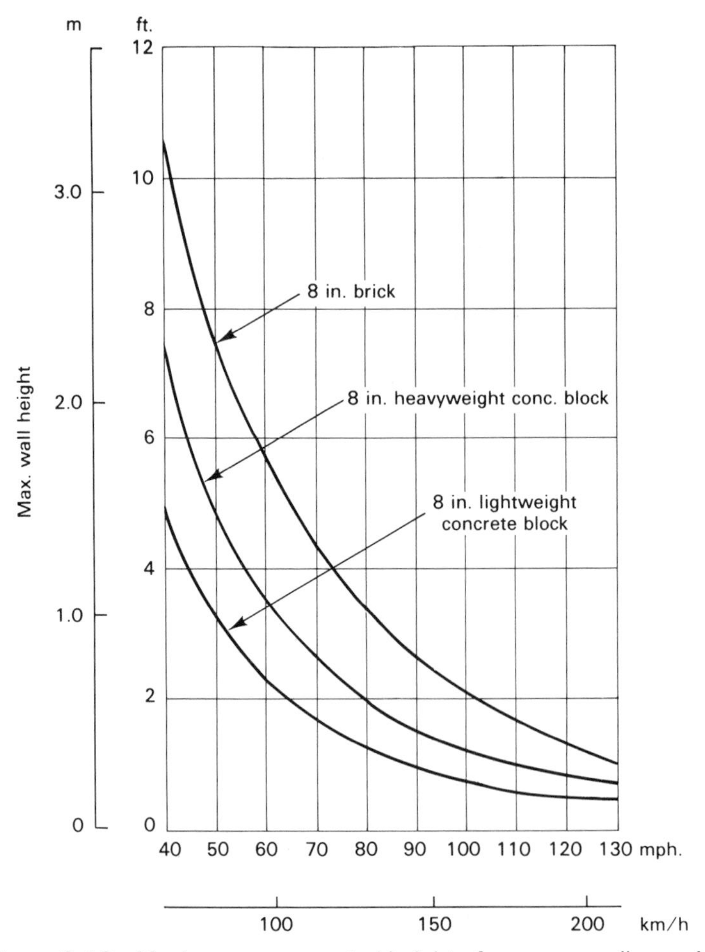

Figure 6-16 Maximum unsupported height of masonry wall vs. wind speed.

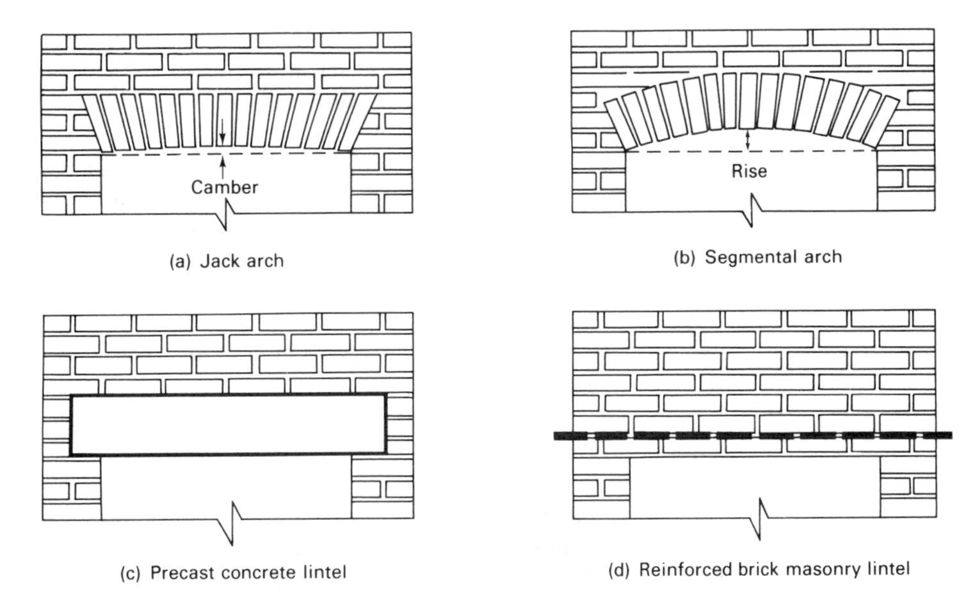

Figure 6-17 Support of masonry over wall openings. (S. W. Nunnally, *Construction Methods and Management.* 2nd ed. © 1987, p. 338. Reproduced by permission of Prentice-Hall, Inc., Englewood Cliffs, NJ).

A reinforced horizontal beam of masonry or concrete used to provide additional strength and reduce cracking of masonry walls is a *bond beam* (Figure 6-18). The requirements for bond beams are established by the designer and building code. Bond beams are commonly used at the foundation level, intermediate floor levels, and roof level.

Intersecting masonry walls must be adequately connected to each other for structural support. Intersecting walls may be rigidly connected or connected by a

Figure 6-18 Concrete bond beam.

flexible anchorage. Rigid connections commonly employ interlocking masonry units set in mortar or reinforcing steel set in mortar joints or grout. They must be carefully constructed in accordance with plans and specifications. *Expansion joints* are designed to allow differential movement of walls or other structural elements due to drying, temperature change, or foundation settlement. Expansion joints are usually connected by flexible metal ties or interlocking masonry placed without mortar in the joints. A flexible sealant must be applied to the exterior of the joint to prevent moisture penetration. *Control joints* are designed to control the location of cracks resulting from longitudinal wall movement due to shrinkage or temperature change. They normally consist of grooved or weakened wall sections, but they may incorporate flexible anchors or interlocking sliding joints like expansion joints. As a result, the terms *expansion joint* and *control joint* are frequently used interchangeably.

Effects of Weather

Masonry units must be protected from the weather during storage as described in Section 2-5. In addition, the top of exposed masonry under construction should be protected from rain. If masonry walls are allowed to become saturated during construction, they may take months to completely dry out and will experience more than normal drying shrinkage. Staining of brick surfaces by dissolved salts, called *efflorescence,* is greatly increased when the brick is allowed to become saturated during construction.

The effects of extreme temperatures on masonry construction are similar to those described in Chapter 5 for concrete construction. Masonry should not be laid on a frozen base, because mortar will not properly bond to a frozen surface. When masonry must be placed at temperatures below 40°F (5°C), heat the masonry materials before laying and keep the masonry structure warm during curing. Other recommendations for cold-weather masonry construction are provided in Reference 3. Extremely hot weather may require shading the work, cooling of masonry materials, and damp-curing for satisfactory results.

PROBLEMS

1. Explain the meaning of the following masonry terms:
   (a) Course
   (b) Header
   (c) Collar joint
   (d) Stretcher
   (e) Wythe
2. What type of concrete masonry unit is most commonly used in small building construction? Describe this unit.
3. **(a)** What is the minimum mortar joint thickness when 1/4 in. (6 mm) joint reinforcement is used?
   (b) What minimum mortar cover is required for this reinforcement on an exposed face?
4. **(a)** Describe the difference between a modular and a nonmodular brick.
   (b) What is the purpose of tooling mortar joints?
5. Describe the differences between a cavity wall and a masonry bonded hollow wall. Which type of wall is more moisture-resistant?
6. What is reinforced brick masonry and how is it constructed? When is it used?
7. Describe the major steps in the construction of a brick masonry wall.

8. **(a)** Explain the difference between a masonry veneer and a composite wall.
   (b) Describe how a brick veneer is placed on a wood frame house.
9. Find the maximum safe unsupported height of a heavyweight 8 in. (20 cm) concrete block wall under construction subject to a peak wind velocity of 80 mph. (128 km/h).
10. Explain why the top of a brick masonry wall under construction should be protected from rain.

REFERENCES

1. Brick Institute of America. *Building Code Requirements for Engineered Brick Masonry.* McLean, VA, 1969.
2. Gross, James G., and Harry C. Plummer. *Principles of Clay Masonry Construction.* Brick Institute of America, McLean, VA, 1973.
3. International Masonry Industry All-Weather Council. *Guide Specifications for Cold Weather Masonry Construction,* 1970.
4. Nunnally, S.W. *Construction Methods and Management.* 2nd ed. Englewood Cliffs, NJ: Prentice-Hall, 1987.
5. Randall, Frank A., Jr., and William C. Panarese. *Concrete Masonry Handbook.* Portland Cement Association, Skokie, IL, 1985.
6. Brick Institute of America. *Technical Notes on Brick Construction* Series. McLean, VA, various dates.

chapter 7

Frame Construction I

7-1 FRAMING SYSTEMS

Frame Construction

Frame construction employs studs, joists, and rafters to form the building frame. Some 90 percent of the homes built in the United States employ wood frame construction. Many low-rise multifamily residences as well as small commercial and industrial structures also utilize this type of construction. Some of the advantages of wood frame construction include low cost, ease of construction, and energy efficiency (when properly constructed and insulated). Similar framing systems employing metal framing also exist (see Chapter 11).

In conventional *wood frame construction,* framing members normally consist of 2 in. (5 cm) nominal thickness lumber, with studs spaced 16 or 24 in. (40 or 60 cm) on center. The minimum size and maximum spacing of framing members as well as minimum construction requirements are often specified in building codes. After completion of the building floor frame, wall frame, and roof frame, the walls and roof are sheathed, and roofing and exterior covering are applied (Chapter 9). Windows, doors, and exterior trim are installed (Chapter 10) to complete the building exterior.

The two principal forms of wood frame construction, platform frame construction and balloon frame construction, are described in following paragraphs. A third method of framing, *plank-and-beam* framing, is illustrated in Figure 7-1. Floor and roof plank [usually nominal 2 in. (50 mm) lumber] are supported by posts and beams spaced up to 8 ft. (2.4 m) apart. Vertical support may be provided by posts or conventional wall framing. When posts are used, framing similar to that described below is placed between posts to frame windows, doors, and other openings. Several advantages are claimed for plank-and-beam framing, the principal one that labor cost is reduced because of the smaller number of framing members used. A number of people find the architectural effect produced by this method of

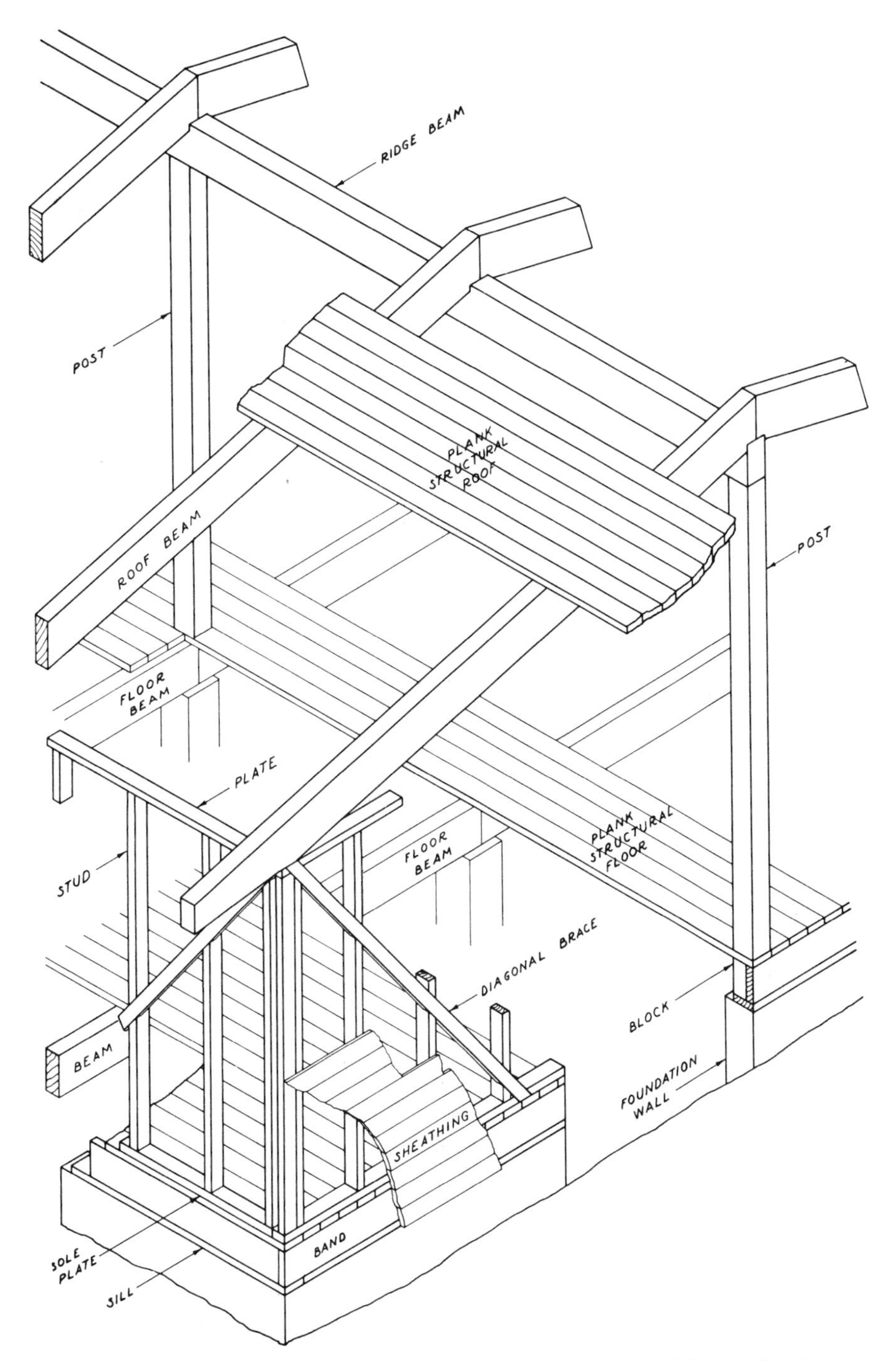

Figure 7-1 Plank-and-beam framing (courtesy National Forest Products Association, Washington, DC).

framing attractive. Recommended nailing requirements for framing are described in Section 8-5.

Platform Frame Construction

Platform frame construction is the most widely used wood framing system in the United States. In this method of framing, the subfloor of each story extends to the outside edge of the building frame and forms a platform on which to erect the walls

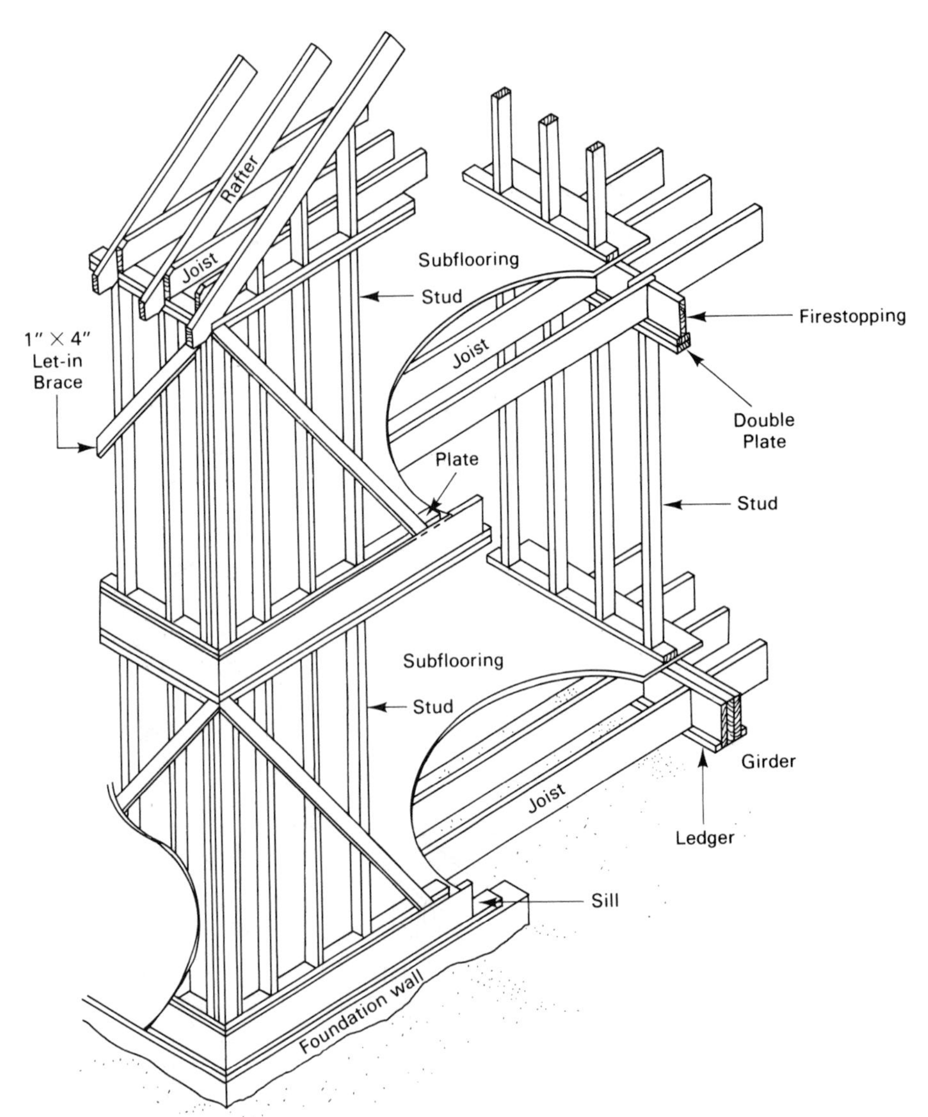

Figure 7-2 Platform frame construction (courtesy National Forest Products Association, Washington, DC).

of that story (Figure 7-2). Note in the figure that the bottom plate of each story's walls rests on the joists and subfloor (platform) for that story. This method of construction lends itself well to prefabricated construction techniques. Walls, for example, are often preassembled and tilted into place on the floor platform.

Balloon Frame Construction

Balloon frame construction (Figure 7-3) is primarily employed in two-story wood frame buildings whose exterior is to be covered with stucco or masonry veneer. As you see in the figure, the exterior wall studs extend all the way from the foundation sill to the top of the second floor walls. As a result, the possible differential movement between the building frame and the exterior covering is reduced, resulting in less cracking of the veneer. Because there is no platform for the second story walls to rest on, second floor joists are supported by strips ("ribbons") notched (or "let in") into the exterior wall studs.

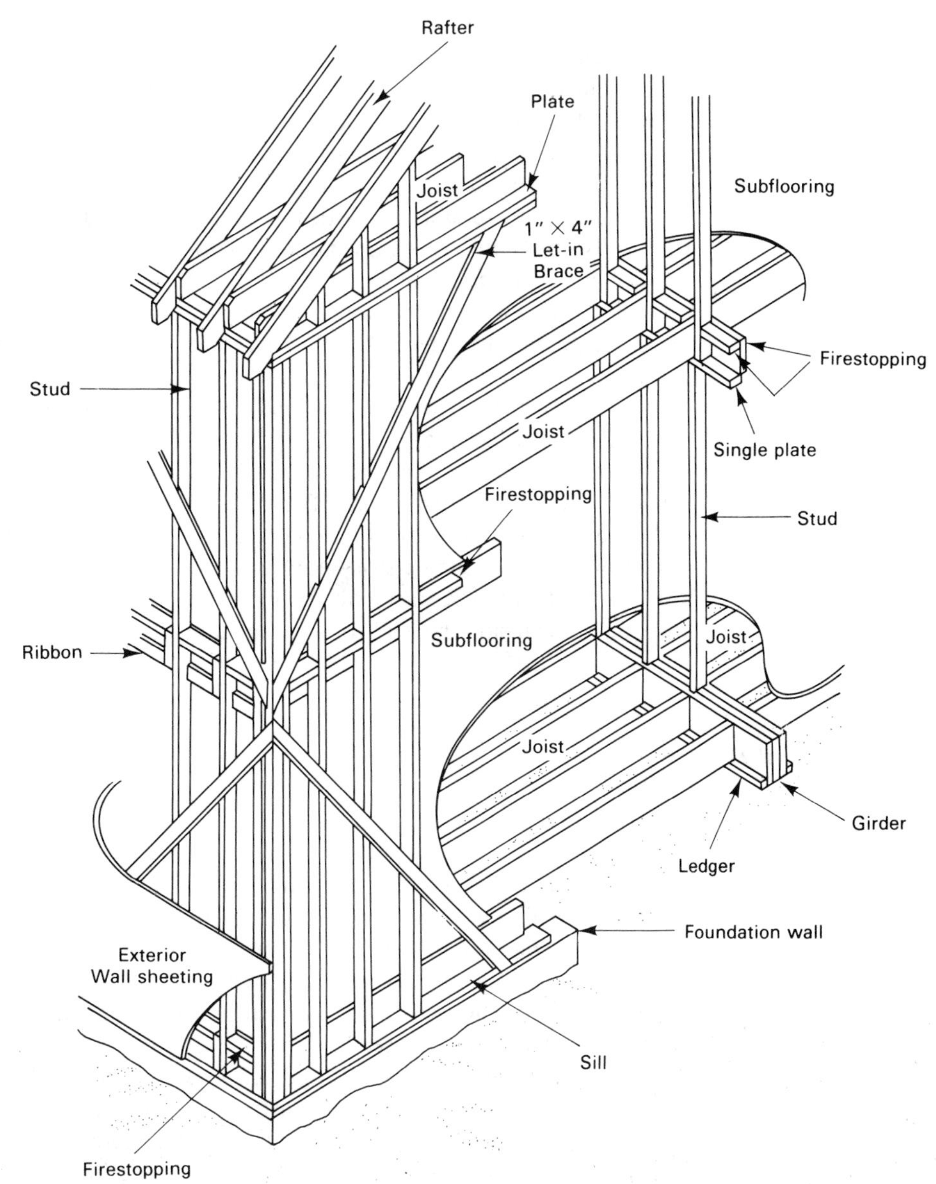

Figure 7-3 Balloon frame construction (courtesy National Forest Products Association, Washington, DC).

Modular Dimensions

In the United States, construction lumber, except for studs, is normally available in lengths of 10 ft. (3 m) to 20 ft. (6 m) in 2-ft. (60 cm) increments. Stud lumber is available in 8 ft. (2.4 m) lengths. The accepted length tolerance is −0 in. to +3 in. (0 to +76 mm). Actual length tolerance is usually at least +1/2 in. (13 mm) to allow for cutting into shorter lengths. As a result, significant savings in lumber cost can be realized by the use of modular building dimensions. Such modular dimensioning greatly reduces the lumber waste and cost of cutting associated with fitting framing members into the required dimensions. Studies conducted by the National Association of Home Builders (NAHB) Research Foundation have found that savings as much as 9 percent in material and labor cost for a house were achieved by the use of modular dimensioning alone (Reference 5, Numbers 1 to 4).

In applying modular dimensioning, the National Forest Products Association recommends a basic module of 4 in. [When using metric dimensions, this corresponds closely to a 100 mm module (actually 3.94 in.). The metric equivalents presented in the following discussion are not exact conversions but are based on this 100 mm module.] Larger dimensions should use increments of 16, 24, and 48 in. (400, 600, and 1200 mm), with a major module of 48 in. (1200 mm) and a minor module of 24 in. (600 mm). With the commonly used 16 in. (400 mm) spacing of wall studs, joists, and rafters, window and door spacing should also use this module. Overall building dimensions should be based on the 48 in. (1200 mm) module where possible; otherwise, on the 24 in. (600 mm) module. For maximum economy in joist material, the building depth should be a multiple of 4 ft. (1200 mm); i.e., 24 ft., 28 ft., or 32 ft. (7.2, 8.4, or 9.6 m). Cost savings in joist material have amounted to as much as 17 percent when a building's exterior dimensions were converted to this 4 ft. (1200 mm) module.

It has also been found that the use of 24 in. (600 mm) center-to-center spacing for wall studs, floor joists, and rafters further reduces framing costs. Most building codes now permit 24 in. (600 mm) center-to-center spacing of 2 × 4 in. (50 × 100 mm) wall studs for single-story homes. However, the use of 2 × 6 in. (50 × 150 mm) studs at this spacing facilitates increased energy efficiency of the building by creating space inside the wall for additional insulation (5 1/2 in. vs. 3 1/2 in. or 140 mm vs. 89 mm).

7-2 THE FLOOR FRAME

Description

The floor frame of a wood frame building is that portion of the structure above the foundation wall and below the flooring. It normally consists of sills, beams or girders, posts or piers, and subfloor. Some of the methods of constructing a typical floor frame for platform frame construction are illustrated in Figure 7-4. In platform frame construction, joists support all floor loads including exterior walls and interior partitions. Joists are, in turn, supported by sills resting on the foundation walls and by beams or girders supported by foundation walls, piers, or posts.

Note that in balloon frame construction (Figure 7-3), the ground floor joists do not support the exterior walls because the exterior wall studs rest directly on the sills. In slab-on-grade construction, described in Section 4-2, there is no floor frame for the ground floor because the exterior walls rest directly on the foundation, and the concrete slab acts as a subfloor.

The construction procedures described in this section reflect generally accepted construction practice. However, the requirements of the plans and specifications and of the local building code must always be observed.

Sills, Beams, and Girders

As shown in Figure 7-4, the exterior ends of floor joists rest on *sills* (also called "sill plates") while interior joist support is provided by beams or girders. The sill is usually a nominal 2 in. (5 cm) thick member anchored to the foundation wall. Anchors should consist of 1/2 in. (13 mm) diameter bolts placed 6 to 8 ft. (1.8 to 2.4 m) on center with at least two bolts in each sill piece. Anchor bolts should be embedded 7 in. (18 cm) in concrete walls or 15 in. (38 cm) in masonry unit walls. Approved light metal sill plate anchors may be substituted for anchor bolts. A sill sealer as shown is placed between the sill and foundation wall to reduce air infiltration.

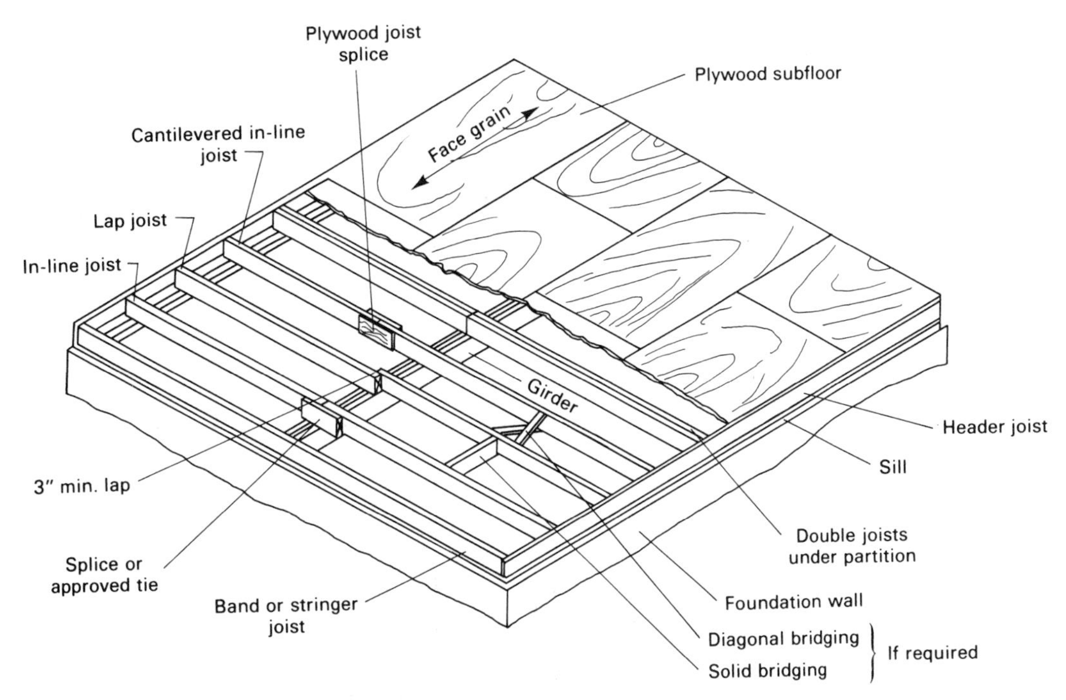

Figure 7-4 Methods of constructing a floor platform.

A *beam* is a horizontal member that provides major structural support. A *girder,* frequently a built-up member, performs a similar function. Thus, the terms beam and girder are frequently used interchangeably. Beams and girders supporting joists may consist of glued laminated timber (glulam); solid timber; built-up lumber; trusses of wood, wood and plywood, steel, or other material; or rolled sections of steel or aluminum.

Built-up wood girders (Figure 7-5) are usually fabricated from nominal 2 in. (5 cm) lumber nailed together. Notice that the individual pieces of lumber making up the girder are nailed together at the girder ends and at each splice, as well as near the top and bottom edges. For a built-up girder composed of three pieces of lumber as illustrated, each outside piece is nailed to the center piece. The end joints of individual pieces of lumber making up the girder are staggered and should fall over a support. However, if the girder is continuous over three or more supports, joints may fall either over a support or at a location one-quarter of the distance between supports.

The ends of beams and girders may be supported by placing them into notches in the foundation walls as illustrated in Figure 7-5. In this case, the top of the girder should be flush with the top of the sill. For protection against decay and insects, a 1/2 in. (13 mm) air space should be provided between the girder and the concrete on the end and sides of the girder unless pressure-treated wood is used. Alternatively, girders may rest on top of the sill with the girder top level with the joist tops. In this case, the joists are framed into the side of the girder. The minimum bearing of the girder on supports should be 4 in. (10 cm). If used, metal bearing plates must be of adequate size and strength to safely distribute the design load at the support. Note that girders should have a minimum clearance of 12 in. (30 cm) above the ground unless treated wood is used.

The maximum girder span, as well as the minimum size of supporting columns and footings, required by one building code is shown in Table 7-1.

TABLE 7-1

Allowable Span for Girders and Required Size of Columns and Footings to Support Roofs, Interior Bearing Partitions and Floors*

Size of girder required		Spacing of girder[2] "S"	Type of loading[3]			Size of columns required[4]		Size of plain concrete footing required[4]
Wood[1]			*A*	*B*	*C*	*Steel*	*Wood*	
4″ × 12″	6″ × 10″	10′	5′-6″	—	—	3″ Steel pipe[5]	4″ × 4″	2′ × 2′ × 8″
		15′	4′-0″	—	—			
		20′	—	—	—			
	6″ × 12″	10′	8′-6″	5′-0″	—			
		15′	6′-0″	4′-0″	—			
		20′	4′-6″	—	—			
		10′	12′-0″	9′-0″	8′-0″		6″ × 6″	4′ × 4′ × 16″[6]
		15′	10′-0″	8′-0″	7′-0″			
		20′	8′-0″	7′-0″	6′-0″			
		10′	16′-0″	12′-6″	11′-0″		8″ × 8″	4′-3″ × 4′-3″ × 17″[6]
		15′	13′-6″	10′-6″	10′-0″			
		20′	12′-0″	9′-6″	8′-0″			
		10′	20′-0″	16′-0″	13′-6″			
		15′	17′-0″	13′-6″	11′-6″			
		20′	15′-0″	12′-0″	10′-0″			

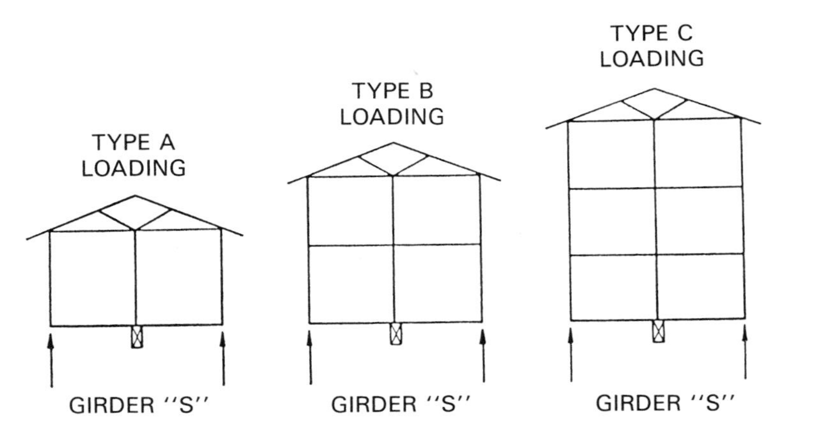

*Reproduced from the 1986 edition of the *CABO One and Two Family Dwelling Code,* Copyright 1986, and 1987 Amendments, Copyright 1987, with permission of the publishers—Building Officials and Code Administrators International, International Conference of Building Officials and Southern Building Code Congress International, Inc.

[1]Spans for wood girders are based on No. 2 grade lumber. No. 3 grade may be used with appropriate design.

[2]The spacing ''S'' is the tributary load in the girder. It is found by adding the unsupported spans of the floor joists on each side which are supported by the girder and dividing by 2.

[3]Figures under ''type of loading'' columns are the allowable girder spans.

[4]Required size of columns is based on girder support from two sides. Size of footing is based on allowable soil pressure of 2,000 pounds per square foot.

[5]Standard weight.

[6]Footing thickness is based on the use of plain concrete with a specified compressive strength of not less than 2,500 pounds per square inch at 28 days. If approved, the footing thickness may be reduced based on an engineering design utilizing higher-strength concrete and/or reinforcement.

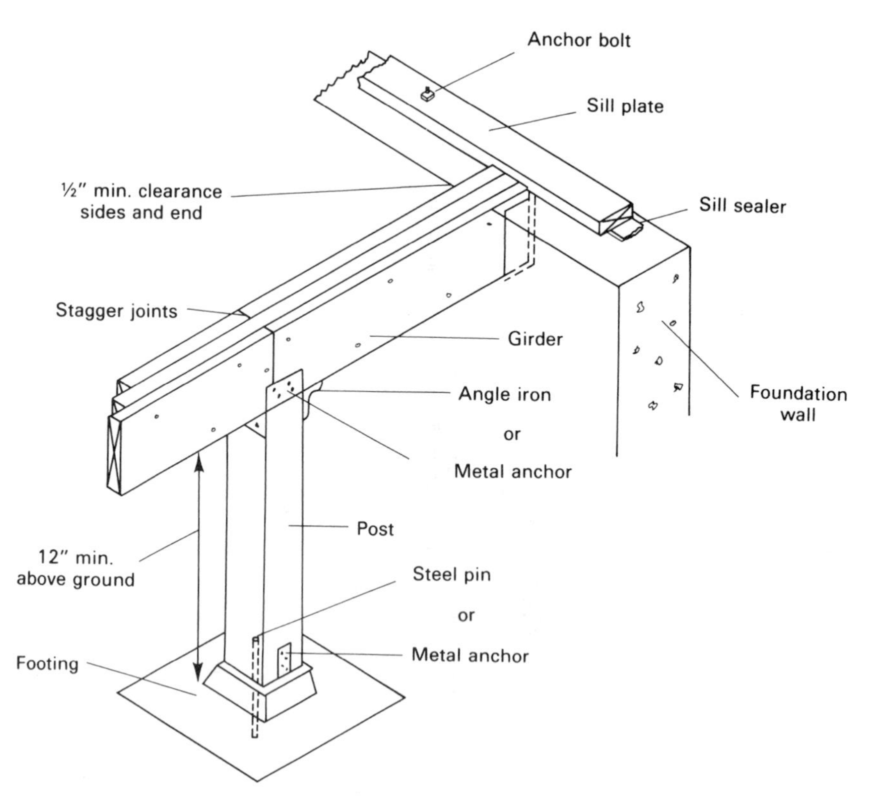

Figure 7-5 Built-up wood floor girder (adapted from U. S. Department of Agriculture).

Posts, Columns, and Piers

Intermediate support for beams and girders is provided by posts, columns, or piers. Wood *posts,* as illustrated in Figure 7-5, are often used. Note that the post is anchored to the girder by an angle iron and lag screws or by a light metal framing anchor nailed in place. At the bottom end, the post is anchored to the footing by a steel pin or metal anchor. Steel *columns* may also be used. When used, they are usually anchored to the girder and to the footing by bolts.

Piers of concrete or masonry may also be used to provide intermediate support of beams or girders. Girders should be anchored to piers by 1/2 in. (13 mm) diameter bolts embedded 6 in. (15 cm) in concrete or 15 in. (38 cm) in masonry. Concrete block piers should have the cores filled with grout if the pier height is more than four times the pier's least dimension. Ungrouted concrete block piers should be capped with 4 in. (10 cm) of solid masonry or concrete. Alternatively, the cores of the top course of hollow block may be filled with grout. The unsupported height of nonreinforced concrete piers, solid masonry piers, and grouted concrete block piers should not exceed ten times the pier's least dimension.

Floor Joists

Ground floor *joists* are supported by sills and girders, while upper floor joists are supported by wall studs. The maximum allowable span of Southern Pine floor joists supporting the usual floor live load of 40 lbs./sq. ft. (1.9 kPa) are given in Table 7-2. Similar span tables for other woods and floor loads can be found in Reference 8.

The methods by which joists are framed into sills are shown in Figure 7-6. Lateral support of joists at each end is required. In platform frame construction,

TABLE 7-2

Maximum Span for Southern Pine Floor Joists (ft-in) (Courtesy Southern Forest Products Association)

Size in.	Spacing in. o.c.	Dense Sel Str KD and No. 1 Dense KD	Dense Sel Str, Sel Str KD, No. 1 Dense and No. 1 KD	Sel Str, No. 1 and No. 2 Dense KD	No. 2 Dense, No. 2 KD and No. 2	No. 3 Dense KD	No. 3 Dense	No. 3 KD	No. 3
2 × 5	12.0	9-3	9-1	8-11	8-9	**8-3**	**8-0**	**7-8**	**7-4**
	13.7	8-11	8-9	8-7	8-5	**7-9**	**7-6**	**7-2**	**6-11**
	16.0	8-5	8-3	8-2	8-0	**7-2**	**6-11**	**6-7**	**6-5**
	19.2	7-11	7-10	7-8	7-6	**6-6**	**6-4**	**6-0**	**5-10**
	24.0	7-4	7-3	7-1	7-0[1]	**5-10**	**5-8**	**5-5**	**5-3**
2 × 6	12.0	11-4	11-2	10-11	10-9	**10-1**	**9-9**	**9-4**	**9-0**
	13.7	10-10	10-8	10-6	10-3	**9-5**	**9-2**	**8-9**	**8-5**
	16.0	10-4	10-2	9-11	9-9	**8-9**	**8-6**	**8-1**	**7-10**
	19.2	9-8	9-6	9-4	9-2	**8-0**	**7-9**	**7-4**	**7-1**
	24.0	9-0	8-10	8-8	8-6[1]	**7-1**	**6-11**	**6-7**	**6-4**
2 × 8	12.0	15-0	14-8	14-5	14-2	**13-3**	**12-11**	**12-4**	**11-11**
	13.7	14-4	14-1	13-10	13-6	**12-5**	**12-1**	**11-6**	**11-1**
	16.0	13-7	13-4	13-1	12-10	**11-6**	**11-2**	**10-8**	**10-3**
	19.2	12-10	12-7	12-4	12-1	**10-6**	**10-2**	**9-9**	**9-5**
	24.0	11-11	11-8	11-5	11-3[1]	**9-5**	**9-1**	**8-8**	**8-5**
2 × 10	12.0	19-1	18-9	18-5	18-0	**16-11**	**16-5**	**15-8**	**15-2**
	13.7	18-3	17-11	17-7	17-3	**15-10**	**15-5**	**14-8**	**14-2**
	16.0	17-4	17-0	16-9	16-5	**14-8**	**14-3**	**13-7**	**13-1**
	19.2	16-4	16-0	15-9	15-5	**13-5**	**13-0**	**12-5**	**12-0**
	24.0	15-2	14-11	14-7	14-4[1]	**12-0**	**11-8**	**11-1**	**10-9**
2 × 12	12.0	23-3	22-10	22-5	21-11	**20-7**	**20-0**	**19-1**	**18-5**
	13.7	22-3	21-10	21-5	21-0	**19-3**	**18-9**	**17-10**	**17-3**
	16.0	21-1	20-9	20-4	19-11	**17-10**	**17-4**	**16-6**	**16-0**
	19.2	19-10	19-6	19-2	18-9	**16-3**	**15-10**	**15-1**	**14-7**
	24.0	18-5	18-1	17-9	17-5[1]	**14-7**	**14-2**	**13-6**	**13-0**

Floor joists—40 psf live load. All rooms except sleeping rooms and attic floors. (Spans shown in light face type are based on a deflection limitation of $l/360$. Spans shown in bold face type are limited by the recommended extreme fiber stress in bending value of the grade and includes a 10 psf dead load.)

[1]The span for No. 2 grade, 24 inches o.c. spacing is: 2×5, 6-10; 2×6, 8-4; 2×8, 11-0; 2×10, 14-0; 2×12, 17-1.

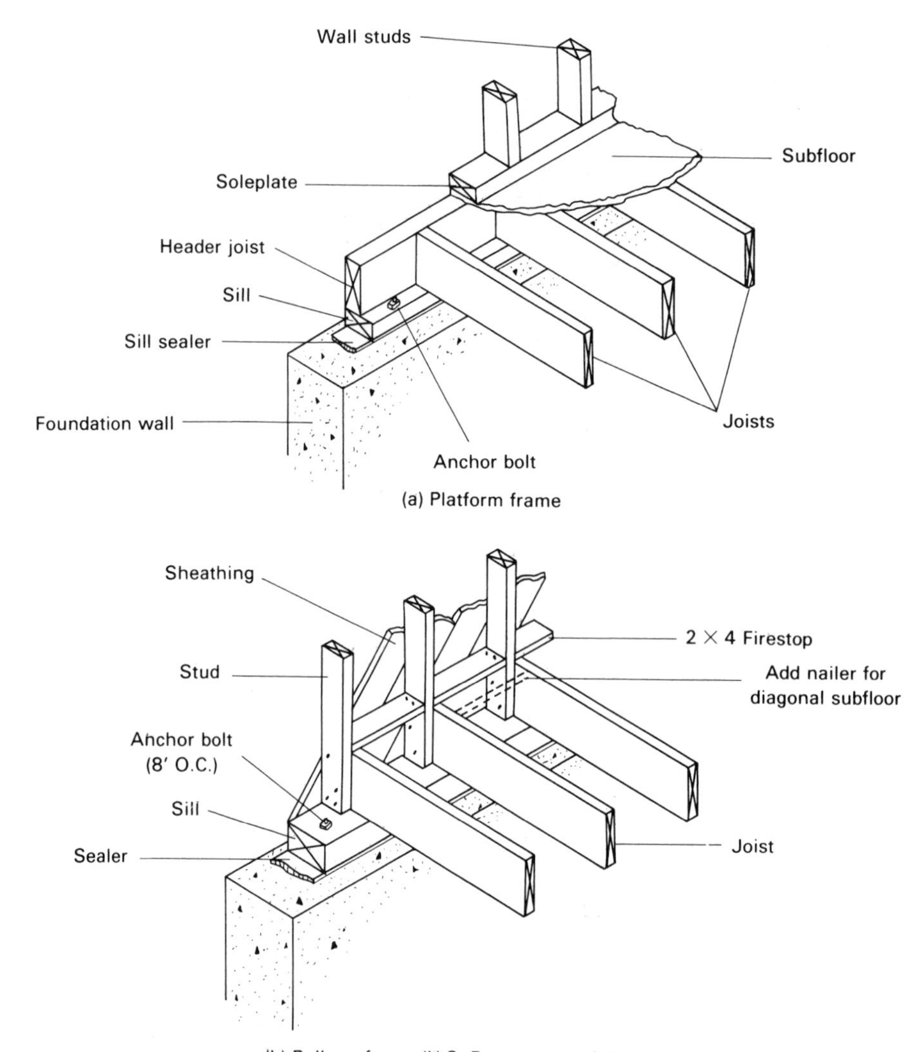

Figure 7-6 Joist framing at sills.

lateral support at the sill is normally provided by nailing the ends of joists to a *header joist*. Exterior floor joists parallel to the line of joists are *stringer joists*. However, all exterior joists are sometimes referred to as *band joists* or *rim joists*. Joists are toenailed to supporting sills and girders. If the joist depth-to-thickness ratio exceeds six, intermediate *bridging* between joists must be provided at intervals not exceeding 8 ft. (2.4 m). Bridging (see Figure 7-4) may consist of solid nominal 2 in. (5 cm) wood, diagonal wood strips, or prefabricated metal braces nailed between adjacent joists. Alternatively, 1 × 3 in. (25 × 76 mm) wood strips may be nailed across the bottom joist edges. The bearing of joist ends on supports should not be less than 1 1/2 in. (3.8 cm) on wood or metal or 3 in. (7.6 cm) on masonry or concrete. Joists should have a minimum clearance of 18 in. (46 cm) above the ground unless treated wood is used.

In balloon frame construction, the ground floor exterior joist ends also rest on sills. Joist ends are nailed to the studs as shown. However, notice that solid blocking between studs or joist ends is also required to act as a firestop [Figure 7-6(b)]. For upper floors, the exterior ends of floor joists are supported by 1 × 4 in. (25 × 102 mm) ribbon strips notched ("let in") into the studs (Figure 7-7), and the joists are nailed to the studs.

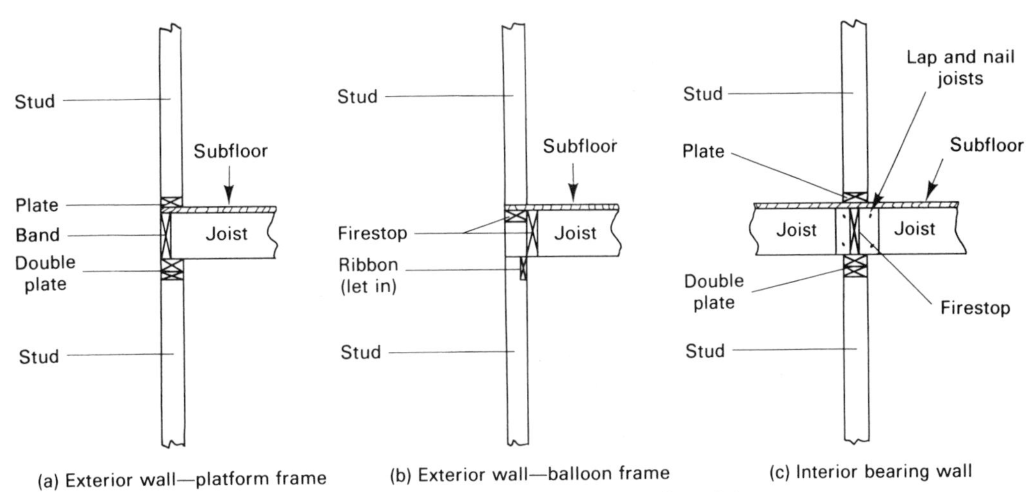

Figure 7-7 Framing of upper floor joists.

For platform frame construction, second floor joists rest on the top plates of the walls (Figure 7-7). Note that double plates are usually used for bearing walls supporting joists or rafters. When studs and joists are spaced 24 in. (61 cm) on center and double 2 × 4 in. (50 × 100 mm) plates are used, the joists must be placed within 5 in. (13 cm) of the center of the stud below it. Alternatively, the size of top plates may be increased to 2 × 6 in. (50 × 150 mm). Some codes permit the use of a single top plate to support joists or rafters when the joists are centered over the studs with a tolerance of 1 in. (25 mm). In this case, the top plates must also be tied together at joints, corners, and intersecting walls by an approved galvanized metal tie.

Joist support at a girder can be provided in either of two ways: resting the joists on the top of the girder (Figure 7-4) or framing the joists into the side of the girder (Figure 7-8). When joists rest on top of the girder, a line of joists may overlap (lap joists) or butt together (in-line joists). For lap joists, a minimum overlap of 3 in. (7.6 cm) should be provided and the overlapping joists nailed together. Joists that butt together over a girder should be tied together by a nominal 2 in. (50 mm) splice, a metal tie, or other approved method. In-line joists can also be connected using the cantilevered splice system shown in Figure 7-4. Since one joist is cantilevered over the girder, this technique stiffens the floor system and often permits the use of joists one size smaller than usual. In addition, it may permit more economical use of lumber by allowing standard lumber lengths to be more fully utilized. The ends of the cantilevered joists are spliced with nailed plywood splice plates.

When using joists that meet over a support, more economical use of joist lumber can be made by using a house depth based on a 4 ft. (1.2 m) module; that is, a depth of 24, 28, or 32 ft. (7.3, 8.5, or 9.8 m). The length of each joist will then be a multiple of 2 ft. (61 cm), corresponding to a standard length of framing lumber. The thickness of the header joist provides a lap over the support of at least 1 1/2 in. (38 mm) on each side for a minimum lap of 3 in. (76 mm). The effective joist span (clear span) can be reduced by using sill plates and girder bearing plates wider than the usual 2 × 4 in. (50 × 100 mm). For example, the use of 2 × 6 in. (50 × 150 mm) sills and bearing plates reduces the joist clear span by 8 1/4 in. (21 cm) compared with the usual 2 × 4 in. (50 × 100 mm) sill and bearing plates. This reduction in clear span will often permit the use of a smaller joist or lower grade of lumber. The use of a field-glued plywood subfloor (see the following section) also increases floor frame stiffness and may permit the use of a smaller or lower grade joist.

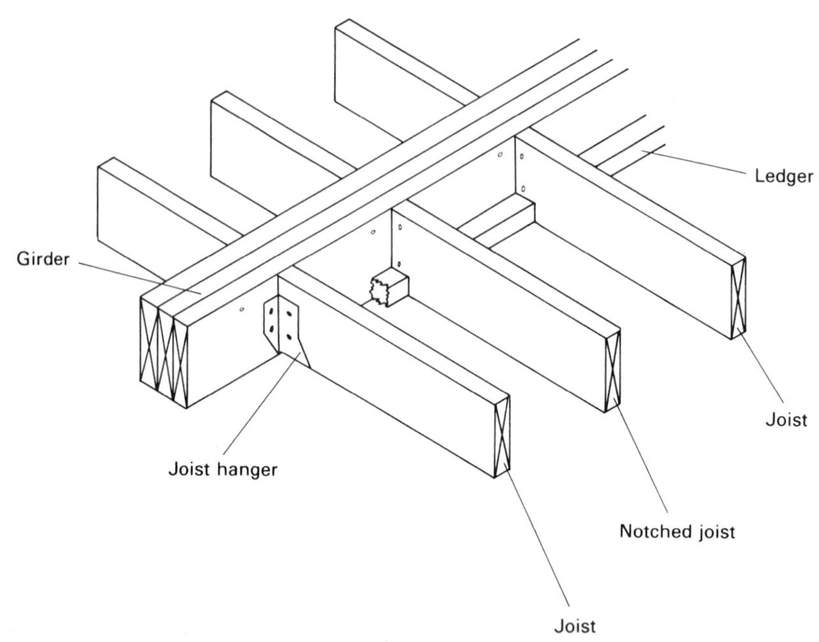

Figure 7-8 Joists framed into girder.

Some common methods for framing joists into girders are illustrated in Figure 7-8. Joists are attached to girders by *joist hangers* or rest on *ledger strips* nailed to the girder. Ledgers should be at least 2 × 2 in. (50 × 50 mm) in size.

When joist ends must be notched as shown in Figure 7-8, the depth of the notch should not exceed one-quarter of the joist depth. Notching of floor joists for pipes or cables should not exceed one-sixth of the joist depth. Notches should not be located within the center one-third of the joist span. The diameter of any holes drilled for pipes or cables should not exceed one-third of the joist depth. Such holes should not be located closer than 2 in. (5 cm) to the top or bottom of the joist.

Floor joists should be doubled under bearing partitions that run parallel to the floor joists. When utility lines run vertically through the partition, the two members of the doubled floor joist should be separated by solid blocking spaced not more than 4 ft. (1.2 m) on center to provide a passage for the utility lines. Bearing walls or partitions running perpendicular to the floor joists should be located directly over the underlying foundation wall or girder. In most cases, however, they may be offset as much as the depth of the joist if necessary.

Framing Floor Openings

The framing for typical floor openings required for stairs, chimneys, and fireplaces is illustrated in Figure 7-9. Note that the opening is framed by *trimmer joists* and *headers.* A single header the same size as floor joists may be used for spans of 4 ft. (1.2 m) or less. Double headers should be used for spans greater than 4 ft. (1.2 m). If the span exceeds 6 ft. (1.8 m), the headers should be supported by joist hangers or other approved anchors. Double trimmer joists should be used at the sides of openings.

Floor Beams and Trusses

Floor beams and trusses are finding increasing use for floor support in place of conventional floor joists. Floor beams have higher load-carrying capacity than do floor joists. As a result, they are able to span larger openings or be spaced further

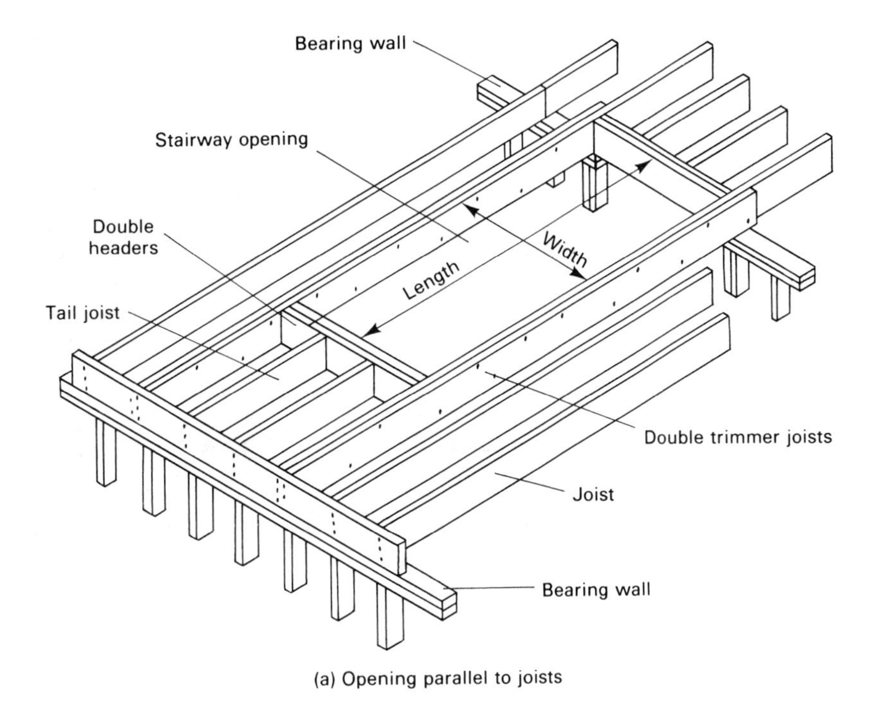

(a) Opening parallel to joists

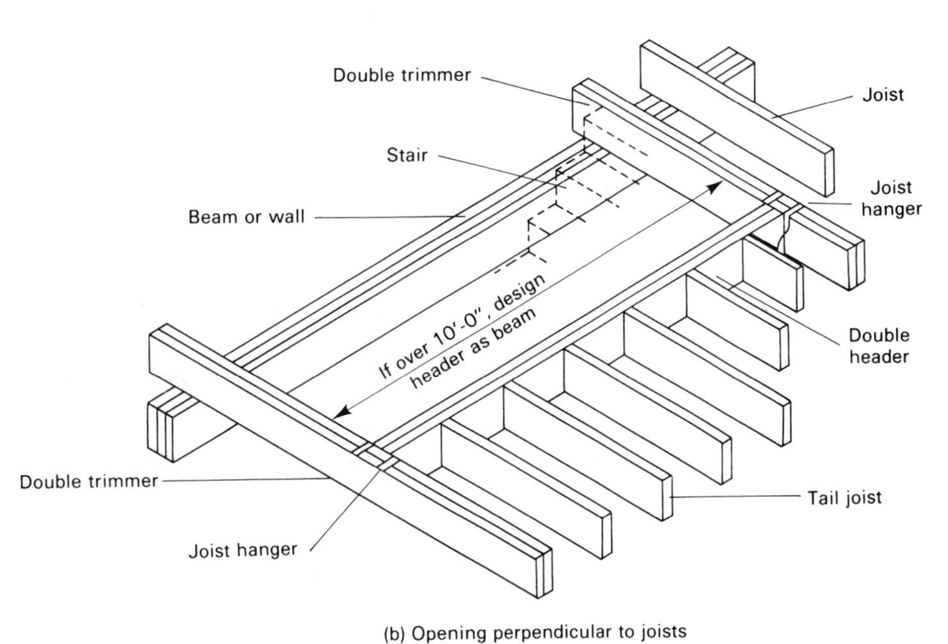

(b) Opening perpendicular to joists

Figure 7-9 Framing for floor openings (adapted from U. S. Department of Agriculture).

apart than floor joists. While solid timber or glued laminated beams may be used, built-up wood beams are lighter for equal load-carrying ability. As a result, wood I-beams are most often used in place of joists (see Figure 7-10). The wood I-beam shown uses 2 × 4 in. (50 × 100 mm) top and bottom flanges bonded to a 3/8 in. (10 mm) plywood web.

In addition to their light weight and high load capacity, wood I-beams can tolerate relatively large holes for passage of ducts and utility lines. Figure 7-11 shows maximum hole size and minimum spacing permitted by one beam manufacturer. The 2

Figure 7-10 Wood I-beams used for floor support (courtesy Alpine Engineered Products, Inc.).

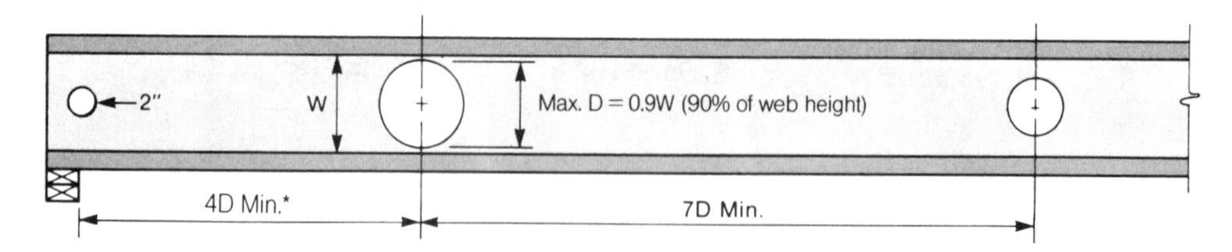

Figure 1. Round hole maximum size and minimum spacing.
Maximum round hole size is 0.9 times the web depth. Any round hole must be located at least 4 diameters from inside face of bearing. Adjacent round holes must be at least 7 diameters of the larger hole apart, center-to-center. A 2" hole may be cut anywhere in the web (pre-punched knockouts are provided 24" o.c.).

Figure 7-11 Openings in wood I-beams (courtesy Alpine Engineered Products, Inc.).

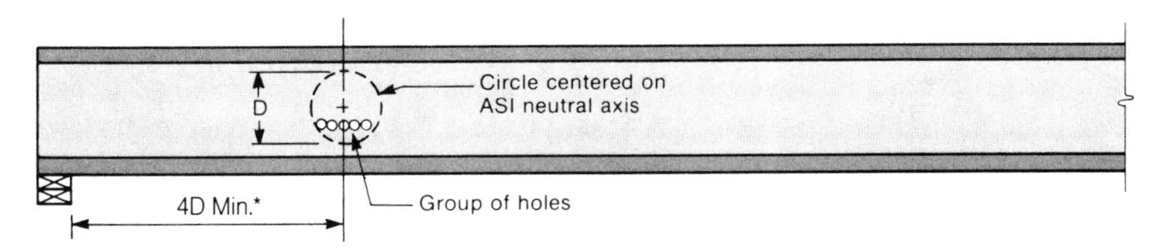

Figure 2. Group of small holes.
A group of small holes must fit inside a circle meeting the limitations for round holes (see Figure 1).

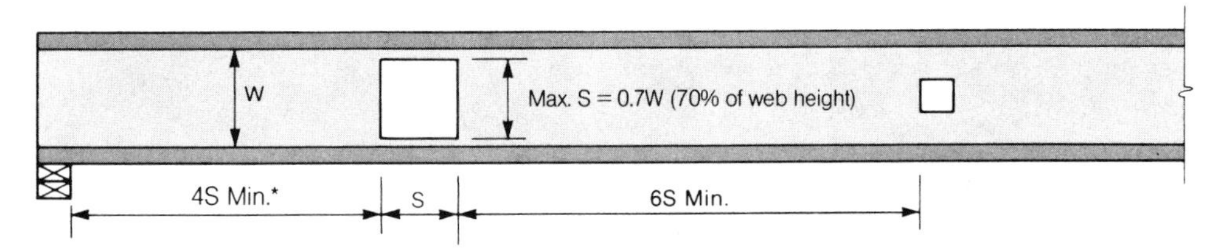

Figure 3. Square hole maximum size and minimum spacing.
Maximum square hole size is 0.7 times the web depth. Any square hole must be located at least 4 times the length of a side from inside face of bearing. Adjacent square holes must be at least six times the length of a side of the larger hole apart.

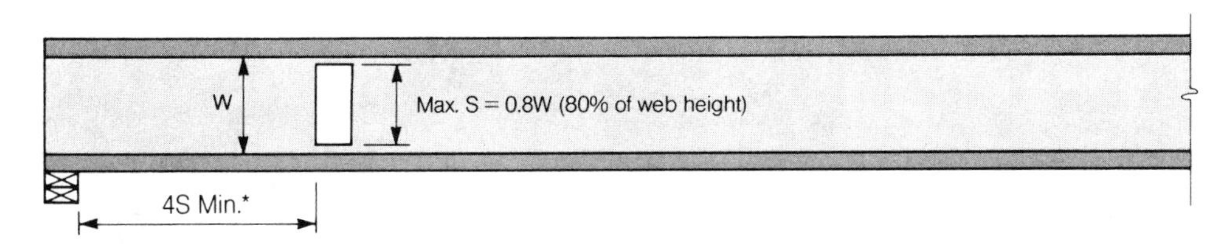

Figure 4. Vertical rectangular hole maximum size and minimum spacing.
Vertical rectangular holes follow the rules for square holes using the longer side for all calculations.

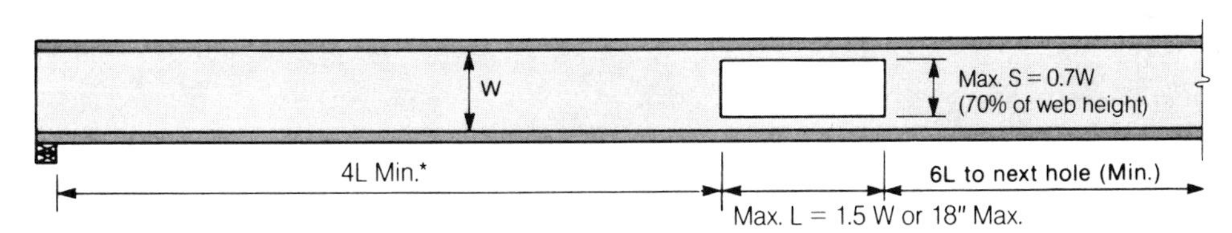

Figure 5. Rectangular hole maximum size and minimum spacing.
Maximum rectangular hole size is 0.7 times the web depth for the shorter side, with the longer side 1.5 times the web depth or 18" at most. Any rectangular hole must be located at least 4 times the length of the longer side from the inside face of a bearing. Adjacent rectangular holes must be at least 6 times the length of the longer long side apart.

Legend

W = Inside flange dimension.

D = Diameter of round opening.

S = Vertical dimension of square or rectangular opening.

L = Horizontal dimension of rectangular opening.

Notes: If adjacent holes have different shapes, the spacing between them is the greater of the two obtained by following the rules for the shapes involved.

Recommendations in figures 1 thru 5 apply only to simple support, uniform load situations. Exceptions to the criteria shown in figures 1 thru 5 may be possible. Make special inquiry to your ASI distributor.

Figure 7-11 *(continued)*

× 4 in. (50 × 100 mm) flanges provide a wide support surface and facilitate nailing of subfloor and ceiling panels.

Flat trusses, which provide advantages similar to wood I-beams, are increasingly used as replacements for joists (see Figure 7-12). The characteristics of flat floor trusses are similar to those of flat roof trusses, described in Chapter 8. However, the trusses used for floor support are commonly manufactured with 2 × 4 in. (50 × 100 mm) chords and webs placed horizontally to produce a truss 3 1/2 in. (89 mm) wide. The wide openings between web members facilitate the installation of ducts and utility lines.

Figure 7-12 Wood floor trusses.

Safety precautions that should be taken during the erection of all types of trusses are described in Chapter 8. However, since floors supported by trusses provide a convenient surface for the storage of construction materials, precautions that should be taken in the storage of materials on such floor are illustrated in Figure 7-13.

Subfloors

The *subfloor* forms the upper surface of the floor frame and provides a smooth, level surface for installing partitions and finish flooring. Either plywood or board subflooring may be used. When board subflooring is used, it is usually placed at a 45° angle to the floor joists. This provides additional stiffness to the floor structure and permits finish wood flooring to be installed either parallel or perpendicular to the joists. The end joints of boards and plywood, as well as the edge joints of plywood, should be supported by a joist or blocking.

In the United States, plywood is most commonly used for subflooring. Plywood subflooring should be laid with the long dimension perpendicular to the floor joists as illustrated in Figure 7-14. Since the face grain of plywood runs parallel to the long dimension, the face grain will also be perpendicular to the joists. Note the recommended spacing between plywood panels. The edge support for plywood

Proper weight distribution of construction materials is a must during installation. Do not stack materials on unbraced trusses under any circumstances. Care should be taken to avoid stacking heavy materials in mid-span. Stacking is generally permitted against outside loadbearing wall or placed directly over loadbearing wall (see illustration below).

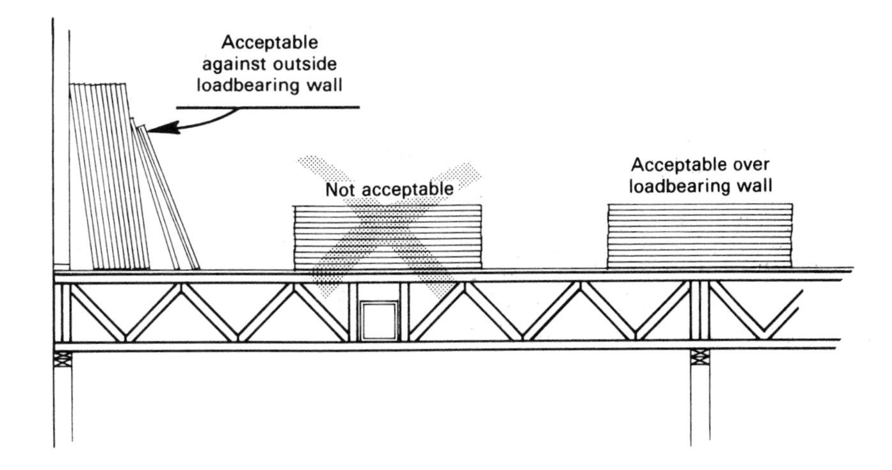

Similar precautions should be taken in stacking decking materials on roof truss systems. Even when materials have been properly located as shown above, be sure to position stack so as to have as many trusses carrying the load as possible. For comprehensive guidelines on erection of trusses, please consult Truss Plate Institute publication BWT-76 – **BRACING WOOD TRUSSES: COMMENTARY and RECOMMENDATIONS.**

Figure 7-13 Safety precautions for floor trusses (courtesy Alpine Engineered Products, Inc.).

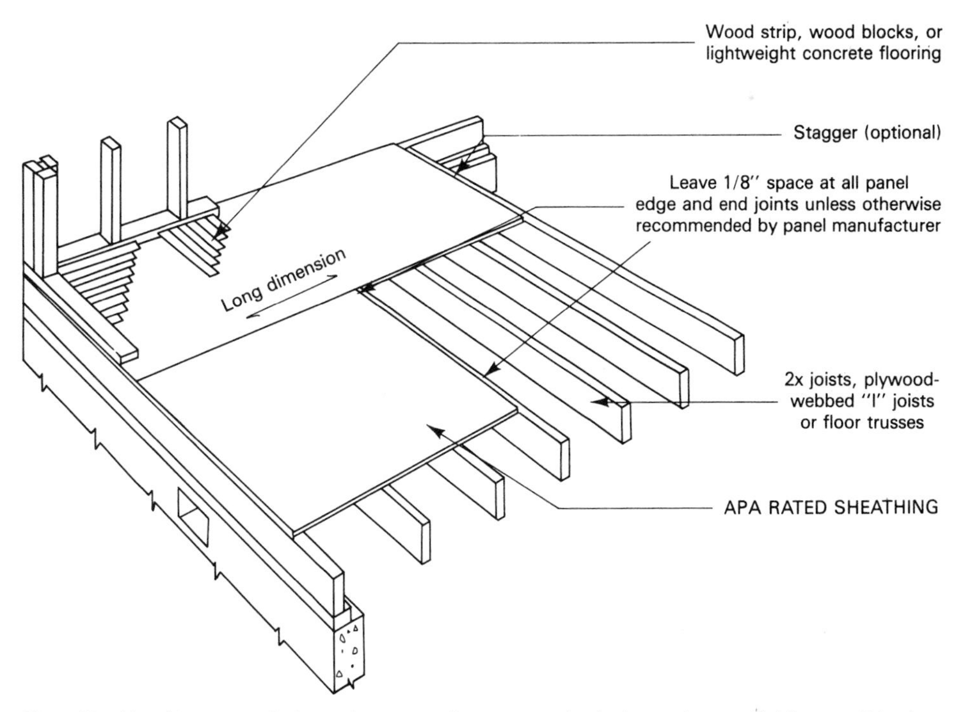

Note: Provide adequate ventilation and use ground cover vapor barrier in crawl space. Subfloor must be dry before applying subsequent layers.

Figure 7-14 Construction of a plywood subfloor (courtesy American Plywood Association).

subflooring may be omitted when using tongue-and-groove plywood panels or when a separate plywood underlayment at least 1/4 in. (6 mm) thick or a wood finish flooring at least 25/32 in. (20 mm) thick is applied over the subfloor. A plywood subfloor plus a plywood underlayment should be used below carpet or other nonstructural flooring unless a combination subfloor-underlayment (single-layer floor) is applied. The minimum plywood thickness and nailing requirements for plywood subflooring recommended by the American Plywood Association (APA) are given in Table 7-3.

Plywood may also be used as a combination subfloor-underlayment (also called a "single-layer" floor). This functions as both a subfloor and a base for nonstructural finish flooring such as carpet and vinyl flooring. APA RATED STURD-I-FLOOR is a proprietary plywood panel developed by the APA for use as a combination subfloor-underlayment for spans up to 48 in. (122 cm). The recommended minimum panel thickness and nailing requirements for this paneling are presented in Table 7-4.

The use of glue as well as nails for fastening the subfloor to joists, beams, or trusses has become popular in recent years. This technique offers several advantages over a conventional nailed plywood subfloor. It significantly increases floor system stiffness and also reduces later floor squeaking and nail-popping. It may also permit the use of smaller or lower grade floor joists. The APA Glued Floor System is a glued plywood floor construction system developed by the APA. Although a single-layer floor is recommended, the technique can be applied to a double-layer floor (subfloor plus underlayment). Panel recommendations for this system are shown in Table 7-5, and fastening requirements are given in Table 7-4. The recommended spacing between glued plywood panels is 1/16 in. (1.6 mm) at both end and edge joints. An approved adhesive should be applied to joist tops and panel tongue and groove edges before each glued panel is placed.

TABLE 7-3

Use of Plywood Subfloor (Courtesy American Plywood Association)

APA Panel Subflooring[a]
(APA-RATED sheathing)

Panel span rating (or group number)	*Panel thickness (in.)*	*Maximum span (in.)*	*Nail size & type*[g]	*Nail spacing (in.)*	
				Supported panel edges	*Intermediate supports*[f]
24/16	7/16	16	6d common	6	12
32/16	15/32, 1/2, 5/8	16[b]	8d common[c]	6	12
40/20	9/16, 19/32, 5/8, 3/4, 7/8	20[d]	8d common	6	12
48/24	23/32, 3/4, 7/8	24	8d common	6	12
1-1/8" Groups 1 & 2[e]	1 1/8	48	10d common	6	6

(a) For subfloor recommendations under ceramic tile, refer to Table 11 (reference 3). For subfloor recommendations under gypsum concrete, contact manufacturer of floor topping.

(b) Span may be 24 inches if 3/4-inch wood strip flooring is installed at right angles to joists.

(c) 6d common nail permitted if panel is 1/2 inch or thinner.

(d) Span may be 24 inches if 3/4-inch wood strip flooring is installed at right angles to joists, or if a minimum 1 1/2 inches of lightweight concrete is applied over panels.

(e) Check dealer for availability.

(f) Applicable building codes may require 10" oc nail spacing at intermediate supports for floors.

(g) Other code-approved fasteners may be used.

TABLE 7-4

Use of Plywood Subfloor-Underlayment Panels (Courtesy American Plywood Association)

APA-RATED STURD-I-FLOOR[a]

Span rating (maximum joist spacing) (in.)	Panel thickness[b] (in.)	Fastening: glue-nailed[c]			Fastening: nailed-only		
		Nail size and type	Spacing (in.)		Nails size and types	Spacing (in.)	
			Supported panel edges	Intermediate supports		Supported panel edges	Intermediate supports[g]
16	19/32, 5/8, 21/32	6d ring- or screw-shank[d]	12	12	6d ring- or screw-shank	6	12
20	19/32, 5/8, 23/32, 3/4	6d ring- or screw-shank[d]	12	12	6d ring- or screw-shank	6	12
24	11/16, 23/32, 3/4	6d ring- or screw-shank[d]	12	12	6d ring- or screw-shank	6	12
	7/8, 1	8d ring- or screw-shank[d]	6	12	8d ring- or screw-shank	6	12
32	7/8, 1	8d ring- or screw-shank[d]	6	12	8d ring- or screw-shank	6	12
48	1 1/8	8d ring- or screw-shank[e]	6	(f)	8d ring- or screw shank[e]	6	(f)

(a) Special conditions may impose heavy traffic and concentrated loads that require construction in excess of the minimums shown. See page 23 (reference 3) for heavy duty floor recommendations.

(b) As indicated above, panels in a given thickness may be manufactured in more than one Span Rating. Panels with a Span Rating greater than the actual joist spacing may be substituted for panels of the same thickness with a Span Rating matching the actual joist spacing. For example, 19/32-inch-thick Sturd-I-Floor 20 oc may be substituted for 19/32-inch-thick Sturd-I-Floor 16 oc over joists 16 inches on center.

(c) Use only adhesives conforming to APA Specification AFG-01, applied in accordance with the manufacturer's recommendations. If non-veneered panels with sealed surfaces and edges are to be used, use only solvent-based glues; check with panel manufacturer.

(d) 8d common nails may be substituted if ring- or screw-shank nails are not available.

(e) 10d common nails may be substituted with 1-1/8-inch panels if supports are well seasoned.

(f) Space nails 6 inches for 48-inch spans and 12 inches for 32-inch spans.

(g) Applicable building codes may require 10″ oc nail spacing at intermediate supports for floors.

TABLE 7-5
Recommendations for APA Glued Floor System (Courtesy American Plywood Association)

Panel Recommendations for APA Glued Floor System[a]

Joist spacing (in.)	*Flooring type*	*APA panel grade and span rating*	*Possible thickness (in.)*
16	Carpet and pad	STURD-I-FLOOR 16 oc, 20 oc	19/32, 5/8, 21/32
	Separate underlayment or structural finish flooring	RATED SHEATHING 24/16, 32/16, 40/20, 48/24	7/16, 15/32, 1/2, 19/32, 5/8, 23/32, 3/4
19.2	Carpet and pad	STURD-I-FLOOR 20 oc, 24 oc	19/32, 5/8, 23/32, 3/4
	Separate underlayment or structural finish flooring	RATED SHEATING 40/20, 48/24	19/32, 5/8, 23/32, 3/4
24	Carpet and pad	STURD-I-FLOOR 24 oc, 32 oc	11/16, 23/32, 3/4, 7/8, 1
	Separate underlayment or structural finish flooring	RATED SHEATING 48/24	23/32, 3/4
32	Carpet and pad	STURD-I-FLOOR 32 oc, 48 oc	7/8, 1, 1 1/8
48	Carpet and pad	STURD-I-FLOOR 48 oc	1 3/32, 1 1/8

(a) For panel recommendations under ceramic tile, refer to Table 11 (reference 3).

7-3 THE WALL FRAME

Exterior Walls—Platform Frame

The wall frame for a platform frame building consists of lower horizontal members (called *soleplates* or *soles),* vertical members (called *studs),* and upper horizontal members (called *top plates* or *plates).* The corner of a wall frame for a typical platform frame building is illustrated in Figure 7-15.

Studs are normally spaced 16 or 24 in. (41 or 61 cm) on center with their wide face perpendicular to the wall. The maximum spacing usually allowed for 2 × 4 in. (50 × 100 mm) studs is 24 in. (61 cm) for a one-story building and 16 in. (41 cm) for the first story of a two-story building. Two by six inch (50 × 150 mm) studs may be spaced at 24 in. (61 cm) on center for one- or two-story buildings and 16 in. (41 cm) on center for the first floor of a three-story building. The width of soleplates and top plates should match the width of studs. Single soleplates and double top plates are commonly used. When studs are spaced 24 in. (61 cm) on center with double top plates, supported floor joists, floor trusses, or roof trusses must be placed within 5 in. (13 mm) of the center of the stud below it unless 2 × 6 in. (50 × 150 mm) top plates are used. Joints in members of double top plates should be staggered by 4 ft. (1.2 m) along the wall. At corners and wall intersections, one member of the top plate should overlap the other as shown in Figure 7-15.

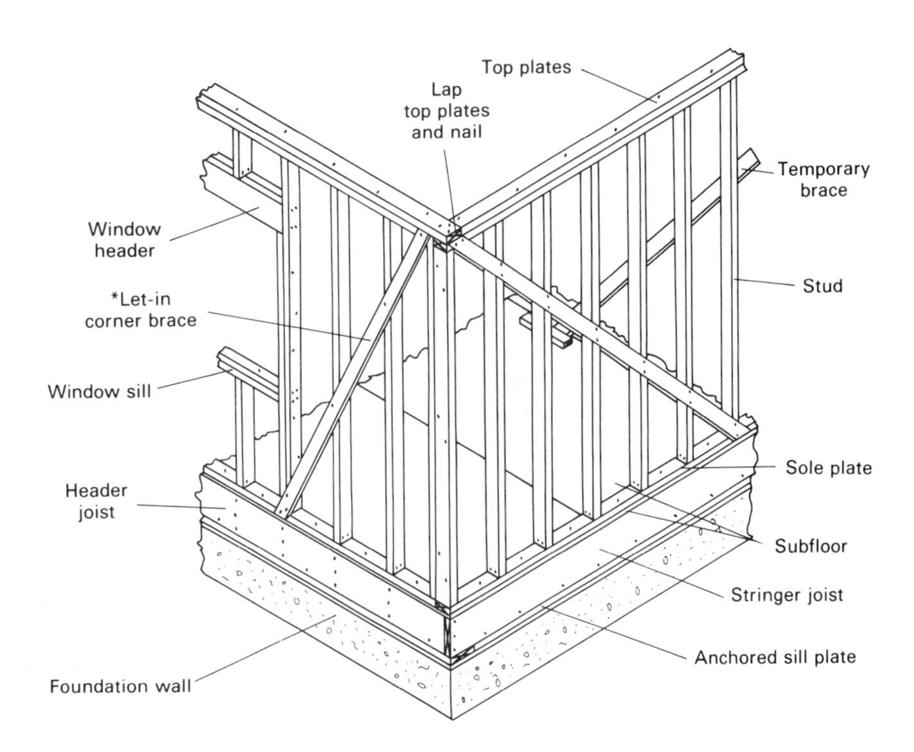

*See text for horizontal bracing requirements.

Figure 7-15 Exterior wall frame—platform construction (adapted from U. S. Department of Agriculture).

Some building codes permit the use of a single top plate for bearing walls when rafters or joists are centered over studs with a tolerance of 1 in. (25 mm). In this case, the top plates must also be tied together at joints, corners, and intersecting walls by an approved metal tie.

A ceiling height of 8 ft. (244 cm) is widely used. To allow space for ceiling finish and finish flooring, wall studs 7 ft. 9 in. (236 cm) long may be used with nominal 2 in. (50 mm) sole plates and double top plates to yield an actual wall frame height (subfloor to top of the top plate) of 8 ft. 1 1/2 in. (248 cm). Many codes require a minimum ceiling height of 7 ft. 6 in. (229 cm). This translates to a wall frame height of about 7 ft. 7 1/2 in. (232 cm) to allow for ceiling finish and finish flooring.

Studs should be located such that they provide clearance for vertical utility pipes. When top plates must be notched or cut for utility pipes, an 18-gauge metal tie at least 1 1/2 in. (38 mm) wide should be fastened across the opening with 4-16d nails at each end.

Horizontal bracing is required for all exterior walls. Single-story buildings should be braced at each end (corner) and at every 25 ft. (7.6 m) of wall length. In general, bracing may consist of:

1. Diagonal (making a 45° angle with studs) 1 × 4 in. (25 × 100 mm) braces let in to studs, soleplates, and top plates.
2. Diagonal wood sheathing at least 5/8 in. (16 mm) in actual thickness.
3. Plywood sheathing panels at least 48 × 96 in. (122 × 144 cm) in size and 15/32 in. (12 mm) thick.
4. Other approved structural sheathing.
5. Approved metal strap ties.

In Figure 7-15, the corner is braced by a 1 × 4 in. (13 × 100 mm) let-in brace. When plywood or other approved structural sheathing is used (see Section 8-1), no additional wall bracing is required. When nonstructural sheathing is used, one sheet of 15/32 or 1/2 in. (12 or 13 mm) plywood applied to each side of the corner serves as corner bracing. In areas subject to high seismic or wind loads, the first story of two-story buildings and the second story of three-story buildings should have at least 25 percent of the wall length braced by structural sheathing. In these areas, the first story of three-story buildings should have at least 40 percent of the wall length braced by structural sheathing.

Since metal strap ties provide resistance only in tension, they must be applied in an X-pattern on each side of the corner. This may interfere with window placement. However, a rigid T-shaped metal brace is available, which requires only one brace on each side of the corner. A saw cut is required to permit insertion of the brace's web into the wall studs.

Exterior wall sections must be securely tied together at corners and must provide an adequate nailing surface for exterior sheathing and interior finish. Four common methods for framing exterior wall corners are illustrated in Figure 7-16. Method (a) is recommended as the least expensive in terms of time and material, when permitted by the building code. Notice that a nailer strip or drywall clip is used to support the drywall at the interior corner. By comparison, method (b) requires the installation of a third stud, while method (c) requires a fourth stud, and method (d) requires extra labor for installing blocking and insulation between two of the studs. Note also that all four methods provide support and a nailing base for exterior sheathing and interior drywall.

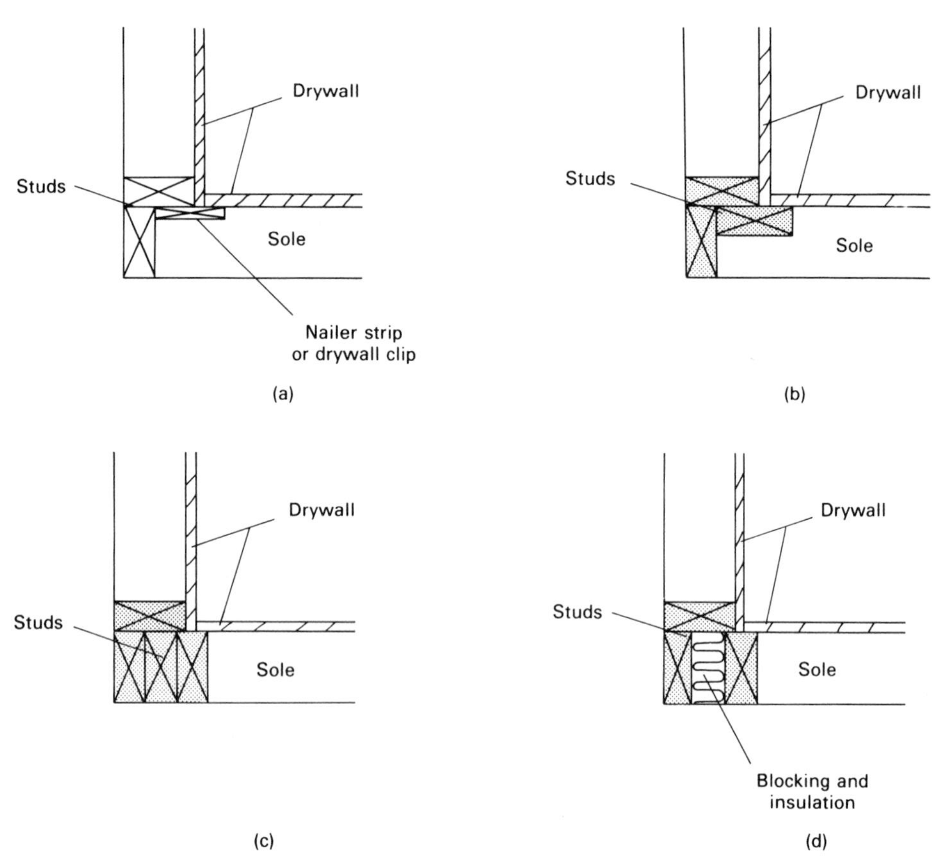

Figure 7-16 Framing exterior wall corners.

For ease of assembly in platform frame construction, wall frames are often assembled in a horizontal position and tilted into place on the floor frame. In Figure 7-15, note the use of a temporary brace to hold a section of wall frame in alignment until the entire exterior wall frame has been erected and tied together.

Openings in Exterior Walls

Framing for a typical window and door opening in an exterior wall is illustrated in Figure 7-17. The *header* serves to transfer vertical loads to the supporting studs at each end of the header. Double pieces of nominal 2 in. (5 cm) lumber are commonly used to form the header, although beams or trusses of wood or metal may be used. Spacers [usually 1/2 in. (13 mm) plywood] are used between the members of double headers so that the header is flush with each side of the wall. When a single piece of nominal 4 in. (10 cm) lumber is used for the header, the width of the header will match the width of 2 × 4 in. (50 × 100 mm) studs. Table 7-6 shows the maximum header span permitted by the CABO code.

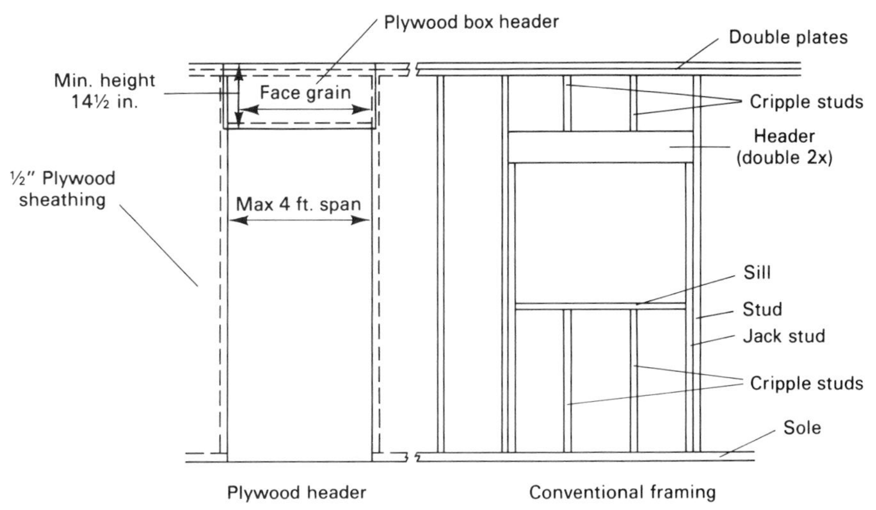

Figure 7-17 Framing wall openings.

When the clear width of the opening does not exceed 3 ft. (0.9 m), the header may be fastened to a single stud at each end by face nailing or by use of a framing anchor. For clear spans of 3-6 ft. (0.9–1.8 m), double studs (one *jack stud* placed under the header and one full-length stud) should be used at each end of the header as illustrated in Figure 7-17. For clear spans exceeding 6 ft. (1.8 m), the header should be supported by three studs (two jack studs and one full length stud) at each end.

Research by the NAHB Research Foundation indicates that the conventional header for an opening up to 4 ft. (1.2 m) wide can be replaced by a one-sided box beam formed by nailing 1/2 in. (13 mm) plywood sheathing at least 14 1/2 in. (37 cm) high to studs, plates, and other framing (see Figure 7-17). The plywood should be fastened along the edges and supports with 8d common nails spaced 4 in. (10 cm) on center.

The *sill* or *window sill* shown in Figure 7-17 forms the bottom member of window openings. Although double sills are often used, as shown in Figure 7-15, the sill and the *cripple studs* below it are actually not loadbearing. The wall's vertical loads are resisted by the top plates, header, and the studs. Therefore, the use of single sills reduces framing costs. Another method of reducing framing costs is to

TABLE 7-6
Maximum Header Span*

Header Design Chart
(Douglas Fir-Larch, Southern Pine No. 2,
Roof Load = 20 LL + 10 DL; Floor Load = 40 LL + 10 DL)
F_b = 1200   E = 1,600,000 F_v = 90

Header supporting:	*Header size:*	*Maximum allowable header span (ft.-in.)*								
		Design procedure								
		Nonstruct. sheath.			*½" Insul. board sheath.***			*½" plywood sheath.‡*		
		House depth (ft.)			*House depth (ft.)*			*House depth (ft.)*		
		24′	*28′*	*32′*	*24′*	*28′*	*32′*	*24′*	*28′*	*32′*
Roof	2-2 × 4	4-7	4-6	4-3	4-11	4-8	4-6	5-7	5-4	5-1
	2-2 × 6	6-8	6-4	5-11	6-11	6-7	6-4	7-7	7-3	7-0
	2-2 × 8	8-3	8-0	7-7	8-6	8-2	7-11	8-11	8-7	8-4
	2-2 × 10	9-10	9-6	9-3	10-0	9-8	9-4	10-4	10-0	9-9
	2-2 × 12	11-4	10-11	10-7	11-5	11-1	10-9	11-10	11-5	11-1
Roof plus one story (bearing)	2-2 × 4	5-2	4-11	4-9	5-4	5-1	4-10	5-8	5-5	5-2
	2-2 × 6	6-9	6-5	6-0	6-11	6-7	6-3	7-4	7-0	6-8
	2-2 × 8	8-0	7-5	7-0	8-1	7-8	7-3	8-5	8-2	7-9
	2-2 × 10	9-3	8-9	8-3	9-4	9-0	8-5	9-7	9-3	8-11
	2-2 × 12	10-5	10-1	9-7	10-6	10-2	9-9	10-10	10-5	10-1

Roof plus one story (no bearing)	2-2 × 4	4-8	4-5	4-3	4-9	4-6	4-3	5-1	4-10	4-8
	2-2 × 6	5-10	5-5	5-1	6-0	5-7	5-3	6-7	6-2	5-9
	2-2 × 8	6-9	6-3	5-11	7-0	6-6	6-1	7-6	7-0	6-7
	2-2 × 10	8-0	7-5	6-11	8-2	7-7	7-1	8-8	8-1	7-7
	2-2 × 12	9-3	8-7	8-1	9-5	8-9	8-3	9-11	9-3	8-8
Roof plus two stories (bearing)	2-2 × 4	4-8	4-5	4-3	4-9	4-6	4-3	5-1	4-10	4-8
	2-2 × 6	5-10	5-5	5-1	6-0	5-7	5-3	6-7	6-2	5-9
	2-2 × 8	6-9	6-3	5-11	7-0	6-6	6-1	7-6	7-0	6-7
	2-2 × 10	8-0	7-5	6-11	8-2	7-7	7-1	8-8	8-1	7-7
	2-2 × 12	9-3	8-7	8-1	9-5	8-9	8-3	9-11	9-3	8-8
Roof plus two stories (no bearing)	2-2 × 4	3-11	3-8	3-5	4-1	3-9	3-7	4-5	4-2	4-0
	2-2 × 6	4-8	4-4	4-0	4-10	4-5	4-2	5-3	4-11	4-7
	2-2 × 8	5-5	5-0	4-8	5-7	5-2	4-10	6-0	5-7	5-3
	2-2 × 10	6-4	5-11	5-6	6-6	6-0	5-8	6-11	6-5	6-0
	2-2 × 12	7-5	6-10	6-5	7-6	7-0	6-6	7-11	7-4	6-11

*Reproduced from the 1986 edition of the *CABO One and Two Family Dwelling Code,* Copyright 1986, and 1987 Amendements, Copyright 1987, with permission of the publishers—Building Officials and Code Administrators International, International Conference of Building Officials and Southern Building Code Congress International, Inc.

**Sheathing or combined sheathing/siding having a minimum density of 18 pcf.

‡Minimum ½-inch plywood sheathing or combined sheathing/siding applied between the bottom of the header, the top of the top plate and between the center lines of the broken vertical studs at the ends of the header and nailed to the header, top plates, cripples and studs—6 inches o.c. at the edges and 12 inches o.c. at intermediate framing.

Note: Linear interpolation for house widths not in table are permitted. For example, assume a 26-foot-wide house with ½-inch plywood sheathing—roof load, 2 × 6 header: allowable header span = 7 feet 5 inches.

Tables based on maximum 1½-foot overhangs and band joists used at floors.

■ symbol represents supporting beam or structural bearing wall below floor.

● symbol represents location of header.

Header spans identified as having "no bearing" construction apply to both interior and exterior load-bearing walls which have tributary areas equal to one-half the house depth. Header spans identified with "bearing" construction apply only to exterior bearing walls with tributary areas equal to one fourth of the house depth.

Nominal 4-inch size single headers may be substituted for nominal 2-inch double headers.

move the header up to replace one of the top plates. In this case, cripple studs would be required between the header and the top of the opening only if needed for support of sheathing or interior finish.

Interior Walls—Platform Frame

Interior walls are constructed like exterior walls except that horizontal bracing is not normally required. Interior bearing walls commonly use a double top plate to support floor and ceiling joists, but a single top plate may be used under the same conditions prescribed for exterior walls.

Nonbearing interior walls commonly use a single top plate. In this case, the top plate should provide overlapping or be tied at corners and intersections with other walls. Joints in single top plates should be tied together with an approved metal tie or a nominal 2 in. (5 cm) splice at least 16 in. (41 cm) long. Studs of interior nonbearing walls are sometimes placed with their wide face parallel to the wall in order to provide clearance for utility lines.

For interior nonbearing walls, a single 2 × 4 in. (50 × 100 mm) member placed with its wide dimension horizontal can serve as a header for openings up to 8 ft. (2.4 m) wide when the vertical distance to a parallel nailing surface (usually a top plate) is not more than 24 in. (61 cm).

Interior walls and partitions must be tied into exterior walls and must provide an adequate nailing surface for the interior finish. Four common methods for framing interior walls into exterior walls are illustrated in Figure 7-18. Method (a) is recommended when permitted by the local building code. In this method, support

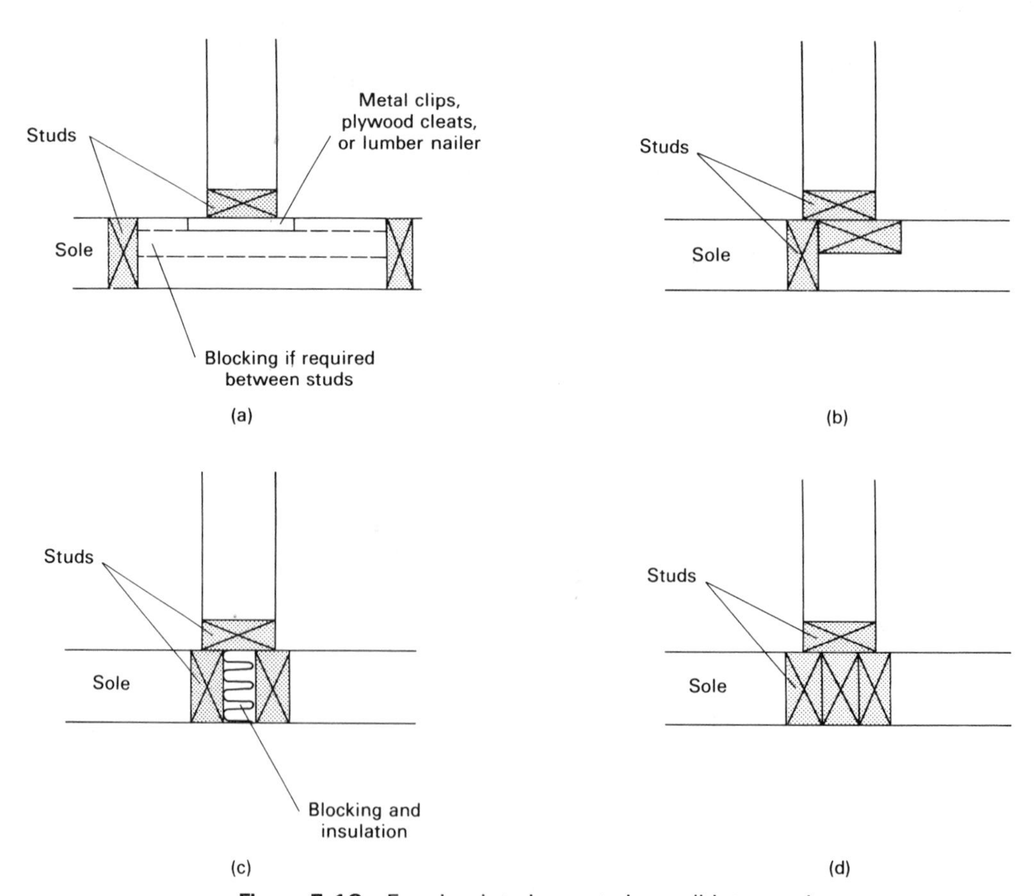

Figure 7-18 Framing interior–exterior wall intersections.

for the interior finish is provided by metal clips or cleats of 3/8 in. (10 mm) plywood or 1 in. (25 mm) lumber. If a backup stud must be used, method (b) is least expensive in terms of material and labor. Methods (c) and (d) are similar to methods previously discussed for framing exterior wall corners.

Interior Walls—Balloon Frame

Balloon frame construction is illustrated in Figure 7-3. As noted earlier, in this method of construction the exterior wall studs extend all the way from the sill to the roof frame. Studs and joists are toenailed to the sill and face-nailed to each other. The exterior ends of upper floor joists are supported by a ribbon strip let in to the wall studs and face-nailed to the studs. Exterior walls must be braced in the same manner as for platform frame construction.

Interior bearing walls support upper floor joists on the walls' top plate as in platform frame construction (see Figure 7-7). A single top plate is commonly used, and the upper floor joist is placed directly over the supporting stud. Note the use of blocking for firestopping of the upper wall. Note, also, that the second-floor interior wall studs rest directly on the top plate of the first-floor wall and are toenailed to the top plate and face-nailed to the second-floor joists. Alternatively, interior bearing walls may be framed in the same manner as for platform frame construction.

PROBLEMS

1. Describe the major differences between platform frame construction and balloon frame construction.
2. The following questions pertain to the use of modular dimensions in wood frame buildings.
   (a) Explain the advantages of using modular dimension in planning the building's dimensions and the location of window and door openings.
   (b) What major module should be used for the building's exterior dimensions?
3. A single-story building with a central interior bearing partition and central girder is 24 ft. (7.3 m) deep. Using Table 7-1, determine the following requirements for the central floor girder.
   (a) If No. 2 grade lumber is used, what minimum size girder is required when girder supports are spaced at 6 ft. (1.8 m) intervals?
   (b) What minimum size footing is required for girder supports?
   (c) Describe the acceptable columns or posts for girder support.
4. Find the minimum size of floor joist required for the building of Problem 3 with a live floor load of 40 lbs./sq. ft. (1.9 kPa). Joists are No. 2 grade Southern Pine spaced 16 in. (41 cm) on center.
5. **(a)** When is bridging required for floor joists?
   (b) Describe acceptable types of bridging for floor joists.
6. What is the maximum depth that the end of a nominal 2 × 8 in. (50 × 200 mm) floor joist should be notched?
7. What limitations should be observed when boring holes in floor joists for the passage of pipe and electrical conduit?
8. Describe how the upper plates of an exterior bearing wall should be cut for the passage of a plumbing vent pipe.
9. An exterior wall of a building 28 ft. (8.5 m) deep supports only the roof and is sheathed with 1/2 in. (13 mm) plywood.
   (a) What size header is required for a window opening 6 ft. (1.8 m) wide?
   (b) Describe the stud support required for the header.

10. Explain how plywood sheathing might be used to eliminate the conventional header over a window opening in an exterior bearing wall.

REFERENCES

1. American Institute of Timber Construction. *Timber Construction Manual.* 3rd ed. New York: Wiley, 1985.
2. Anderson, L. O. *Wood-Frame House Construction.* Agriculture Handbook No. 73. U.S. Department of Agriculture, Washington, DC, 1975.
3. American Plywood Association. *APA Design/Construction Guide: Residential and Commercial.* Tacoma, WA, 1988.
4. Council of American Building Officials. *CABO One and Two Family Dwelling Code.* Falls Church, VA.
5. National Forest Products Association. *Construction Cost Saver* Series. Washington, DC, various dates.
6. National Forest Products Association. *Manual for House Framing.* Wood Construction Data No. 1. Washington, DC, 1970.
7. Southern Forest Products Association. *Southern Pine Maximum Spans for Joists and Rafters.* New Orleans, LA, 1978.
8. National Forest Products Association. *Span Tables for Joists and Rafters.* Washington, DC, 1977.

chapter 8

Frame Construction II

8-1 THE ROOF FRAME

Roof Types

The roof frame provides horizontal support for the building walls as well as a waterproof covering for the building. The roof frame also provides a base for the installation of sheathing, roofing, and ceilings.

Common types of roofs include flat roofs, shed roofs, gable roofs, and hip roofs, as illustrated in Figure 8-1. *Flat roofs* are simple to frame but difficult to insulate, ventilate, and waterproof. They also provide no attic space for storage or other uses. Although called flat, their upper surface is usually slightly sloped so that water does not pond on the roof surface. *Shed roofs* combine the ease of framing of flat roofs with sufficient slope to provide good drainage. They can also provide limited attic space. *Gable roofs* are probably the most common type of roof for homes and small buildings. They are easy to ventilate and insulate and provide considerable attic space. When combined with dormers as shown in Figure 8-1, they can provide additional living space at low cost. *Hip roofs* eliminate the gable ends and simplify maintenance of exterior walls and fascia. However, they are somewhat more difficult to frame than gable roofs and provide less attic space.

Framing Flat and Shed Roofs

The framing for flat and shed roofs is similar to floor framing. However, since roofs commonly include an eave or overhang to provide shade and protection from the weather, the roof frame must project beyond the wall frame. Typical roof framing for a flat roof is illustrated in Figure 8-2. Note the use of a double header (or trimmer joist) to support the cantilevered *lookout rafters,* which are perpendicular to the line of roof joists. The maximum span for roof joists can be determined from span tables for floor joists with the same design loads (see Table 7-2).

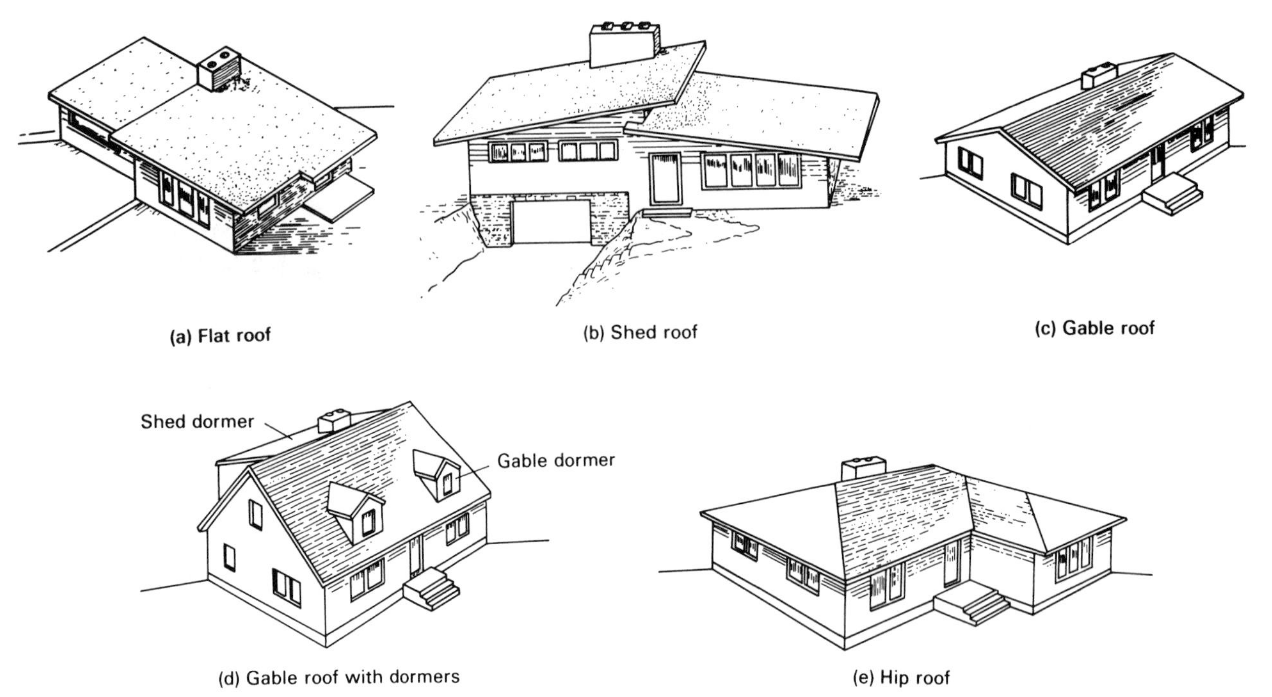

Figure 8-1 Common roof types (U. S. Department of Agriculture).

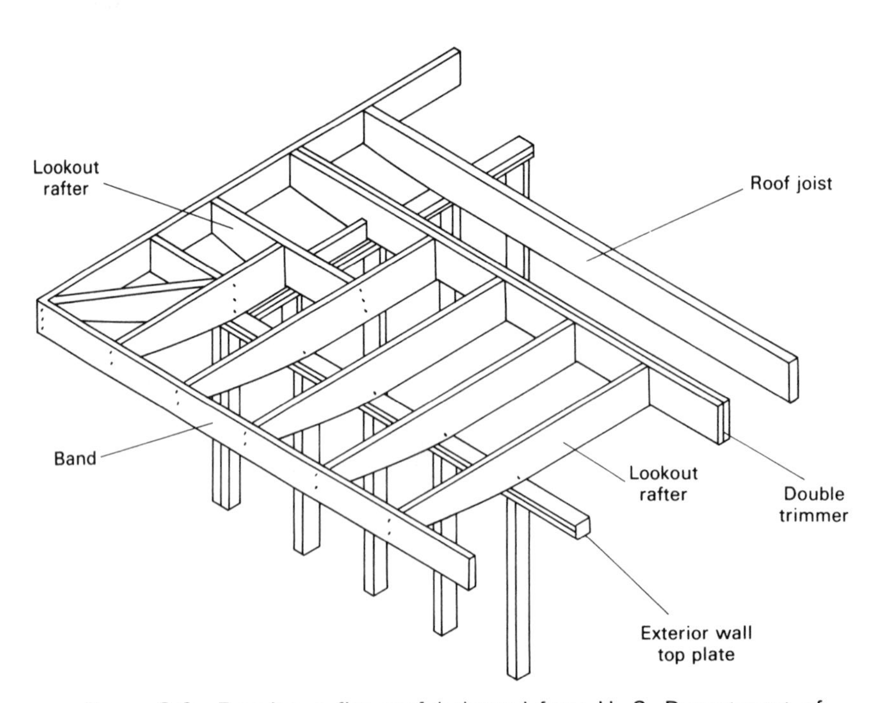

Figure 8-2 Framing a flat roof (adapted from U. S. Department of Agriculture).

Access and Ventilation

Building codes commonly require that attic space be accessible through an opening inside the building. A minimum opening size of 22 × 30 in. (56 × 76 cm) is often required. Requirements for roof and attic ventilation are discussed in Chapter 12.

8-2 CONVENTIONAL RAFTER FRAMING

The Roof Frame

The framing for a typical gable roof is illustrated in Figure 8-3. *Rafters* form the upper (or compression) chords of the triangular roof frame composed of rafters and ceiling joists. The *ceiling joists* serve as the lower (tension) chords of the frame and support the ceiling below. The *ridge board* serves as horizontal bracing between rafters and provides a base for installing sheathing and roofing. The *collar beams* strengthen the rafter frame and help to resist horizontal wind loads. Rafters may be notched as shown to rest on the wall plates. This notch is sometimes called a ''birdsmouth.'' Notice the framing for the gable end.

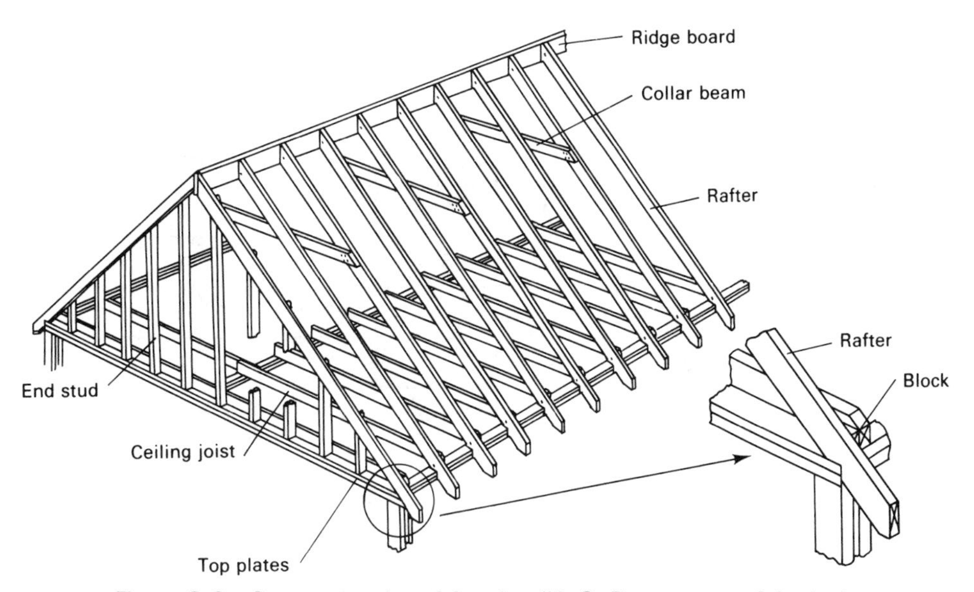

Figure 8-3 Conventional roof framing (U. S. Department of Agriculture).

The maximum allowable span of Southern Pine rafters and ceiling joists for typical design loads are given in Tables 8-1 and 8-2. Span tables for other wood species and loads can be found in Reference 9.

Framing Details

A typical rafter and joist roof frame is shown in Figure 8-4. Notice that the *slope* of the roof is expressed as the ratio of the roof's rise (vertical distance) to its run (horizontal distance). A roof's *rise* is the difference in elevation between the roof at the outer edge of the wall plate (building line) and at the ridge. The *run* is the horizontal distance from the center of the ridge to the outer edge of the wall plate.

TABLE 8-1

Maximum Span for Southern Pine Rafters (ft-in) (Courtesy Southern Forest Products Association)

Size in. (Grade →)	and spacing in. o.c.	Dense Sel Str KD	Sel Str KD and Dense Sel Str	No. 1 Dense KD	Sel Str	No. 1 Dense	No. 1 KD	No. 2 Dense Kd	No. 1
	12.0	11-6	11-3	11-6	11-1	11-3	11-3	11-1	11-1
	13.7	11-0	10-9	11-0	10-7	10-9	10-9	10-7	10-7
2 × 4	16.0	10-5	10-3	10-5	10-0	10-3	10-3	10-0	10-0
	19.2	9-10	9-8	9-10	9-5	9-8	9-8	9-5	9-5
	24.0	9-1	8-11	9-1	8-9	8-11	8-11	8-9	**8-7**
	12.0	14-9	14-6	14-9	14-3	14-6	14-6	14-3	14-3
	13.7	14-1	13-10	14-1	13-7	13-10	13-10	13-7	**13-6**
2 × 5	16.0	13-5	13-2	13-5	12-11	13-2	**13-2**	12-11	**12-6**
	19.2	12-7	12-5	12-7	12-2	**12-4**	**12-0**	**11-10**	**11-5**
	24.0	11-8	11-6	**11-6**	**11-3**	**11-1**	**10-9**	**10-7**	**10-3**
	12.0	18-0	17-8	18-0	17-4	17-8	17-8	17-4	17-4
	13.7	17-3	16-11	17-3	16-7	16-11	16-11	16-7	**16-6**
2 × 6	16.0	16-4	16-1	16-4	15-9	16-1	**16-1**	15-9	**15-3**
	19.2	15-5	15-2	15-5	14-10	**15-1**	**14-8**	**14-5**	**13-11**
	24.0	14-4	14-1	**14-1**	**13-8**	**13-6**	**13-1**	**12-11**	**12-6**
	12.0	23-9	23-4	23-9	22-11	23-4	23-4	22-11	22-11
	13.7	22-9	22-4	22-9	21-11	22-4	22-4	21-11	**21-9**
2 × 8	16.0	21-7	21-2	21-7	20-10	21-2	**21-2**	20-10	**20-2**
	19.2	20-4	19-11	20-4	19-7	**19-11**	**19-4**	**19-0**	**18-5**
	24.0	18-10	18-6	**18-7**	**18-1**	**17-10**	**17-3**	**17-0**	**16-5**
	12.0	30-4	29-9	30-4	29-2	29-9	29-9	29-2	29-2
	13.7	29-0	28-6	29-0	27-11	28-6	28-6	27-11	**27-9**
2 × 10	16.0	27-6	27-1	27-6	26-6	27-1	**27-0**	26-6	**25-8**
	19.2	25-11	25-5	25-11	25-0	**25-5**	**24-8**	**24-3**	**23-6**
	24.0	24-1	23-8	**23-8**	**23-1**	**22-9**	**22-1**	**21-8**	**21-0**

Rafters—High slope (over 3 in 12)—with no finished ceiling—20 psf live load + 7 psf dead load—light roofing. (Spans shown in light face type are based on a deflection limitation of ℓ 180. Spans shown in bold face are limited by the recommended extreme fiber stress in bending value of the grade and includes a 7 psf deadload.)

No. 2 Dense	*No. 2 KD*	*No. 2*	*No. 3 Dense KD*	*No. 3 Dense*	*No. 3 KD*	*No. 3*	*Construction KD*	*Construction*	*Standard KD*	*Standard*
10-10	10-10	10-10	9-4	9-0	8-7	8-3	9-9	9-4	7-4	7-1
10-4	10-4	10-4	8-9	8-5	8-1	7-8	9-2	8-9	6-11	6-7
9-10	9-10	9-7	8-1	7-9	7-5	7-1	8-6	8-1	6-5	6-1
9-3	9-2	8-9	7-4	7-1	6-10	6-6	7-9	7-4	5-10	5-7
8-6	8-3	7-10	6-7	6-4	6-1	5-10	6-11	6-7	5-3	5-0
13-11	13-8	13-2	11-3	10-11	10-5	10-0				
13-3	12-9	12-3	10-6	10-2	9-9	9-5				
12-3	11-10	11-4	9-9	9-5	9-0	8-8				
11-3	10-10	10-5	8-10	8-7	8-3	7-11				
10-0	9-8	9-3	7-11	7-8	7-4	7-1				
17-0	16-8	16-1	13-8	13-4	12-8	12-3				
16-3	15-8	15-0	12-10	12-5	11-10	11-6				
15-0	14-6	13-11	11-10	11-6	11-0	10-7				
13-8	13-3	12-8	10-10	10-6	10-0	9-8				
12-3	11-10	11-4	9-8	9-5	9-0	8-8				
22-5	22-0	21-2	18-1	17-7	16-9	16-2				
21-5	20-7	19-10	16-11	16-5	15-8	15-1				
19-10	19-1	18-4	15-8	15-2	14-6	14-0				
18-1	17-5	16-9	14-3	13-10	13-3	12-9				
16-2	15-7	15-0	12-9	12-5	11-10	11-5				
28-7	28-1	27-0	23-1	22-5	21-4	20-7				
27-3	26-3	25-3	21-7	20-11	20-0	19-3				
25-3	24-4	23-5	20-0	19-5	18-6	17-10				
23-1	22-3	21-4	18-3	17-8	16-10	16-4				
20-7	19-10	19-1	16-4	15-10	15-1	14-7				

TABLE 8-2

Maximum Span for Southern Pine Ceiling Joists (ft-in) (Courtesy Southern Forest Products Association)

Size in.	Grade / spacing in. o.c.	Dense Sel Str KD and No. 1 Dense KD	Dense Sel Str, Sel Str KD, No. 1 Dense and No. 1 KD	Sel Str, No. 1 and No. 2 Dense KD	No. 2 Dense	No. 2 KD	No. 2	No. 3 Dense KD	No. 3 Dense	No. 3 KD	No. 3	Con-struc-tion KD	Con-struc-tion	Stand-ard KD	Stand-ard
2 × 4	12.0	10-5	10-3	10-0	9-10	9-10	9-10	**8-10**	**8-6**	**8-2**	**7-9**	**9-3**	**8-10**	**7-0**	**6-9**
	13.7	10-0	9-9	9-7	9-5	9-5	9-5	**8-3**	**8-0**	**7-8**	**7-3**	**8-8**	**8-3**	**6-7**	**6-3**
	16.0	9-6	9-4	9-1	8-11	8-11	8-11	**7-8**	**7-4**	**7-1**	**6-9**	**8-0**	**7-8**	**6-1**	**5-10**
	19.2	8-11	8-9	8-7	8-5	8-5	**8-3**	**7-0**	**6-9**	**6-5**	**6-2**	**7-4**	**7-0**	**5-6**	**5-4**
	24.0	8-3	8-1	8-0	7-10	**7-9**	**7-5**	**6-3**	**6-0**	**5-9**	**5-6**	**6-7**	**6-3**	**4-11**	**4-9**
2 × 5	12.0	13-5	13-2	12-11	12-8	12-8	**12-6**	**10-8**	**10-4**	**9-10**	**9-6**				
	13.7	12-10	12-7	12-4	12-1	12-1	**11-8**	**9-11**	**9-8**	**9-3**	**8-11**				
	16.0	12-2	11-11	11-9	11-6	**11-3**	**10-9**	**9-3**	**8-11**	**8-6**	**8-3**				
	19.2	11-5	11-3	11-0[2]	**10-8**	**10-3**	**9-10**	**8-5**	**8-2**	**7-9**	**7-6**				
	24.0	10-8	10-5[1]	10-3[2,3]	**9-6**	**9-2**	**8-10**	**7-6**	**7-4**	**7-0**	**6-9**				

	12.0	16-4	16-1	15-9	15-6	15-6	**15-3**	**13-0**	**12-8**	**12-0**	**11-8**				
	13.7	15-8	15-5	15-1	14-9	14-9	**14-3**	**12-2**	**11-10**	**11-3**	**10-11**				
2 × 6	16.0	14-11	14-7	14-4	14-1	**13-9**	**13-2**	**11-3**	**10-11**	**10-5**	**10-1**				
	19.2	14-0	13-9	13-6[2]	**13-0**	**12-6**	**12-0**	**10-3**	**10-0**	**9-6**	**9-2**				
	24.0	13-0	12-9[1]	12-6[2,3]	**11-8**	**11-2**	**10-9**	**9-2**	**8-11**	**8-6**	**8-3**				
	12.0	21-7	21-2	20-10	20-5	20-5	**20-1**	**17-2**	**16-8**	**15-10**	**15-4**				
	13.7	20-8	20-3	19-11	19-6	19-6	**18-9**	**16-0**	**15-7**	**14-10**	**14-4**				
2 × 8	16.0	19-7	19-3	18-11	18-6	**18-1**	**17-5**	**14-10**	**14-5**	**13-9**	**13-3**				
	19.2	18-5	18-2	17-9[2]	**17-2**	**16-6**	**15-10**	**13-7**	**13-2**	**12-7**	**12-1**				
	24.0	17-2	16-10[1]	16-6[2,3]	**15-4**	**14-9**	**14-2**	**12-1**	**11-9**	**11-3**	**10-10**				
	12.0	27-6	27-1	26-6	26-0	26-0	**25-7**	**21-10**	**21-3**	**20-3**	**19-7**				
	13.7	26-4	25-10	25-5	24-11	24-11	**24-0**	**20-6**	**19-10**	**18-11**	**18-4**				
2 × 10	16.0	25-0	24-7	24-1	23-8	**23-1**	**22-2**	**18-11**	**18-5**	**17-6**	**16-11**				
	19.2	23-7	23-2	22-8[2]	**21-10**	**21-1**	**20-3**	**17-3**	**16-9**	**16-0**	**15-6**				
	24.0	21-10	21-6[1]	21-1[2,3]	**19-7**	**18-10**	**18-1**	**15-6**	**15-0**	**14-4**	**13-10**				

Ceiling joists—Drywall ceiling—20 psf live load. No future sleeping rooms but limited storage available. (Spans shown in light face type are based on a deflection limitation of $l/240$. Spans shown in a bold face type are limited by the recommended extreme fiber stress in bending value of the grade and includes a 10 psf dead load.)

[1]The span for No. 1 KD grade, 24 inches o.c. is: 2×5, 10-2; 2×6, 12-5; 2×8, 16-5; 2×10, 20-11.

[2]The span for No. 1 grade is: 2×5, 19.2 o.c., 10-10; 24 o.c., 9-8; 2×6, 19.2 o.c., 13-3; 24 o.c., 11-10; 2×8, 19.2 o.c., 17-5; 24 o.c., 15-7; 2×10, 19.2 o.c., 22-3; 24 o.c.; 19-11.

[3]The span for No. 2 Dense KD grade, 24 inches o.c. is: 2×5, 10-0; 2×6, 12-3; 2×8, 16-2; 2×10, 20-7.

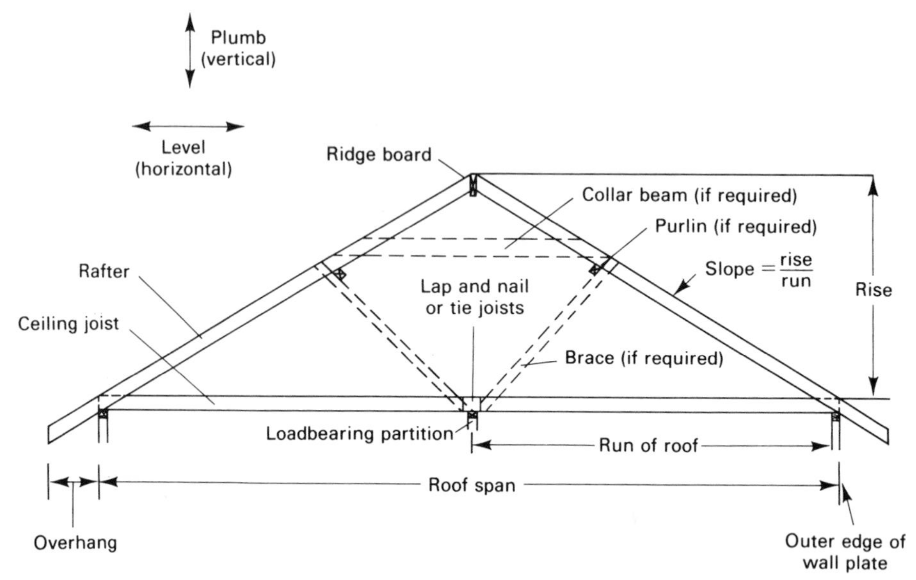

Figure 8-4 Typical rafter frame.

However, the roof *pitch* equals the rise divided by the building width or roof span. Thus, pitch is numerically equal to one-half the roof slope.

In the United States, roof slope is expressed as the rise (in.) in a 12 in. (30 cm) run. For example, a roof that rises 3 in. (7.6 cm) per foot (30 cm) of horizontal distance has a slope of 3 in 12 or 3/12. If the building's width is 24 ft. (7.3 m), the rise would be 3 ft. (0.9 m) in a run of 12 ft. (3.7 m) for a slope of 3/12. However, the roof's pitch equals the 3 ft. rise divided by the 24 ft. roof span or 1/8 pitch.

A vertical cut, such as that at the ridge end of the rafter, is called a *plumb cut,* and a horizontal cut, such as that at the wall plate, is called a *level cut.* The level cut in the rafter where the rafter rests on the wall plate is often called a *heel cut* or *seat cut.*

Nominal 2 in. (50 mm) lumber is commonly used for rafters and ceiling joists. Ceiling joists should be installed before rafter installation is begun. The ceiling joists should be fastened so that they form a continuous tie between opposing wall plates. This holds the walls in alignment while rafters are installed. When ceiling joists are parallel to rafters, the joists also act to resist the lateral thrust induced by the rafters. However, when ceiling joists are not parallel to rafters, *rafter ties* should be installed between the ends of rafter pairs to serve as the tension member of the roof frame. When used, rafter ties should be spaced not more than 4 ft. (1.2 m) on center and located as close to the wall plate as possible.

The ridge board should be made of nominal 1 in. (25 mm) or thicker lumber. The depth of the ridge board should be at least equal to the depth of the cut end of the rafter fastened to it. Rafters are placed directly opposite each other and nailed to the ridge board. *Collar beams,* usually nominal 1 × 6 in. (25 × 150 mm) lumber, are often installed on every third rafter pair. When used, collar beams should be located in the upper third of the roof height. Some codes require that bracing tied together with purlins be installed between rafters and intermediate supports (bearing walls). *Purlins* are horizontal beams installed between rafters. When used to provide lateral support to rafters, they should be installed at the junction of rafters and braces as illustrated in Figure 8-4.

Rafters are usually notched to bear on wall plates. At least 1 1/2 in. (38 mm) of bearing should be provided by rafters or ceiling joists resting on supports of wood or metal. This should be increased to 3 in. (76 mm) on concrete. However, notches at the end of rafters or joists should not exceed one-fourth of their depth. Other notches in joists should not exceed one-sixth of the joist depth and should not be located in the middle third of the joist span. As an exception, a notch of up to one-third of the rafter or joist depth may be allowed when located in the top of the member within a distance equal to the depth of the member from a support. The diameter of any holes bored in rafters or ceiling joists should not exceed one-third of the depth of the member and should not be located closer than 2 in. (50 mm) to the edge of the member. Cantilevered members should not be notched without performing a structural analysis of the member.

Lateral support of rafters and ceiling joists is required at each support. When the depth-to-width ratio of the member is greater than 6, bridging is required at intervals not exceeding 10 ft. (3.1 m).

When the roof overhangs the gable end of the building more than about 8 in. (20 cm), special framing, called a *ladder,* is required for the gable overhang (Figure 8-5). Notice that the end rafter *(fly rafter)* is supported by cantilever beams called *lookout* beams. Lookouts are usually spaced 16 or 24 in. (41 or 61 cm) apart. Nailing blocks may be used to assist in attaching roof sheathing.

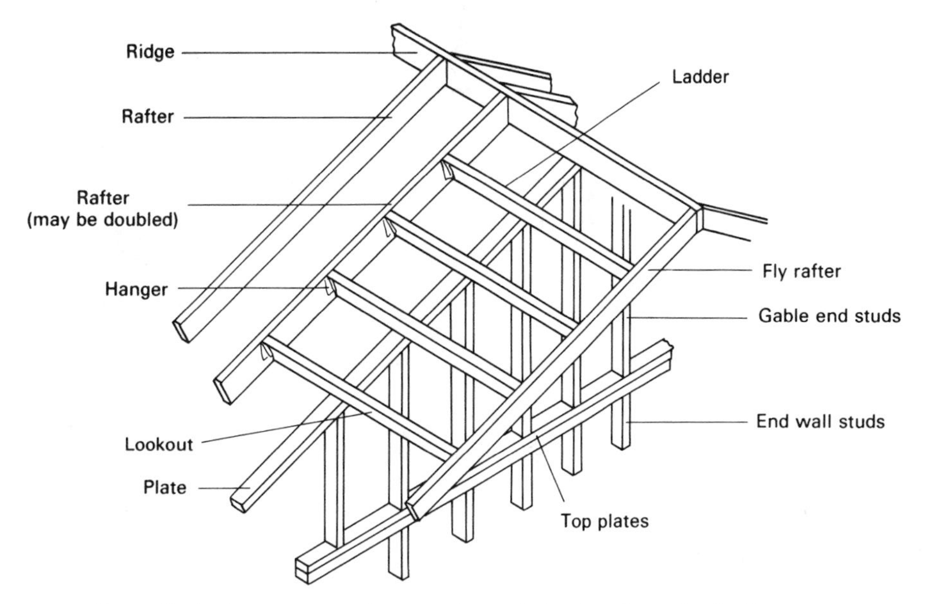

Figure 8-5 Framing a gable overhang (adpated from U. S. Department of Agriculture).

Hip Roofs

The framing for a hip roof is illustrated in Figure 8-6. Notice the use of *hip rafters* and *valley rafters* where roof planes intersect. *Jack rafters* are shorter than common rafters and extend from the fascia to a hip rafter or from the ridge board to a valley rafter. Hip and valley rafters should be nominal 2 in. (50 mm) or thicker lumber with a depth at least as great as the cut end of the attached jack rafters.

Dormer Framing

The framing for a typical gable dormer is illustrated in Figure 8-7. Double trimmer rafters and headers are usually required to surround the dormer opening.

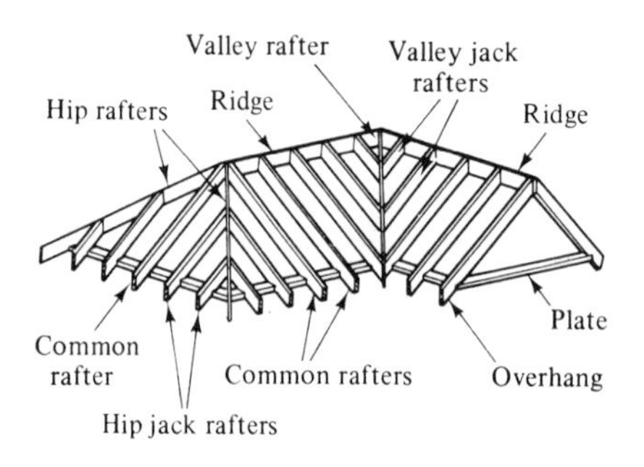

Figure 8-6 Hip roof framing (Paul Calter, *Practical Math Handbook for the Building Trades.* © 1983, p. 210. Reproduced by permission of Prentice-Hall, Inc., Englewood Cliffs, NJ).

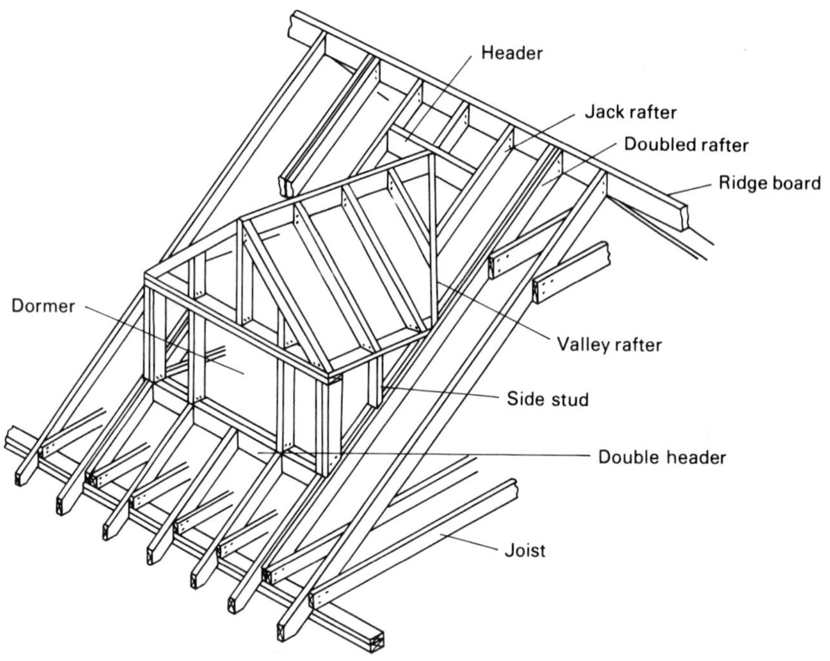

Figure 8-7 Framing a gable dormer (adapted from U. S. Department of Agriculture).

8-3 LAYING OUT RAFTERS

Measuring Rafters

It is important that rafters be accurately measured and cut so that they fit snugly against plates, ridges, and other framing. The principal methods by which the length of a rafter may be determined include:

1. Calculating the length using the properties of right triangles.
2. Using a rafter table (found on many framing squares).
3. Stepping off the length with a framing square.

Although the pocket electronic calculator greatly simplifies mathematical calculations, the geometry of some rafters is complex, and it is easy to make errors when computing rafter length. The method employing rafter tables is widely used and is described in the following paragraphs. Although some carpenters use the third method, it is time consuming and subject to error.

All methods for measuring and marking rafter cuts involve some use of the *framing square,* one of the carpenter's most useful tools. The longer and wider blade of the square (see Figure 8-9) is the *body,* and the smaller blade is the *tongue.* The side of the square containing the manufacturer's trademark is the *face,* and the reverse side is the *back.* The inside and outside edges of the face and back are engraved with measuring scales. In the United States, scale divisions are marked in inches with subdivisions in eighths and sixteenths. Some scales also include subdivisions of tenths and twelfths of an inch as well as tables to convert from hundredths of an inch to sixteenths of an inch. The rafter table described below is often engraved on the face of the body of the square.

Common Rafters

The "line length" of a common rafter (also called the "measuring line length") is defined as the shortest distance between the outer edge of the plate and the center of the ridge (see Figure 8-8). This distance should be measured along the line *A-B* parallel to the rafter edges. Note that the measuring line *(A-B)* forms the hypotenuse of the right triangle *A-B-C* formed by the measuring line *(A-B),* the rise *(B-C),* and the run *(C-A).* Also notice that the distance *D-F* along the rafter edge is equal to the measuring line length *(A-B).*

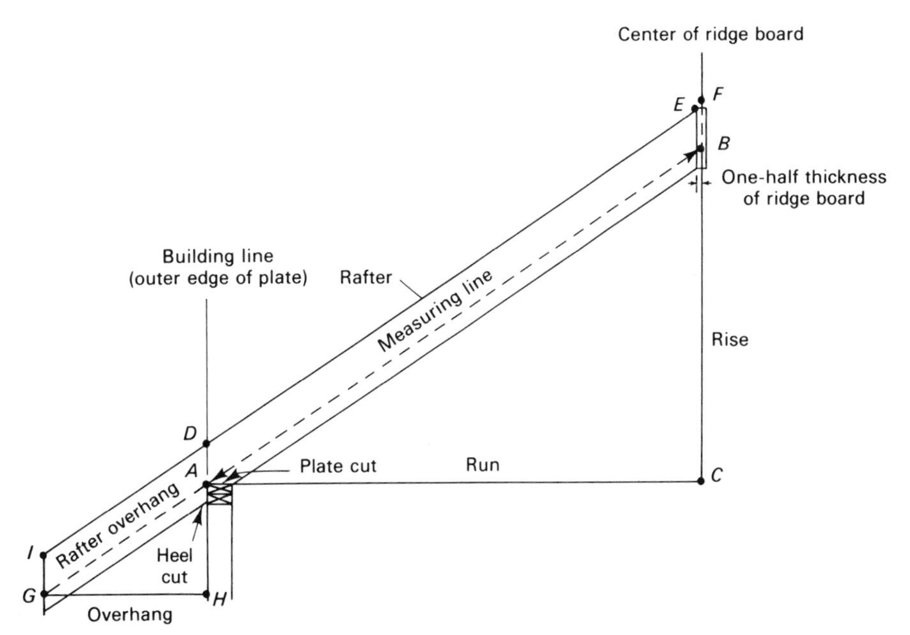

Figure 8-8 Measuring a rafter.

The *rafter table* often found on framing squares is reproduced in Table 8-3. For each roof slope, it indicates the rafter length per foot of run, the length of the first jack rafter (which is also the difference in length between each succeeding jack rafter), and the side cut required for jack, hip, and valley rafters.

To find the line length *(A-B)* in inches for a common rafter, simply multiply the run *(A-C)* in feet by the table value "length per ft. run, common rafter" for the roof slope. The length of the rafter overhang *(A-G = D-I)* is calculated by multiplying the rafter table factor by the run of the overhang *(G-H).* The total rafter length from the center of the ridge to the end as measured along the edge *(F-I)* is the sum of *F-D* and *D-I.* The procedure is illustrated in Example 8-1.

TABLE 8-3
Rafter Table

	Length per ft. of run		*Length of 1st jack & difference in length*		*Side cut*	
Slope in./12	*Common*	*Hip or valley*	*16″ o.c.*	*24″ o.c.*	*Jack*	*Hip or valley*
2	12.16	17.09	16-1/4	24-5/16	11-13/16	11-15/16
3	12.37	17.23	16-1/2	24-3/4	11-5/8	11-13/16
4	12.65	17.44	16-7/8	25-5/16	11-3/8	11-11/16
5	13.00	17.69	17-5/16	26	11-1/16	11-1/2
6	13.42	18.00	17-7/8	28-13/16	10-3/4	11-5/16
7	13.89	18.36	18-1/2	27-3/4	10-3/8	11-1/16
8	14.42	18.76	19-1/4	28-7/8	10	10-7/8
9	15.00	19.21	20	30	9-5/8	10-5/8
10	15.62	19.70	20-13/16	31-1/4	9-1/4	10-3/8
11	16.28	20.22	21-11/16	32-9/16	8-7/8	10-1/8
12	16.97	20.78	22-5/8	33-15/16	8-1/2	9-7/8
13	17.69	21.38	23-9/16	35-3/8	8-1/8	9-5/8
14	18.44	22.00	24-9/16	36-7/8	7-13/16	9-3/8
15	19.21	22.65	25-5/8	38-7/16	7-1/2	9-1/16
16	20.00	23.32	26-11/16	40	7-3/16	8-3/4
17	20.81	24.02	27-3/4	41-5/8	6-15/16	8-1/2
18	21.63	24.74	28-7/8	43-1/4	6-11/16	8-1/4

EXAMPLE 8-1

Problem: Find the line length, rafter overhang length, and total length from the center of the ridge of a common rafter when the roof slope is 4/12, the building width is 24 ft., the roof overhang is 2 ft., and the ridge board is nominal 2 in. lumber.

Solution:

Length/ft. run = 12.65 (Table 8-3)

$$\text{Run} = \frac{\text{span}}{2} = \frac{24}{2} = 12 \text{ ft.}$$

Line length *(A-B = D-F)* = 12.65 × 12 = 151.8 in.
Run of overhang = 2 ft.
Rafter overhang *(A-G = D-I)* = 12.65 × 2 = 25.3 in.
Length from ridge center *(F-I)* = 151.8 + 25.3 = 177.1 in.
= 14 ft. 9 1/8 in.

The usual procedure for laying out a rafter is described below. In this procedure, no calculation is made of the final rafter length *(E-I)*. Rather, this length is measured after the fact. The rafter is laid out from the center of the ridge and later shortened by one-half the thickness of the ridge board. However, it is possible to calculate the distance along the rafter edge from the ridge board to the end of the rafter *(E-I)*. The distance from the ridge board to the building line *(E-D)* can be found by multiplying the rafter table factor by the run less one-half the thickness of the ridge board (in feet). The final rafter length *(E-I)* can then be found by adding the rafter overhang to this distance. That is, *E-I = E-D + D-I*. The procedure is illustrated in Example 8-2.

EXAMPLE 8-2

Problem: Find the actual rafter length along its top edge *(E-I)* for the rafter of Example 8-1.

Solution:

Length/ft. run = 12.65 (Table 8-3)

Run = 12 ft.

$$\text{Run} - 1/2 \text{ ridge board} = 12.0 - \frac{1.5}{2 \times 12}$$

$$= 12.0 - 0.0625 = 11.9375 \text{ ft.}$$

Distance $D\text{-}E = 12.65 \times 11.9375 = 151$ in.

Rafter overhang $(A\text{-}G = D\text{-}I) = 12.65 \times 2 = 25.3$ in.

Rafter length $(E\text{-}I) = 151 + 25.3 = 176.3$ in.

$= 14$ ft. 8 5/16 in.

From the above examples, it can be seen that the amount by which the rafter is shortened (as measured along the rafter edge) to accommodate the ridge board is the difference between the rafter lengths found in Examples 8-1 and 8-2. Thus, the distance $E\text{-}F = 177.1 - 176.3 = 0.8$ in. By comparison, one-half the thickness of the ridge board (measured perpendicular to a plumb line) is 0.75 in. Although the difference in these values is small for a roof slope of 4/12, the difference increases with increasing roof slope.

The procedure for laying out a common rafter with a slope of 4/12 is as follows (see Figure 8-9). First, select the straightest available board of the required size to serve as a pattern rafter. Place the framing square on the rafter in position 1 so that "12 in." on the edge of the body and "4 in." on the edge of the tongue line up with the edge of the rafter. When the square is aligned in this manner, any line drawn along the edge of the tongue will be plumb (vertical), and any line drawn along the body will be level (horizontal) for a roof slope of 4/12.

Move the square to the end of the board and draw the plumb line for the ridge center *(A-B)*. Measure off perpendicular to this line a distance equal to one-half the thickness of the ridge board and draw another plumb line *(C-D)*. This line represents the cutting line for the ridge board end of the rafter.

Now, measure along the rafter edge a distance from point *A* equal to the line length. With the square in position 2, draw a plumb line at this point. This plumb

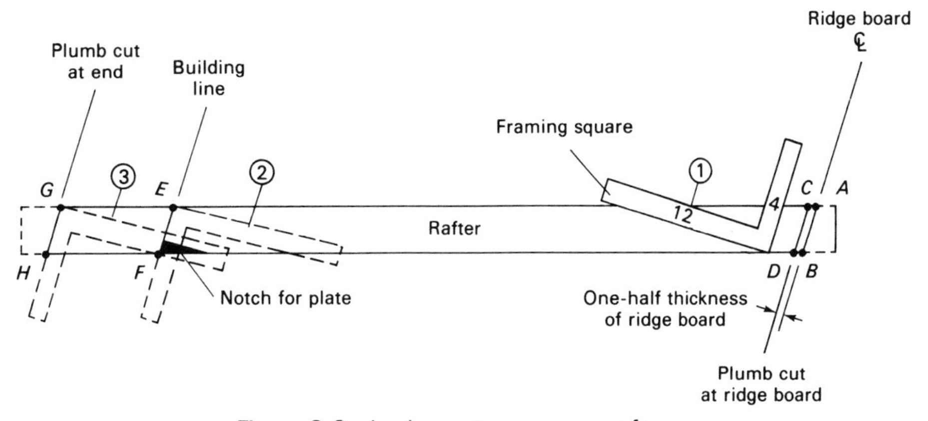

Figure 8-9 Laying out a common rafter.

line *(E-F)* represents the building line. Mark off the desired depth of the seat cut and draw a level line through this point to mark the plate cut.

Next, measure off from point E a distance corresponding to the rafter overhang. With the square in position 3, draw a plumb line at this point representing the cutting line for the outer end of the rafter. If a fascia board is to be attached to the end of the rafter, shorten the rafter overhang by the thickness of the fascia board (measured perpendicular to the plumb line).

When the rafter has been cut as marked, use this rafter as a pattern for cutting all the common rafters for this roof.

Hip and Valley Rafters

As seen in Figure 8-6, hip and valley rafters are placed at a 45° angle to the common rafters. As a result, their run *per 12 in. of common rafter run* is approximately 17 in. (actually $12 \times \sqrt{2} = 16.97$ in.). Using this factor, the line length of a hip or valley rafter could be calculated as the hypotenuse of a right triangle. However, the rafter table (Table 8-3) provides multiplying factors for the line length of a hip or valley rafter per foot of common rafter run for each roof slope. The line length from the center of the ridge to the building line can be found by multiplying this factor by the run of the common rafters. The procedure for calculating rafter length is illustrated by the following example.

EXAMPLE 8-3

Problem: Find the line length and rafter overhang for a hip rafter for the building of Example 8-1.

Solution:

Common rafter run = 12 ft.
Roof slope = 4/12
Roof overhang = 2 ft.
Length/ft. run = 17.44 (Table 8-3)
Line length = $17.44 \times 12 = 209.28$ in.
Overhang length = $17.44 \times 2 = 34.88$ in.

Laying out a hip or valley rafter is basically the same as laying out a common rafter. However, a scale value of "17 in." (rather than "12 in.") must be used on the body of the framing square while the roof slope (in. per 12 in.) is used on the tongue of the square to obtain plumb and level cuts. With the framing square in this position, any line drawn along the tongue of the square is plumb, and any line drawn along the body of the square is level. The procedure for measuring and marking end cuts and the birdsmouth notch are described below and illustrated in Figure 8-10.

The ridge end of the rafter (both hip and valley) must first be shortened to account for the thickness of the ridge board (the horizontal distance between lines *A-B* and *C-D* in Figure 8-10(a). Since the rafter is placed at a 45° angle to the ridge, the amount of shortening is calculated using Equation 8-1.

$$\text{Shortening} = 0.707 \times t \qquad (8\text{-}1)$$

where t = thickness of ridge board

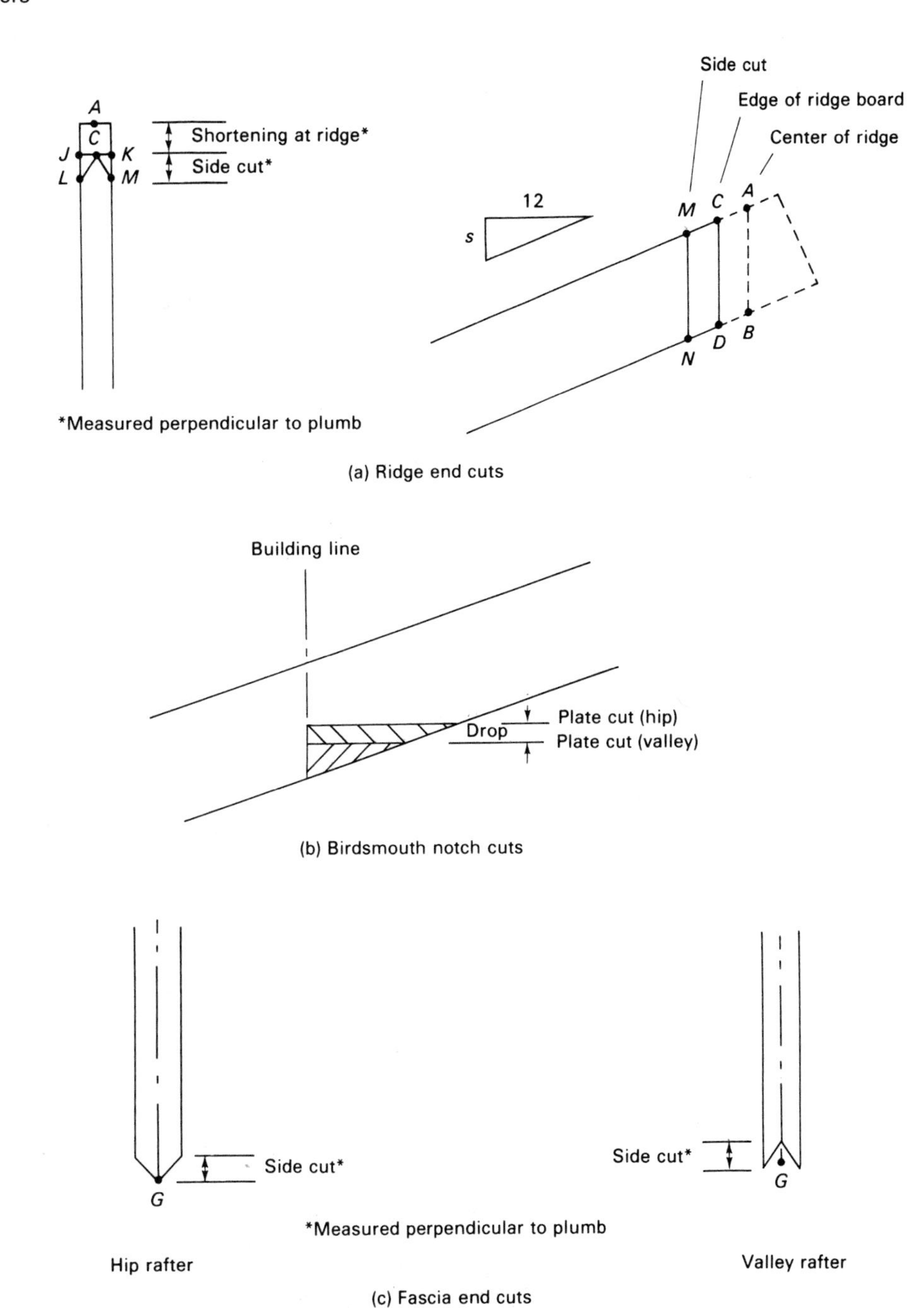

Figure 8-10 Laying out hip and valley rafters.

To locate line *C-D,* draw a plumb line *(A-B)* on the side of the rafter opposite the ridge center (point *A).* Measure off perpendicular to this line the distance calculated using Equation 8-1 (distance *A-C).* Draw the plumb line *C-D* through this point. This line represents the line on which the rafter center line will intersect the side of the ridge board.

The ridge end must now be angled or side cut on each side to fit between the common rafters or ridge boards. Draw plumb lines on each side of the rafter at *J* and *K* opposite point *C.* Measure off perpendicular to the plumb lines at *J* and *K* the distance obtained from Equation 8-2.

$$\text{Side cut} = \frac{t}{2} \qquad (8\text{-}2)$$

where t = rafter thickness

Now draw plumb lines through these points *(L* and *M)* on each side of the rafter. Finally, connect point C with points L and M on the rafter edges. These lines and the plumb lines through L and M represent the cutting lines for the rafter ridge end.

The procedure for making the plumb and level cuts for the birdsmouth notch is similar to that used for common rafters. Again, remember to use "17 in." on the body of the square instead of "12 in." to align the square for plumb and level cuts. In addition, the depth of the birdsmouth notch *in a hip rafter* must be increased by a certain amount ("drop") to permit the top edges of the hip rafter to line up with the tops of the common rafters [see Figure 8-10(b)]. One method of calculating the required drop is the use of Equation 8-3.

$$\text{Drop (in.)} = \frac{s}{17} \times \frac{t}{2} \qquad (8\text{-}3)$$

where s = roof slope per 12 in. (in.)
t = thickness of ridge board (in.)

The cuts for the fascia end of the rafter are illustrated in Figure 8-10(c). A plumb cut through point G represents the end of the rafter at the center line. Hip rafters must be angled by side cuts on each side. Follow the same procedure as described for end cuts at the ridge board except that one-half the rafter thickness is measured off *toward the ridge* on each side. If a fascia board is to be attached to the end of the rafter, the rafter must be shortened by the following amount before making the side cuts.

$$\text{Shortening at fascia (in.)} = 1.414 \times t \qquad (8\text{-}4)$$

where t = thickness of fascia (in.)

For valley rafters, end cuts at the fascia are angled in exactly the reverse of those for hip rafters [see Figure 8-10(c)]. Again, if a fascia board is to be attached to the end of the rafter, the rafter must be shortened by the amount obtained from Equation 8-4 before laying out the side cuts.

There is an alternate method for laying out rafter side cuts employing the framing square and rafter table. To use this method (see Figure 8-11), locate the factor in Table 8-3 under the column headed "Side Cut, Hip or Valley" corresponding to the roof slope. For a roof slope of 4/12, this factor is 11-11/16. Now line up this factor on the edge of the square body and "12 in." on the edge of the tongue with the rafter edge. A line drawn along the tongue will mark the proper side cut angle. Draw a line through point C to one side of the rafter *(C-L),* then reverse the square and draw a similar line to the other side of the rafter. Lines *C-L* and *C-M* [see Figure 8-10(a)] and plumb lines through L and M represent the cutting lines for the side cuts.

Jack Rafters

Jack rafters ("jacks") are common rafters that have been shortened to fit against hip or valley rafters (see Figure 8-6). Jacks extending from the fascia to a hip rafter are *hip jack rafters.* Jacks extending from the ridge board to a valley rafter are

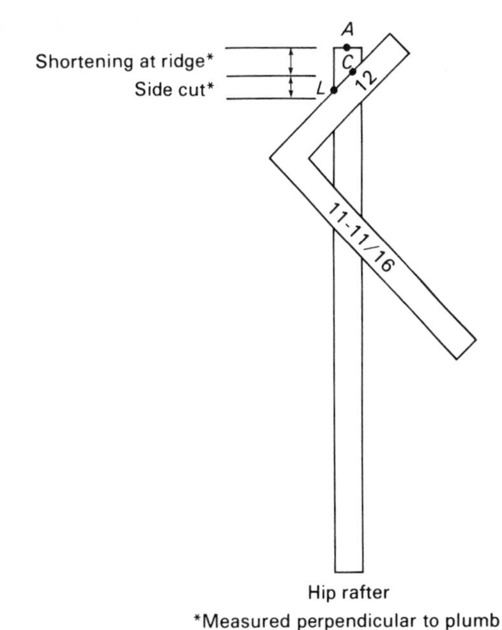

Figure 8-11 Using the framing square to mark a side cut.

valley jack rafters. The procedure for laying out jack rafters is similar to that used for laying out common rafters except for rafter length and for the side cut required for the jack rafter to fit against a hip or valley rafter.

Find the jack length factor in the rafter table (Table 8-3) under the column headed "Length of 1st jack & difference in length" for the specified roof slope and rafter spacing. This factor represents the amount by which the line length of each successive jack rafter must be shortened or lengthened compared to the length of the previous rafter. This calculation is illustrated in Example 8-4.

EXAMPLE 8-4

Problem: Find the line length (center of hip rafter to building line) for the hip jack rafter adjacent to the last common rafter for the roof of Example 8-1.

Solution:

Common rafter line length = 151.8 in. (Ex. 8-1)
= 12 ft. 7 3/4 in.

Roof slope = 4/12

Difference in length = 16 7/8 in. (Table 8-3)

Line length of 1st jack = 12 ft. 7 3/4″ − 16 7/8″
= 11 ft. 2 7/8 in.

A side cut must be made on the end of the jack so it will fit against the side of the hip rafter. The required amount of side cut for each side of the jack rafter is given by Equation 8-2.

The procedure for laying out the side cut is similar to that used for hip and valley rafters except that only a single side cut is made across the rafter. Referring to Figure 8-11, point A represents the end of the line length of the jack, that is, the intersection of the center line of the jack with the center line of the hip rafter. From a plumb line opposite point A, measure off perpendicular to the plumb line the shortening distance determined from Equation 8-1 (distance *A-C)* and draw another plumb line through point *J*. Now measure off perpendicular to this plumb line the side cut distance obtained from Equation 8-2 and draw another plumb line (point *L)*. Draw a line along the rafter edge from point *L* through point *C* to the opposite edge (we will call this point *N)*. Finally, draw a plumb line through point *N*. The line *L-C-N* and the plumb lines through *L* and *N* represent the cutting lines for the jack to fit against the hip rafter.

When a circular saw is available, the procedure is even simpler. After the plumb line has been drawn through point *N* on the long side of the jack, set the saw for a 45° angle of cut and cut along the plumb line.

EXAMPLE 8-5

Problem: Find the side cut required on each side of the jack rafter of Example 8-4 in order for the jack rafter to fit against the hip rafter if the jack rafter is nominal 2 in. lumber.

Solution:

$$\text{Side cut} = \frac{t}{2} = \frac{1.5}{2} = 3/4 \text{ in.} \qquad \text{(Eq. 8-2)}$$

Note that the total side cut for a jack rafter (that is, the horizontal distance measured along the rafter side from the plumb line at one end of the cut to a plumb line opposite the other end of the cut is twice the value obtained from Equation 8-2, or 1 1/2 in. for this example.

The side cut for a jack rafter can also be marked using the framing square and rafter table. The procedure is similar to that described earlier for marking side cuts in hip and valley rafters (see Figure 8-11). In this case, the column headed "Side cut, jack" is used to find the proper value to use on the body of the framing square. Mark the end of the measuring line length of the jack (point *A)* and measure off the shortening distance to locate point *C*. Align "12 in." on the tongue of the square and the table value on the body of the square with the rafter edge. Maintaining this alignment, move the square until point *C* falls along the tongue of the square. Draw a line along the tongue across the rafter through point *C* (we will call this line *L-N)*. This line and the plumb lines through *L* and *N* represent the cutting lines for the side cut.

8-4 ROOF TRUSSES

Characteristics of Trusses

Wood roof trusses are finding increasing use in roof framing in place of conventional rafter and joist construction. Since trusses are often supported only by the exterior walls, interior walls and partitions can be nonbearing. This permits great flexibility in interior floor plan and allows interior walls and partitions to be removed or relocated as desired. Other advantages of roof trusses include:

- *High strength.* Trusses are fabricated of selected components in accordance with an engineered design.

- *Controlled quality.* Units are assembled in a controlled environment under factory-like conditions.
- *Less skilled labor required on-site.* There is no need to lay out, cut, and assemble rafter and joist framing.
- *Open web design.* This permits easy installation of electrical, plumbing, and HVAC systems above the ceiling.
- *Economy.* The installed cost of a truss roof frame is usually less than that of a conventional roof frame.

The components of a common (triangular) roof truss and a flat truss are identified in Figure 8-12.

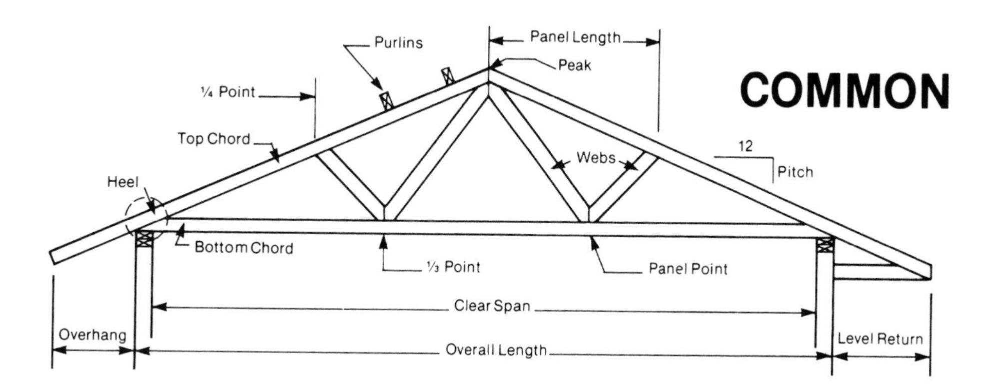

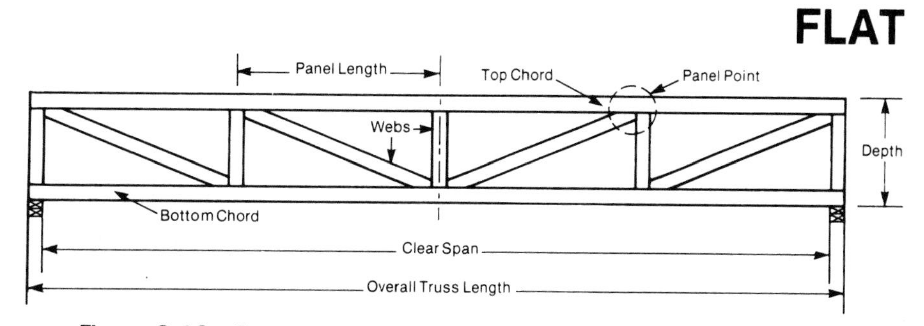

Figure 8-12 Truss components (courtesy Alpine Engineered Products, Inc.)

Types of Trusses

Wood roof trusses are available in many types to accommodate almost any architectural design. Some common types of roof trusses are illustrated and their use described in Figure 8-13. A roof frame utilizing several types of trusses is illustrated in Figure 8-14. Note the method used for constructing hips, valleys, and chimney openings.

Construction Practices

Trusses may be damaged by improper handling during transportation, storage, or erection. Take particular care to avoid excessive lateral bending, which can damage joints and truss members. Trusses are usually delivered to the job site in bundles

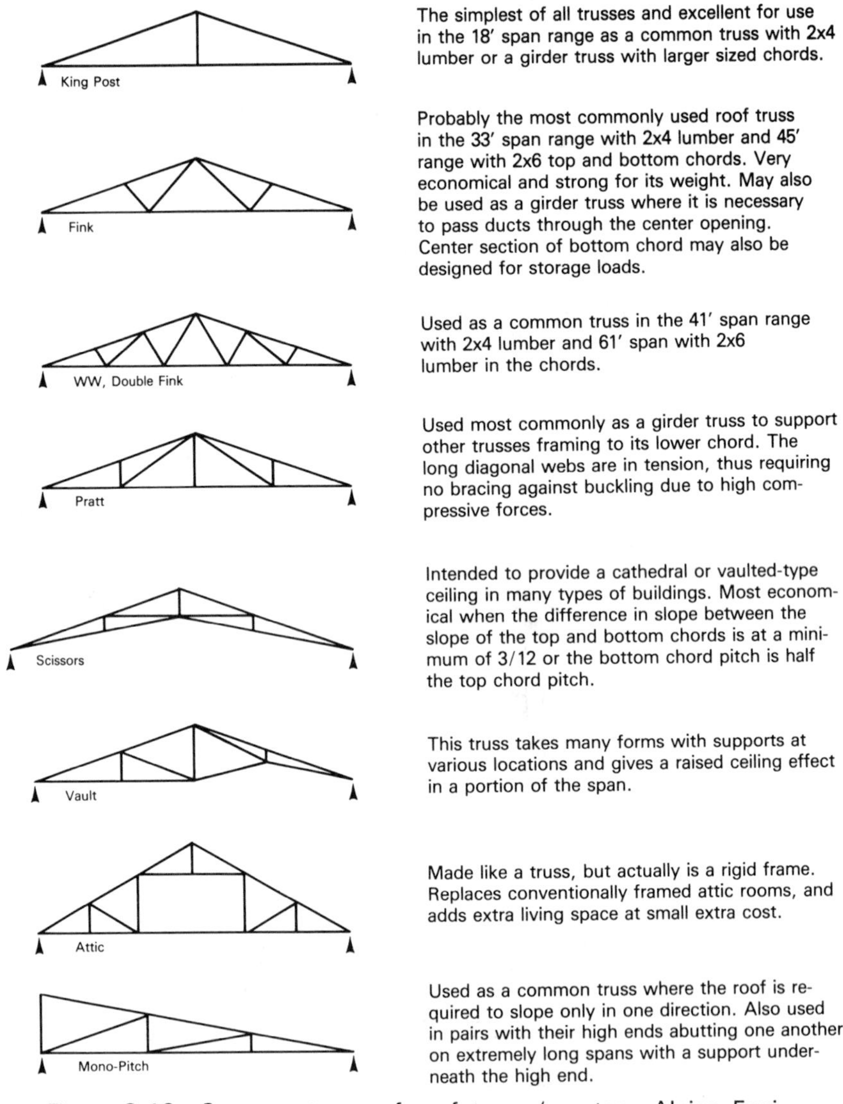

Figure 8-13 Common types of roof truss (courtesy Alpine Engineered Products, Inc.).

banded together by strapping. To minimize damage during storage, trusses should be left bundled until needed. Truss bundles should be stored on blocking, and covered if possible, to minimize water damage. If stored horizontally, trusses should be placed on blocking spaced at intervals not greater than 10 ft. (3 m) on a level surface to prevent excessive lateral bending. If stored vertically, truss bundles should be braced to prevent injury or damage due to overturning.

Trusses shorter than 40 ft. (12 m) can normally be erected satisfactorily by hand. When erected manually, trusses should be raised to the top of the supporting wall in a peak down position. The truss is then rotated into position (peak up) using lifting poles and fastened into place. The use of ramp boards extending from the top of the wall to the ground will allow trusses to be moved to the top of the wall by shoving them up the ramp. This will minimize the lifting required. Larger trusses should be lifted into place by mechanical equipment (crane or forklift) using slings and/or spreader bar. When lifting a truss with a crane, tag lines (see Figure 8-15)

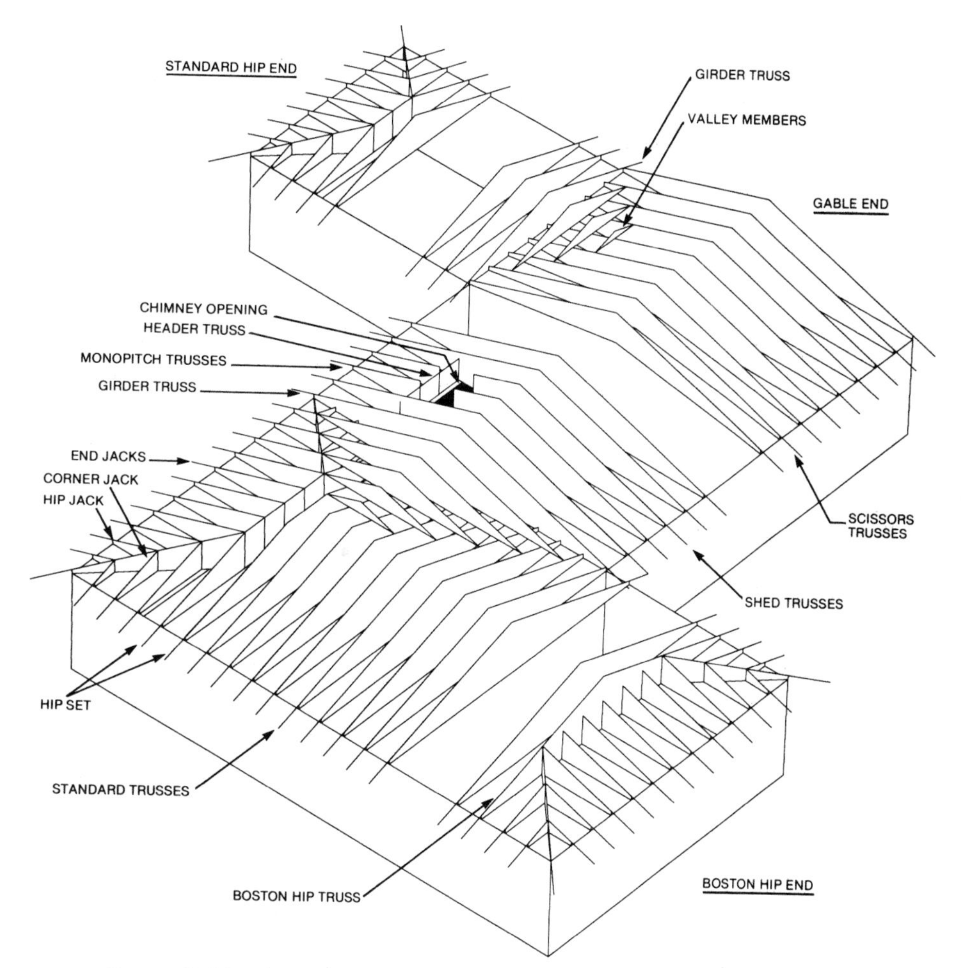

Figure 8-14 A roof frame utilizing trusses (courtesy Alpine Engineered Products, Inc.).

should be utilized to control the swing of the truss and to assist in moving the truss into position. Suggested procedures for lifting trusses using a crane are illustrated in Figure 8-15. Additional guidance on handling and erecting wood trusses is provided in Reference 5.

Lateral bracing of trusses as they are erected is critical to the roof's structural integrity and to the safety of the construction operation. The truss designer will specify the permanent bracing required. However, the builder must provide temporary bracing until the permanent bracing can be put into place. Bracing of the first end truss is critical to the proper alignment of the truss system. Following bracing of the end truss, trusses must be braced in the following three planes:

1. Top chord (sheathing) plane
2. Web (vertical) plane
3. Bottom chord (ceiling) plane

Some suggested methods for bracing trusses are illustrated in Figure 8-16. Additional guidance on bracing trusses is contained in Reference 3. A suggested jobsite inspection checklist is provided in Figure 8-17.

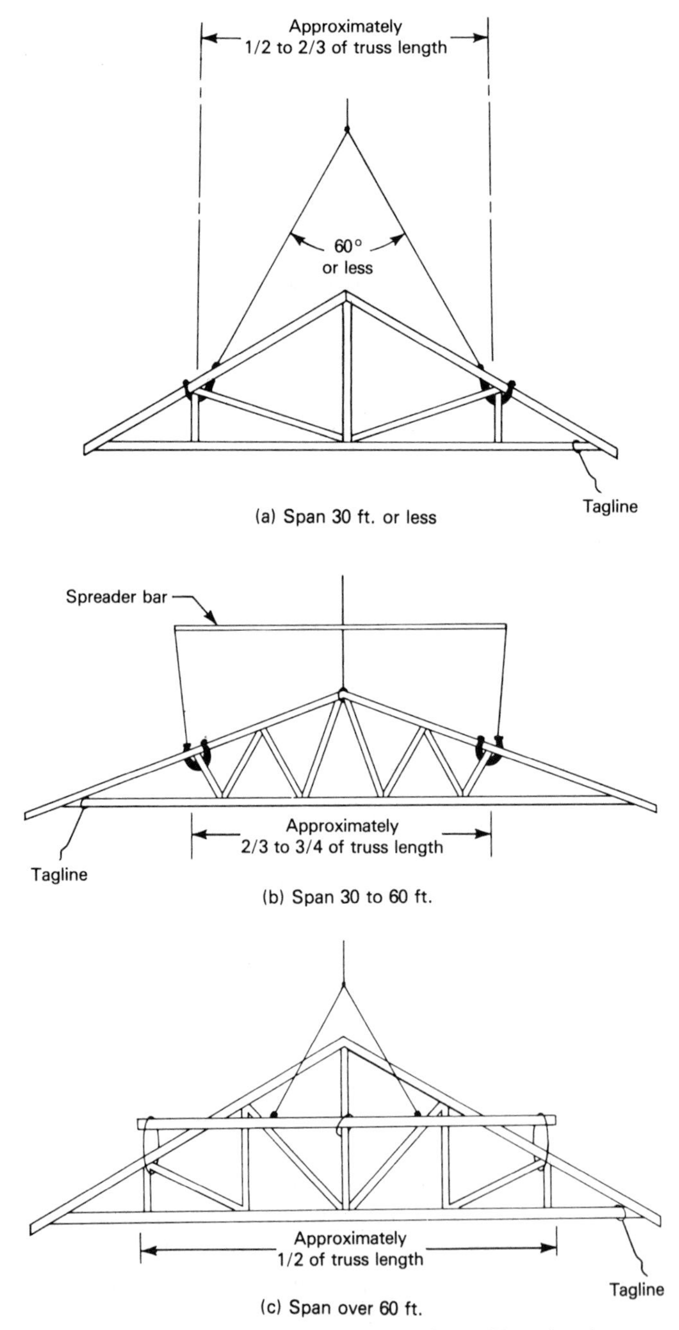

Figure 8-15 Lifting trusses by crane (courtesy Truss Plate Institute, Inc.).

8-5 FASTENERS

A wood structure cannot develop the full strength of its members unless the connections between members are at least as strong as the members themselves. There are a number of different types of fasteners and connectors utilized in frame construction. These include nails, screws, bolts, timber connectors, and light metal framing devices. The allowable load on common wood fasteners can be determined by the methods presented in Reference 7.

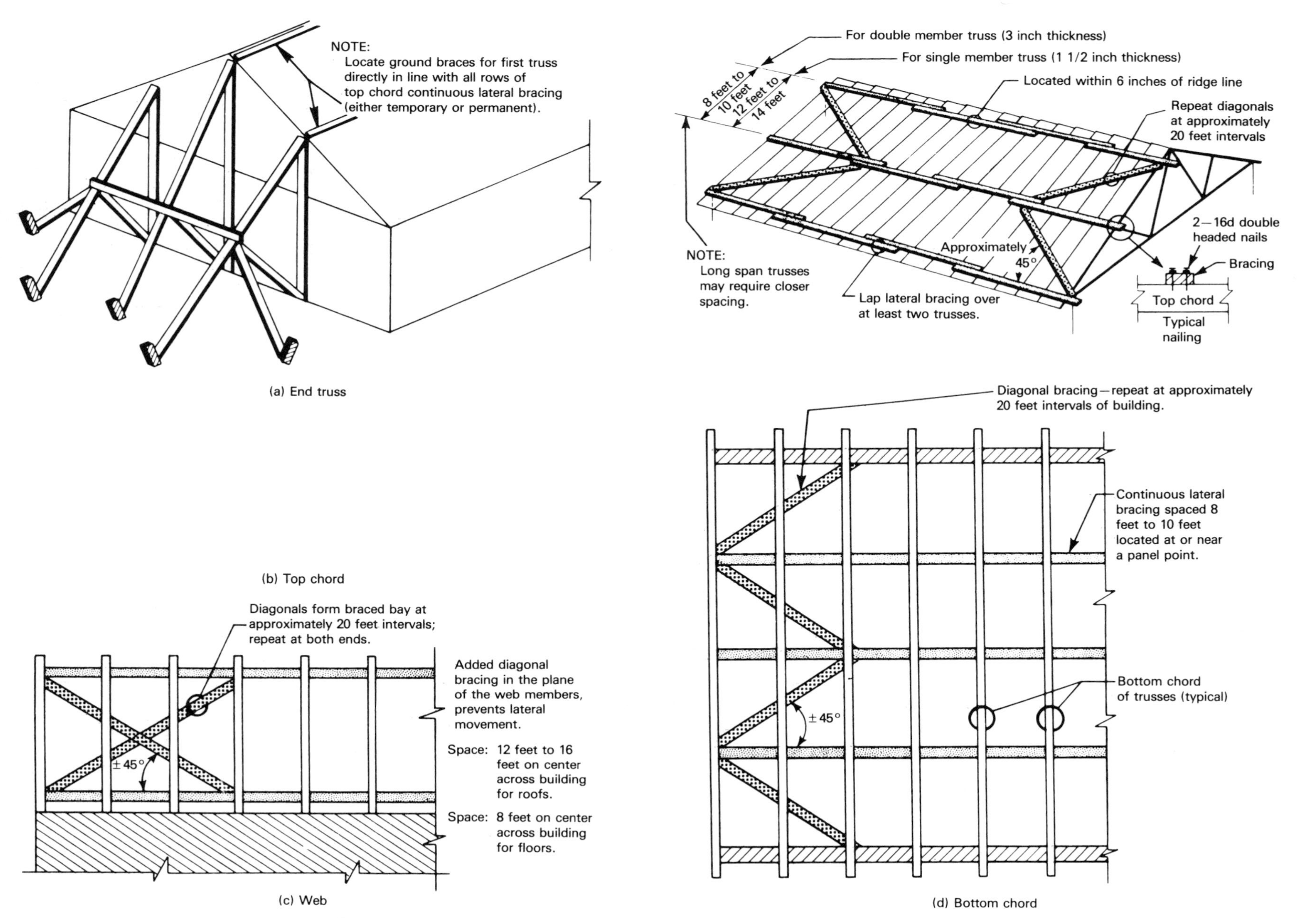

Figure 8-16 Bracing roof trusses (courtesy of Truss Plate Institute, Inc.).

BE SURE TO BRING COPY OF TRUSS DESIGN DRAWINGS FOR READY REFERENCE

		OK	REJECT
A.	**JOBSITE STORAGE (if applicable)**		
1.	Are trusses protected against foul weather?	□	○
2.	Truss bundles intact?	□	○
3.	Trusses supported out of mud, dirt and standing water?	□	○
B.	**BRACING & ERECTION**		
1.	Truss handling techniques proper? (See HET-80)	□	○
2.	Adequate temporary bracing installed during erection? (See BWT-76)	□	○
3.	Are loads being applied to trusses prematurely?	□	○
4.	Is permanent bracing installed as shown on Architect or Engineer's framing plan?	□	○
C.	**CONFIGURATION**		
	Does truss match design drawing?	□	○
		□	○
		□	○
		□	○
		□	○
	If sealed drawings are required has seal been affixed?	□	○
D.	**SOURCE**		
	Is truss manufacturer same as supplier of design drawings?	□	○
	Where code requires approved third-party quality control inspection, does inspector's stamp appear on trusses?	□	○

		GRADE OK	GRADE REJECT	SIZE OK	SIZE REJECT
E.	**LUMBER SIZES & GRADES**				
	Do they match or better design drawing requirements in:				
1.	Top chord?	□	○	□	○
2.	Bottom chord?	□	○	□	○
3.	Web members?	□	○	□	○
4.	Special webs (if required)?	□	○	□	○

		OK	REJECT
F.	**TRUSS CONNECTORS**		
1.	Is connector plate manufacturer the same as specified on drawing?	□	○
2.	Is connector plate size and gauge as specified on all joints?	□	○
3.	Are any joints missing plates?	□	○
4.	Plate position on joint and slotted hole direction in accordance with design?	□	○
G.	**INSTALLATION**		
1.	Cantilevered trusses positioned in correct direction?	□	○
2.	Interior bearing trusses properly positioned?	□	○
3.	Flat trusses right-side up?	□	○
4.	Are all the end walls straight enough to ensure safe and proper bearing?	□	○
5.	Have any of the wall dimensions (thickness of walls) or placement of walls changed to something other than the dimensions called for on the drawing?	□	○
6.	Are trusses being properly nailed to bearing plates, or are the correct type of hangers and fastenings called out in shop drawings being properly applied?	□	○
7.	Are any holes being drilled into the webs or chords?	□	○
8.	Verify location and details of extra trusses required (if any) to handle concentrated loads, stair headers, etc.	□	○
9.	Is truss camber oriented in correct direction (see truss drawings)?	□	○
10.	Is on-center spacing correct?	□	○
H.	**MISHANDLING & ALTERATION**		
1.	Damage due to mishandling?	□	○
2.	Connector plates buckled?	□	○
3.	Missing or broken members?	□	○
4.	Cut, notched, or altered members?	□	○

Wood Truss Jobsite Inspection Check List is only a guide and cannot cover all points and conditions. All points and conditions should comply with sound engineering judgment and construction procedures.

Figure 8-17 Inspection check list for truss installation (courtesy Alpine Engineered Products, Inc.).

Nails

The most widely used type of fastener for frame construction is the nail. Some common types of nails are illustrated in Figure 8-18. For general framing purposes, the *common* nail is most often used. The sizes of common nails are given in Table 8-4. *Finish* nails are used for improved surface appearance. Their heads may be driven flush with the surface or they may be recessed below the surface using a *nail set*. When set, the resulting cavity is filled and painted over to hide the nail's location. *Roofing* nails with their large heads are used to attach soft material such as roofing to sheathing. *Ring shank* and *screw shank* nails have increased resistance to

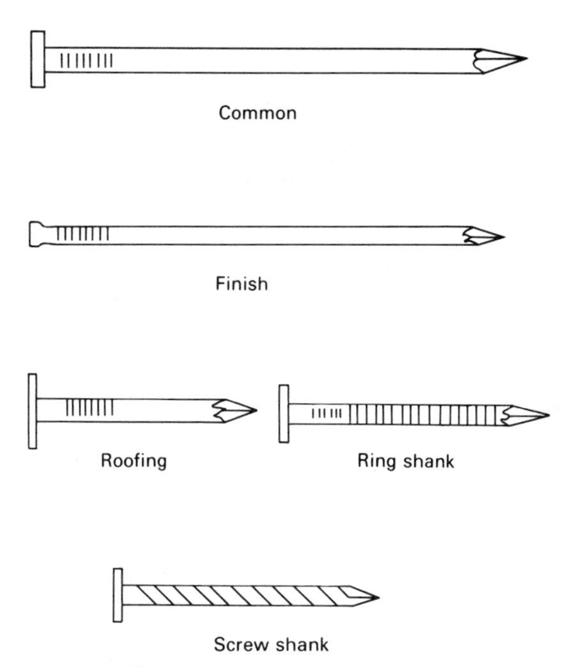

Figure 8-18 Types of nails.

TABLE 8-4
Sizes of Common Nails

Size penney (d)	Length in.	Length cm	Diameter in.	Diameter cm
6	2.00	5.1	0.113	2.9
8	2.50	6.4	0.131	3.3
10	3.00	7.6	0.148	3.8
12	3.25	8.3	0.148	3.8
16	3.50	8.8	0.162	4.1
20	4.00	10.2	0.192	4.9
30	4.50	11.4	0.207	5.3
40	5.00	12.7	0.225	5.7
50	5.50	14.0	0.244	6.2
60	6.00	15.2	0.263	6.7

withdrawal and are often used for attaching sheathing or subfloor to framing. A number of types of nails are available with a galvanized finish to resist rusting.

Suggested fastener schedules for framing and sheathing are provided in Tables 8-5 and 8-6.

Screws

Screws have greater resistance to withdrawal than do nails and are easier to remove without damage. Some common types of wood screws are illustrated in Figure 8-19. The *flathead* screw is most widely used. It is driven into a cone-shaped recess (countersunk) until the head is flush with the surface or recessed below the surface. *Roundhead* screws project above the surface and may be used with washers to increase the effective size of the head. *Oval head* screws are intermediate between flathead and roundhead screws, the head being partly recessed and partly exposed.

TABLE 8-5
Suggested Fastener Schedule—Framing

Members	*Nailing method*	*No. & size fastener*	*Location*
Floor Frame			
Built-up beams & girders	Face	20d	32″ o.c. staggered top & bottom
Ledger to beam or girder	Face	3–16d	Each joist
Header to joist	Face	3–16d	Each joist
Joist to sill or girder	Toe	3–8d	Each joist
Bridging to joist	Toe	2–8d	Each end
Band joist to sill	Toe	10d	16″ o.c.
Joists to studs (balloon)	Face	4–10d	Each joist
Joist lap	Face	3–16d	Each lap
Joist splice, 2″ scab	Face	3–16d	Each joist
Wall Frame			
Sole plate to joist/header	Face	16d	16″ o.c.
Sole plate to stud—horizontal assembly	Face	2–16d	Each stud
Stud to sole place—vertical assembly	Toe	2–16d	Each stud
Double studs	Face	16d	24″ o.c.
Header, 2-pc	Face	16d	16″ o.c. top & bottom
Header to stud	Toe	4–8d	Each stud
Partition stud to ext. stud	Face	16d	16″ o.c.
Top plate to stud	Face	2–16d	Each stud
Upper top plate to lower	Face	16d	16″ o.c.
Upper top plate, laps	Face	2–16d	Each lap
Diagonal brace, 1″	Face	2–8d	Each stud & plate
Built-up corner stud	Face	16d	24″ o.c.
Ceiling & Roof Frame			
Ceiling joist to top plate	Toe	3–8d	Each intersection
Ceiling joist lap	Face	3–16d	Each lap
Rafter to ceiling joist	Face	3–16d	Each joist
Rafter to top plate	Toe	2–16d	Each intersection
Rafter to ridge board, hip, or valley rafter	Toe	4–16d	Each intersection

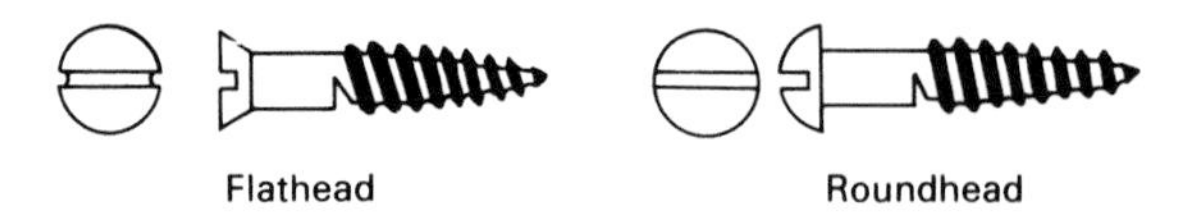

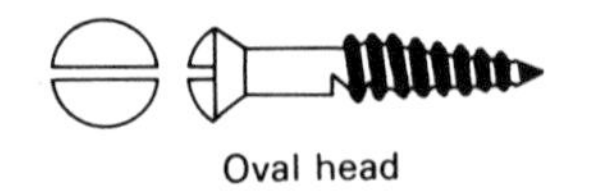

Figure 8-19 Common screws.

TABLE 8-6
Suggested Fastener Schedule—Subfloor and Sheathing

Member	*Nailing method*	*No. & size fastener*	*Location*	
Board Subfloor & Sheathing				
Subfloor to joist:				
1″ × 6″ and smaller	Face	2–8d or 2-1 3/4″ staples	Each joist	
1″ × 8″ and wider	Face	3–8d or 4-1 3/4 staples	Each joist	
2″	Blind & Face	2–16d	Each joist	
Sheathing				
1″ × 6″	Face	2–8d or 2-1 3/4″ staples	Each bearing	
1″ × 8″	Face	2–8d or 3-1 3/4″ staples	Each bearing	
Over 1″ × 8″	Face	3–8d or 4-1 3/4″ staples	Each bearing	
Plywood & Particleboard Combination Subfloor-Underlayment				
			Spacing	
			Edges	*Intermediate*
Nailed only:				
3/4″ or less	Face	6d deformed*	6″	10″
7/8″–1″	Face	8d deformed*	6″	10″
1 1/8″–1 1/4″	Face	8d deformed* or 10d common	6″	6″
Glue-nailed:				
19/32″ to 3/4″	Face	6d deformed* or 10d common	12″	12″
7/8″ and 1″	Face	8d deformed* or 10d common	12″	12″
1 1/8″	Face	8d deformed* or 10d common	6″	6″
Plywood, Particleboard & Other Sheathing				
Plywood & particleboard				
5/16″–1/2″	Face	6d or 16 ga. staples	6″	12″
19/32″–3/4″	Face	8d common or 6d deformed	6″	12″
7/8″–1″	Face	8d	6″	12″
1 1/8″–1 1/4″	Face	10d common or 8d deformed	6″	12″
Other sheathing				
12″ fiberboard	Face	1 1/2″ galv. roof or 6d common or 1 1/8″ staples	3″	6″
25/32″ fiberboard	Face	1 3/4″ galv. roof or 8d common or 1 1/2″ staple	3″	6″
1/2″ gypsum	Face	1 1/2″ galv. roof or 6d common or 1 1/2″ staple	4″	8″

*Ring-shank or screw-shank

They are primarily used for appearance purposes when the head will be exposed. *Sheet metal* screws are designed to be self-tapping in light gauge metal but may also be used in plastics and wood. *One-way* screws are intended to deter removal by vandals. Their head is designed so that the screwdriver blade will slip out of the slot when an attempt is made to remove the screw.

Although most wood screws have a single slot in the head to receive a straight blade screwdriver, other types of head slots are available. The most common of these is the *Phillips* screw, which has a head with a cross-shaped slot designed to receive a Phillips screwdriver.

Some heavy duty screw types are available. The most common is the lag bolt. *Lag bolts* or *lag screws* have square or hex heads designed to be driven by a wrench.

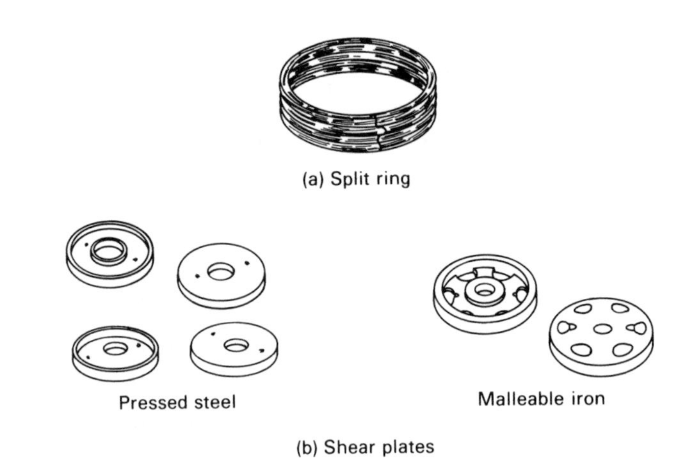

Figure 8-20 Timber connectors (courtesy National Forest Products Associaton, Washington, DC)

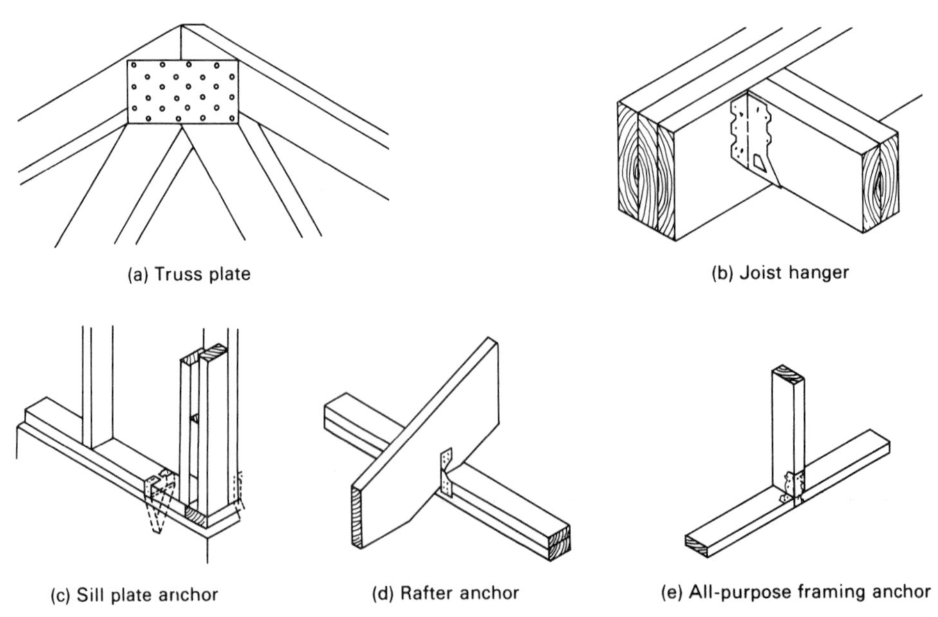

Figure 8-21 Metal framing devices (courtesy TECO Products).

Timber Connectors and Framing Devices

Timber connectors are designed to provide maximum holding power with minimum installation labor. Some common types of connectors are illustrated in Figure 8-20. A bolt or lag screw is used to place the connector under compression between two timber members. *Shear plates* and *split ring* connectors fit into grooves precut into the timber members. The teeth of the *toothed ring* connector are forced into the wood members by the pressure of the bolt joining the members.

Light metal framing devices are also used to increase connection efficiency and reduce installation labor. *Truss plates* are widely used in the construction of the wood trusses described in Section 8-4. Other types of framing devices include *joist hangers, sill plate anchors, rafter anchors,* and *all-purpose framing anchors* (see Figure 8-21).

PROBLEMS

1. Briefly explain the meaning of the following roof terms.
   (a) Gable roof
   (b) Hip roof
   (c) Lookout rafter

2. What is the slope of a roof that has a total rise of 6 ft. (1.8 m) in a building width of 24 ft. (7.3 m)? What is the roof pitch?

3. Explain the function of the following elements of a conventional rafter roof frame.
   (a) Rafter tie
   (b) Purlin
   (c) Collar beam

4. Find the rafter line length and rafter overhang length for a common rafter if the roof slope is 5/12, the building width is 26 ft. (7.9 m), and the roof overhang is 3 ft. (0.9 m).

5. Find the actual length along the center of the top edge of a hip rafter for the roof of Problem 4. No fascia board will be used. The ridge board and all rafters are nominal 2 in. (50 mm) lumber.

6. Calculate the shortening and side cut required for a hip jack rafter to fit against a hip rafter if the hip rafter and jack rafter are both nominal 2 in. (25 mm) lumber. The roof slope is 5/12, and the line length of the jack rafter is 26 in. (66 cm). Sketch the jack rafter cuts and show the cut dimensions.

7. Briefly describe the procedure that should be used to lift a wood roof truss having a span of 44 ft. (13.4 m) and set it into place.

8. How should wood roof trusses be braced as they are erected? Explain the sequence of bracing.

9. Describe how a 1 in. (25 mm) thick glue-nailed plywood subfloor-underlayment should be fastened to the floor frame.

10. Explain how a shear plate is installed to connect two framing members.

REFERENCES

1. Anderson, L. O. *Wood Frame House Construction.* Agriculture Handbook No. 73. U.S. Department of Agriculture, Washington, DC, 1975.
2. American Plywood Association. *APA Design/Construction Guide: Residential and Commercial.* Tacoma, WA, 1988.

3. Truss Plate Institute. *Bracing Wood Trusses: Commentary and Recommendations.* BWT-76. College Park, MD, 1976.
4. Calter, Paul. *Practical Math Handbook for the Building Trades.* Englewood Cliffs, NJ: Prentice-Hall, 1983.
5. Truss Plate Institute. *Handling and Erecting Wood Trusses.* HET '80. College Park, MD, 1980.
6. National Forest Products Association. *Manual for House Framing.* Wood Construction Data No. 1. Washington, DC, 1970.
7. National Forest Products Association. *National Design Specification for Wood Construction.* Washington, DC, 1986.
8. Southern Forest Products Association. *Southern Pine Maximum Spans for Joists and Rafters.* New Orleans, LA, 1978.
9. National Forest Products Association. *Span Tables for Joists and Rafters.* Washington, DC, 1977.

chapter 9

Sheathing, Siding, and Roofing

9-1 SHEATHING

Wall Sheathing

The exterior of the wall frame is covered with sheathing, which provides a base for the installation of siding or other exterior covering. *Structural sheathing* also helps tie the frame together and may serve as horizontal bracing (see Section 7-3). When structural sheathing is used for horizontal bracing, braced panels should be at least 48 in. (1.2 m) wide. *Nonstructural sheathing* helps reduce air infiltration (see Chapter 12) and provides insulation to the wall, but offers no significant structural support. Common structural sheathing materials include board lumber, plywood, particleboard, and structural insulating board (also called ''structural fiberboard''). Nonstructural sheathing includes fiberboard, gypsum board, and rigid foam insulation board. Plywood siding may be applied in a manner that eliminates the need for sheathing as described in Section 9-2. Fastener requirements for sheathing are presented in Table 8-6.

Wood sheathing, usually nominal 1 in. (25 mm) board, may be applied either horizontally (see Figure 7-1) or diagonally. When applied diagonally, no additional horizontal bracing is normally required.

Plywood and particleboard sheathing have largely replaced board sheathing. Plywood sheathing may be applied either horizontally or vertically, as shown in Figure 9-1. Note the recommended spacing between plywood panels. Also note the requirement for blocking behind horizontal joints in panels serving as horizontal bracing. The maximum stud spacing and minimum plywood thickness required by the CABO code are presented in Table 9-1. Both plywood and particleboard panels are usually fastened at 6 in. (15 cm) intervals along panel edges and at 12 in. (30 cm) intervals along intermediate supports.

Particleboard sheathing is applied in a manner similar to plywood sheathing. The maximum allowable span for particleboard wall sheathing allowed by the

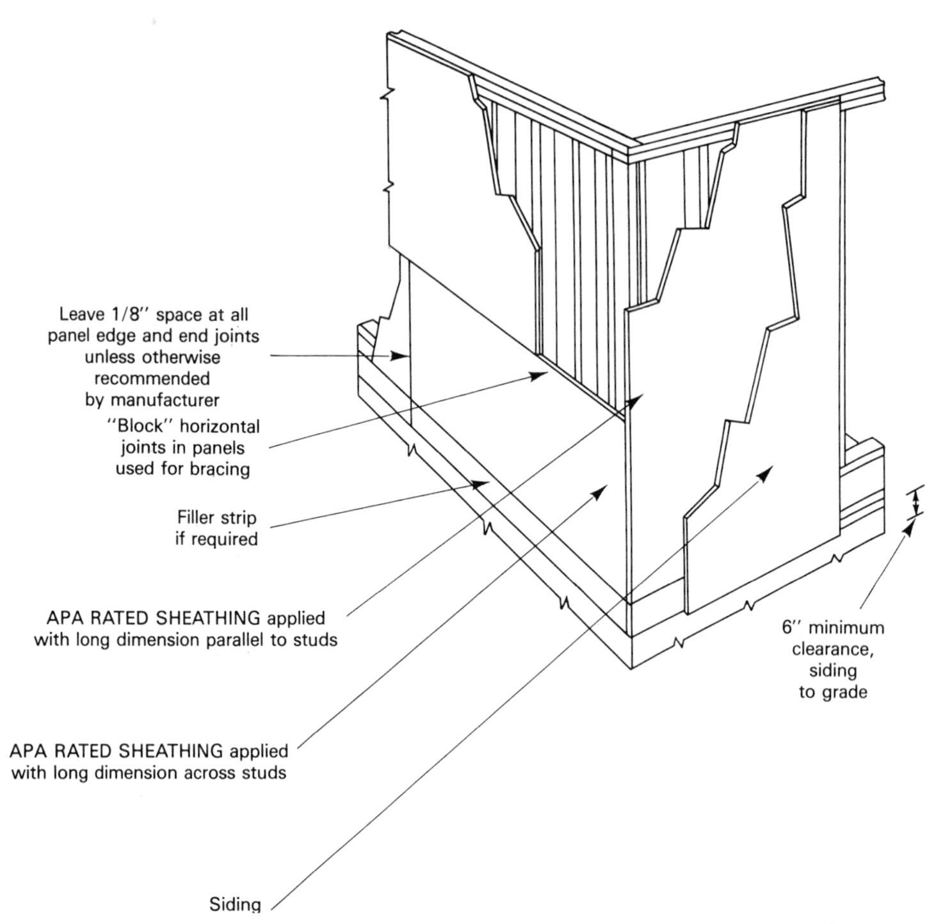

Figure 9-1 Installation of plywood wall sheathing (courtesy American Plywood Association)

TABLE 9-1
Allowable Span of Plywood Wall Sheathing.*

Plywood wall sheathing—face grain parallel or perpendicular to studs[1]

Minimum thickness and construction	*Panel span rating*	*Stud spacing (in.)*		
		Siding nailed to studs	*Siding nailed to sheathing, sheathing parallel to studs*	*Siding nailed to sheathing, sheathing perpendicular to studs*
5/16"	12/0, 16/0, 20/0	16	—	16
3/8", 15/32", and 1/2"—3-ply	16/0, 20/0, 24/0, 32/16	24	16	24
15/32" and 1/2"— 4- and 5-ply	24/0, 32/16	24	24	24

*Reproduced from the 1986 edition of the *CABO One and Two Family Dwelling Code,* copyright 1986, and 1987 amendments, Copyright 1987, with permission of the publishers—Building Officials and Code Administrators International, International Conference of Building Officials and Southern Building Code Congress International, Inc.

[1]Table applies to C-D and C-C grades, including Structural I and II.

TABLE 9-2
Allowable Span of Particleboard Wall Sheathing*

Allowable spans for particleboard wall sheathing[1]
(Not exposed to the weather, long dimension of panel parallel or perpendicular to studs)

Thickness (in.)	*Grade*	*Stud spacing (in.)*	
		When siding is nailed to studs	*When siding is nailed to sheathing*
5/16	2-M-W	16	—
3/8	and	24	16
7/16	2-M-F	24	24
3/8	2-M-1	16	—
	and		
1/2	2-M-2	16	16

*Reproduced from the 1986 edition of the *CABO One and Two Family Dwelling Code,* Copyright 1986, and 1987 amendments, Copyright 1987, with permission of the publishers—Building Officials and Code Administrators International, International Conference of Building Officials and Southern Building Code Congress International, Inc.

[1]The panels may be applied horizontally or vertically. If the panels are applied horizontally, the end joints of the panels shall be offset so that four panel corners will not meet. All panel edges must be supported. Leave a 1/16-inch gap between panels and nail no closer than 3/8 inch from panel edges.

CABO code is presented in Table 9-2. Note the requirements in the table footnote. Panels should be spaced 1/16 in. (1.5 mm) apart, and fasteners should not be placed closer than 3/8 in. (1 cm) to the panel edge. When panels are applied horizontally, panel end joints should be offset so that four panels do not meet at a common joint.

Structural insulating board is applied in a manner similar to plywood and particleboard. However, 4 ft. × 8 ft. (1.2 × 2.4 m) and larger panels should be applied vertically. Fasteners are commonly spaced at 3 in. (8 cm) intervals along panel edges and at 6 in. (15 cm) intervals along intermediate supports.

Gypsum sheathing is usually supplied in panels 2 ft. × 8 ft. (0.6 × 2.4 m) by 1/2 in. (1 cm) thick. Panels are commonly applied horizontally with fasteners spaced at 4 in. (10 cm) intervals along panel edges and 8 in. (20 cm) intervals along intermediate supports.

Rigid foam insulating board sheathing is employed primarily for its insulating value. Panels are usually 4 ft. × 8 ft. (1.2 m × 2.4 m) or larger and are available in several thicknesses depending on the insulating value desired. Panels are applied vertically and joints sealed with tape to provide an air infiltration barrier (see Chapter 12).

Roof Sheathing

The exterior of the roof frame is covered by sheathing to provide a base for the installation of roofing and to transfer roof loads to the roof frame. Materials commonly used for roof sheathing include boards, planks, plywood, particleboard, and fiberboard insulating roof deck.

Board sheathing is commonly applied as closed sheathing placed perpendicular to the roof rafters or trusses as shown in Figure 9-2. However, spaced board sheathing is sometimes used to support wood shingles or shakes, or clay or concrete tiles where additional roofing ventilation is required. Closed board sheathing may be applied diagonally to provide additional resistance to racking (twisting) in the roof frame.

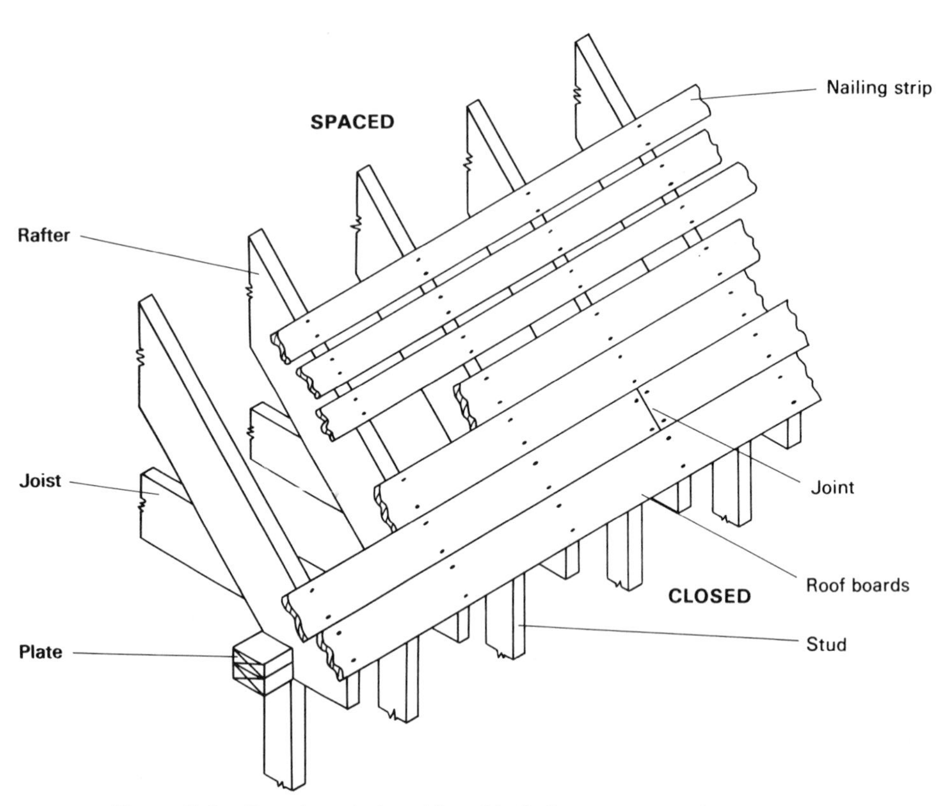

Figure 9-2 Board roof sheathing (U. S. Department of Agriculture).

Plywood and particleboard panels have largely replaced board sheathing for roofs. The proper installation of plywood roof sheathing is illustrated in Figure 9-3. Plywood panels should be installed with the face grain perpendicular to rafters or trusses. Note the spacing between panels and the use of metal panel clips to eliminate the need for blocking between supports. The recommended allowable span for various design loads and panel thicknesses is presented in Table 9-3.

Particleboard roof sheathing may be installed either parallel or perpendicular to supports, but should be continuous over two or more spans. The maximum span and load permitted by the CABO code is presented in Table 9-4. Note that 3/8 in. (1 cm) panels must have tongue and groove edges or have the edges supported by clips or blocking. End joints should be staggered so that four panel corners do not meet at a common joint. A 1/16 in. (1.5 mm) space should be left between panels, and fasteners should not be installed closer than 3/8 in. (1 cm) to the panel edge.

When plank-and-beam framing is used or trusses are widely spaced, tongue and groove planks of nominal 2 in. (50 mm) or thicker lumber may be used as roof decking (sheathing). The underside of such roof decks is often left exposed for architectural effect. Methods for determining the allowable span for plank decking are described in Reference 1. Plywood panels 1 1/8 in. (28 mm) thick may be used to span openings up to 4 ft. (1.2 m) between roof beams, trusses, or purlins. Plywood panels may be used for spans up to 8 ft. (2.4 m) when panels are strengthened by stiffeners spaced 16 or 24 in. (41 or 61 cm) on center.

Fiberboard insulating roof decking is most commonly available in panels 2 ft. by 8 ft. (0.6 × 2.4 m) with thicknesses of 1 1/2 in. (38 mm) or greater. It is applied in a manner similar to wood planking. The maximum span between supports is specified by the manufacturer for various roof loads.

Fastener requirements for board, plywood, and particleboard roof sheathing are given in Table 8-6.

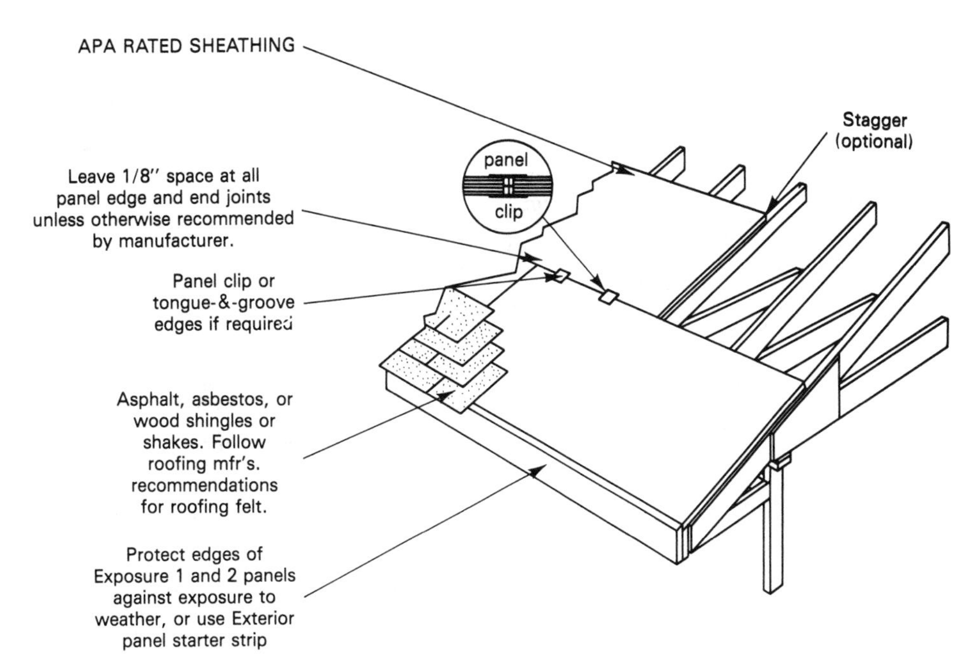

Note: Cover sheathing as soon as possible with roofing felt for extra protection against excessive moisture prior to roofing application.

Figure 9-3 Installation of plywood roof sheathing (courtesy American Plywood Association).

TABLE 9-3
Allowable Loads for Plywood Roof Sheathing (Courtesy of American Plywood Association)

Recommended Roof Loads (psf) for APA RATED SHEATHING with Long Dimension Parallel to Supports[e]
(Nonveneer, composite and 5-ply/5-layer plywood panels unless otherwise noted)

Panel grade	*Thickness (in.)*	*Span rating*	*Maximum span (in.)*	*Load at maximum span*	
				Live	*Total*
APA-STRUCTURAL I-RATED SHEATHING	7/16	24/0, 24/16	24[d]	20	30
	15/32	32/16	24	35[a]	45[a]
	1/2	32/16	24	40[a]	50[a]
	19/32, 5/8	40/20	24	70	80
	23/32, 3/4	48/24	24	90	100
APA-RATED SHEATHING	7/16[b]	24/0, 24/16	16	40	50
	15/32[b]	32/16	24[d]	20	25
	1/2[b]	24/0, 32/16	24[d]	25	30
	19/32	40/20	24	40[c]	50[c]
	5/8	32/16, 40/20	24	45[c]	55[c]
	23/32, 3/4	40/20, 48/24	24	60[c]	65[c]

[a] For 4-ply plywood marked PS 1, reduce load by 15 psf.
[b] Composite panels must be 19/32 inch or thicker.
[c] For composite and 4-ply plywood panels reduce load by 15 psf.
[d] Solid blocking recommended at 24-inch span
[e] When roofing is to be guaranteed by a performance bond, check with roofing manufacturer for panel type, minimum panel thickness, span and edge support requirements.

TABLE 9-4
Allowable Loads for Particleboard Roof Sheathing*

Allowable Loads for Particleboard Roof Sheathing[1 2 3]

Grade	Thickness (in.)	Maximum on-center spacing	Live load (lbs/sq. ft.)	Total load (lbs./sq. ft.)
2-M-W and 2-M-F	3/8[4]	16	45	65
	7/16	16	105	105
	7/16	24	30	40
	1/2	16	110	150
	1/2	24	40	55

*Reproduced from the 1986 edition of the *CABO One and Two Family Dwelling Code,* Copyright 1986, and 1987 Amendments, Copyright 1987, with permission of the publishers—Building Officials and Code Administrators International, International Conference of Building Officials and Southern Building Code Congress International, Inc.

[1]Panels are continuous over two or more spans.

[2]Uniform load deflection limitations: 1/180 of the span under live load plus dead load and 1/240 of the span under live load only.

[3]The panels may be applied parallel or perpendicular or the span of the rafters or joists and shall be continuous over two or more spans. If the panels are applied perpendicular to roof supports, the end joints of the panels shall be offset so that four panel corners will not meet. Cutouts for items such as plumbing and electrical shall be oversized to avoid a force fit. A 1/2 inch gap must be provided between the panel and concrete and masonry walls. Leave a 1/16 inch gap between panels and nail no closer than 3/8 inch from panel edge.

[4]Edges shall be tongue and groove or supported with blocking or edge clips.

Sheathing Paper

Wall. Depending on the type of siding used, it may be necessary to cover the exterior of wall sheathing with *sheathing paper* (also called *building paper*) or a *weather-resistant membrane* to prevent moisture from penetrating the sheathing. Generally, such a barrier is required under stucco, masonry veneer, shingles, and shakes. Sidings that require a moisture barrier are identified in Section 9-2.

When building paper is used, it should consist of an asphalt-saturated felt weighing at least 14 lbs. (6.4 kg) per square [100 sq. ft. (9.3 m^2)]. No. 15 felt weighing 15 lbs. (6.8 kg) per square is widely used. Vertical joints should be lapped at least 6 in. (15 cm) and horizontal joints at least 2 in. (5 cm).

An air infiltration barrier cloth is often used (and may be required by code) over plywood, particleboard, and other types of sheathing. In some cases, the sheathing itself with taped joints may serve as an air infiltration barrier. The requirements for air infiltration barriers are discussed in Chapter 12.

Roof. Sheathing paper for roofs is called *underlayment, roofing felt, roofing paper,* or *tar paper.* Depending on the type of roofing used, underlayment may be required to increase the moisture resistance of the roofing (see Section 9-4). In addition, underlayment serves to waterproof the roof during construction prior to the application of roofing.

Underlayment usually consists of 15 or 30 lbs. (6.8 or 13.6 kg) weight per square asphalt-saturated felt. This material is commonly referred to as No. 15 or No. 30 felt. When applied as a single layer, underlayment should be laid parallel to the roof eave and lapped 2 in. (5 cm) at the top and 4 in. (10 cm) at the ends. When applied as a double layer, underlayment should be lapped 19 in. (48 cm) at the top and 12 in. (30 cm) at the end. End laps of the second layer should be located at least 6 ft. (1.8 m) from the end laps of the underlying layer. Nailing requirements are not usually specified, but underlayment must be nailed sufficiently, using roofing nails, to hold it in place until roofing is applied.

9-2 SIDING

General

Exterior wall coverings protect the building against weather and insects as well as provide an architectural effect. Exterior wall coverings include a number of types of *siding,* masonry veneers, and stucco. Some of the many types of exterior wall coverings available include:

Aluminum boards and panels
Asbestos cement boards and shingles
Brick, stone, and tile veneer
Cement stucco
Fiberboard boards and panels
Hardboard boards and panels
Particleboard panels
Plywood panels and lap siding
Steel boards and panels
Wood boards, shingle, and shakes

The requirements for siding attachment and use of water-resistant membranes set forth in the CABO code are shown in Table 9-5. All fasteners used for attaching siding should be corrosion-resistant (aluminum, galvanized steel, stainless steel, or rust-preventive coated).

Board Siding

Common types of wood board siding are illustrated in Figure 9-4. A recommended method for installing *bevel siding* is illustrated in Figure 9-5. The siding should be lapped a minimum of 1 in. (25 mm). The starting strip at the bottom of the first course of siding causes this row of siding to tilt at the same angle as the other rows. To achieve this effect, the thickness of the starting strip should be the same as the thickness of the top edge of the siding boards.

For appearance purposes, the bottom edge of a horizontal board siding course should fall just above the top of the upper window trim. Likewise, the bottom of window sills should line up with the bottom edge of a row of siding. Depending on window height and width of the siding, it may not be possible to satisfy both requirements. One solution is to use a frieze board (see Figure 11-2) above the window in place of a course of siding.

Drop siding [Figure 9-4(b)] rests flush against the underlying sheathing, with the groove of the upper course fitting over the lip of the course below. Otherwise, it is installed like bevel siding.

Common forms of vertically applied board siding include *board and batten* siding and *tongue and groove* siding, as illustrated in Figure 9-4(c) and 9-4(d). Vertical wood board siding must be fastened to blocking or horizontal nailing strips located not more than 24 in. (61 cm) on center vertically. For wood board and batten siding, boards should be spaced about 1/2 in. (1 cm) apart and the joint covered by a batten. The batten should be fastened by nails spaced along its center line, passing between the two boards.

The installation of aluminum, asbestos cement, fiberboard, hardboard, plywood, vinyl, and other board and lap sidings should be made in accordance with the manufacturer's recommendations and Table 9-5.

TABLE 9-5
Installation of Exterior Wall Coverings*

Weather-Resistant Siding Attachment and Minimum Thickness

Siding material		*Nominal thickness*[1] *(in.)*	*Joint treatment*	*Weather-resistant membrane required*	*Type of supports for the siding material and fasteners*[4,5,9]				
					Wood or plywood sheathing	*Fiberboard sheathing into stud*	*Gypsum sheathing into stud*	*Direct to studs*	*Number or spacing of fasteners*
Horizontal aluminum[8]	Without insulation	.019[10]	Lap	No	.120 nail- 1 1/2″	.120 nail- 2″	.120 nail- 2″	Not allowed	Same as stud spacing
		.024	Lap	No	.120 nail- 1 1/2″ long	.120 nail- 2″ long	.120 nail- 2″ long	Not allowed	
	With insulation	.019	Lap	No	.120 nail- 1 1/2″	.120 nail- 2 1/2″	.120 nail- 2 1/2″	.120 nail- 1 1/2″	
Horizontal asbestos cement Boards Shingles[7]		5/32 1/8	(2) Lap	(2) Yes	.113 nail- 1 1/2″	.113 nail- 2″	.113 nail- 1 3/4″	.113 nail- 1 3/8″	2 nails per shingle
Brick veneer Clay tile veneer Concrete veneer		2 1/4 to 1 2	Sec. 503	Yes	See Sec. 503 and Figure No. R-503.4				
Horizontal fiberboard[3]		1/2	Sec. 503	No	.099 nail- 2″ Staple- 1 3/4″	.113 nail- 2 3/4″ Staple- 2 1/2″	.113 nail- 2 1/2″ Staple- 2 1/4″	.099-nail- 2″ Staple- 1 3/4″	Same as stud spacing
Hardboard[3] Board and batten-vertical		1/4	(2)	(2)	.099 nail- 2″	.099 nail- 2 1/2″	.099 nail- 2″	.099 nail- 1 3/4″	6″ panel edges 8″ inter. sup.
Hardboard[3] Lap-siding-horizontal		7/16	(2)	(2)	.099 nail- 2″	.099 nail- 2 1/2″	.099 nail- 2 1/4″	.099 nail- 2″	Same as stud spacing 2 per bearing

Vertical panel siding	7/16	(2)	(2)	.099 nail-2″	.099 nail-2 1/4″	.099 nail-2″	.080 nail-1 3/4″	6″ panel edges 12″ inter. sup.
Steel[3]	29 ga.	Lap	No	.113 nail-1 3/4″ Staple-1 3/4″	.113 nail-2 3/4″ Staple-2 1/2″	.113 nail-2 1/2″ Staple-2 1/4″	Not allowed	Same as stud spacing
Stone veneer	2	Sec. 503	Yes	See Sec. 503 and Figure No. R-503.4				
Particleboard panels	3/8	(2)	(2)	.113 nail-2″ Staple-1 3/8″	.113 nail-2 1/2″ Staple-2 1/4″	.113 nail-2″ Staple-2″	Not allowed	6″ on edges 8″ inter. sup.
	5/8	(2)	(2)	.113 nail-2″ Staple-1 7/8″	.113 nail-2 1/2″ Staple-2 1/2″	.113 nail-2 1/2″ Staple-2 1/4″	.113 nail-2″ Staple-1 5/8″	6 ″ on edges 8″ inter. sup.
Plywood panels[11] (exterior grade)	3/8	(2)	(2)	.099 nail-2″ Staple-1 3/8″	.113 nail-2 1/2″ Staple-2 1/4″	.099 nail-2″ Staple-2″	.099 nail-2″ Staple-1 3/8″	6″ on edges 12″ inter. sup
Wood[12] Rustic, drop Shiplap	3/8 Minimum 19/32 Average	Lap	No	Fastener penetration into stud—1″			.113 nail-2 1/2″ Staple-2″	Face nailing up to 6″ widths, 1 nail per bearing; 8″ widths and over, 2 nails per bearing
Bevel	7/16	Lap	No					
Butt tipp	3/16	Lap	No					

(*continued*)

TABLE 9-5 (*Continued*)

Weather-Resistant Siding Attachment and Minimum Thickness

Siding material	Nominal thickness[1] (in.)	Joint treatment	Weather-resistant membrane required	Type of supports for the siding material and fasteners[4,5,9]				
				Wood or plywood sheathing	Fiberboard sheathing into stud	Gypsum sheathing into stud	Direct to studs	Number or spacing of fasteners
Shakes[7]	3/8	Lap	Yes	.0915 nail-2"		Staple 2"		
Shingles[7]	3/8	Lap	Yes	16" and 18" shingles		.076 nail-1 1/4"		2-fasteners per shin-gle or shake
						Staple-1 1/4"		
				24" shingles		.080 nail-1 1/2"		
						Staple-1 1/2"		

*Reproduced from the 1986 edition of the *CABO One and Two Family Dwelling Code*, copyright 1986, and 1987 Amendments, copyright 1987, with permission of the publishers—Building Officials and Code Administrators International, International Conference of Building Officials and Southern Building Code Congress International, Inc.

[1]Based on stud spacing of 16 inches o.c. Where studs are spaced 24 inches, siding may be applied to sheathing approved for that spacing.

[2]If boards are applied over sheathing or weather-resistant membrane, joints need not be treated. Otherwise vertical joints must occur at studs and be covered with battens or be lapped.

[3]Shall be of approved type.

[4]Nail is a general description and may be T-head, modified round head, or round head with smooth or deformed shanks.

[5]Staples shall have a minimum crown width of 7/16 inch O.D. and be manufactured of minimum No. 16 gauge wire.

[6]All attachments shall be coated with a corrosion-resistive coating.

[7]Shingles and shakes applied over regular density fiberboard or gypsum sheathing shall be fastened to horizontal wood nailers or fiberboard shingle backer.

[8]Aluminum nails shall be used to attach aluminum siding.

[9]Nails or staples must be aluminum, galvanized, or rust-preventive coated and shall be driven into the studs for fiberboard or gypsum backing.

[10]Aluminum (.019 inch) may be unbacked only when the flat areas are 5 inches or less in the narrow dimension.

[11]Three-eighths inch plywood may be applied directly to studs spaced 16 inches on center. One-half inch plywood may be applied directly to studs spaced 24 inches on center.

[12]Wood board sidings applied vertically shall be nailed to horizontal nailing strips or blocking set 24 inches o.c. Nails shall penetrate 1 1/2 inches into studs, studs and wood sheathing combined, or blocking. A weather-resistant membrane shall be installed weatherboard fashion under the vertical siding unless the siding boards are lapped or battens are used.

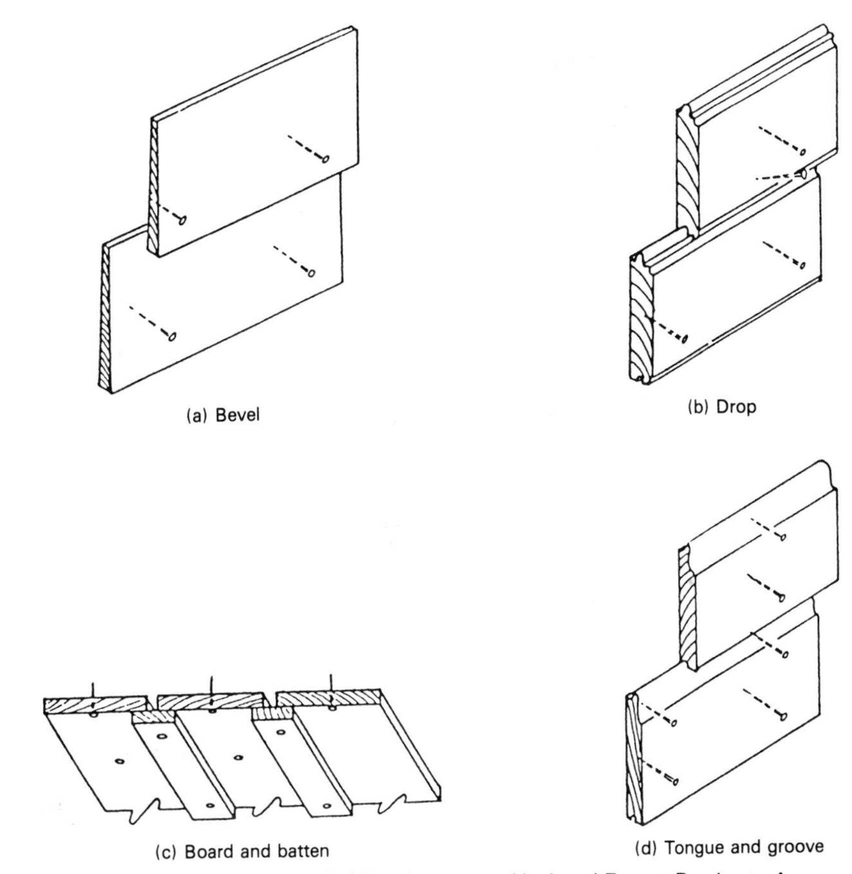

(a) Bevel (b) Drop

(c) Board and batten (d) Tongue and groove

Figure 9-4 Wood board siding (courtesy National Forest Products Association, Washington, DC).

Shingles and Shakes

Wood *shingles* (Figure 9-6) are manufactured in lengths of 16 in., 18 in., and 24 in. (41, 46, and 61 cm). *Shakes* are thick wood shingles that have been hand split and resawn. Wood shingles and shakes may be installed in a single course (that is, one layer thick) or a double course (two layers thick). The installation of single-course wood shingles is illustrated in Figure 9-6. Note that a double course of shingles is used as a starting course at the bottom of the wall to provide the proper tilt to this row of shingles. Vertical joints should be offset as illustrated. The maximum exposure for single-course 16 in. (41 cm) shingles should be 7 1/2 in. (19 cm) as illustrated. The maximum exposure for other shingles is as follows:

	Maximum Exposure [in. (cm)]	
Shingle length	*Single course*	*Double course*
16″ (41 cm)	7 1/2 (19)	12 (30)
18″ (46 cm)	8 1/2 (22)	14 (36)
24″ (61 cm)	11 1/2 (29)	16 (41)

Shingles may be fastened directly to plywood, particleboard, or nail-base-type fiberboard sheathing. When applied over regular density fiberboard, gypsum, or other nonstructural sheathing, shingles should be fastened to horizontal nailing strips.

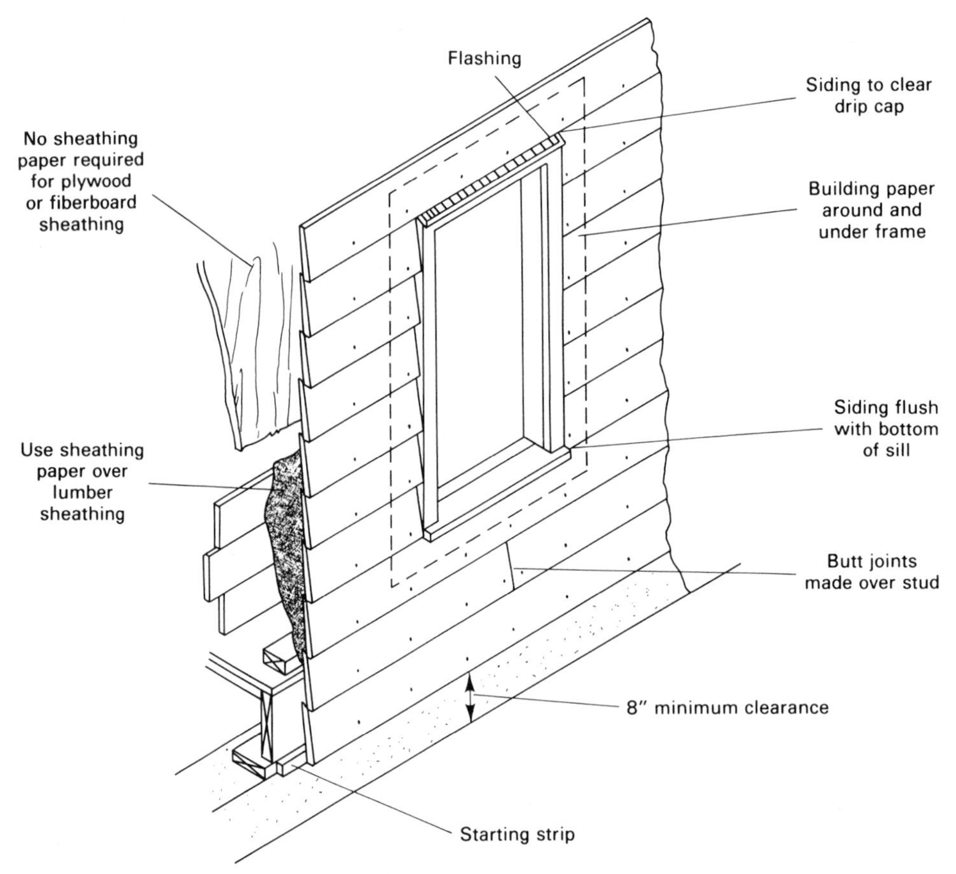

Figure 9-5 Installation of bevel siding (U. S. Department of Agriculture).

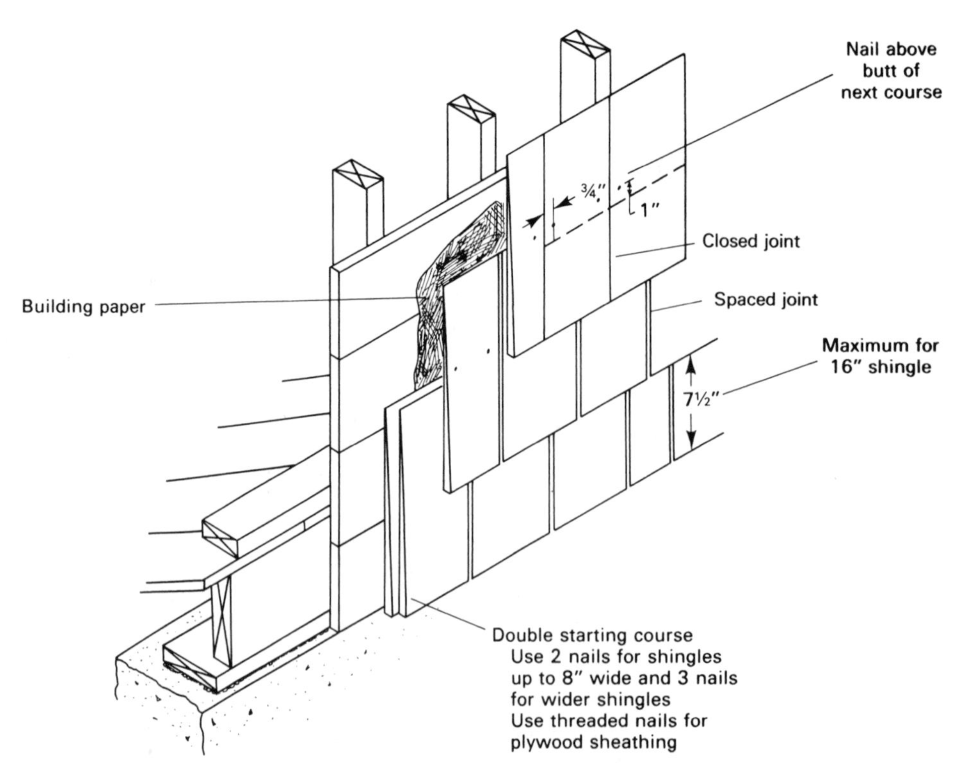

Figure 9-6 Installation of wood shingle siding (adapted from U. S. Department of Agriculture).

Shingles of asbestos cement or other materials should be installed in accordance with the manufacturer's recommendations and Table 9-5.

Plywood and Other Panels

Plywood panel siding may be applied over nailable sheathing with the face grain either vertical (Figure 9-1) or horizontal. No weather-resistant membrane is required under the siding. However, plywood panel siding may also be applied directly over studs or over nonstructural sheathing as shown in Figure 9-7. Note that in this system, no weather-resistant membrane is required when joints fall directly over framing members, and vertical joints are lapped [minimum 1 1/2 in. (38 mm)] or battened. The maximum spacing of studs in this system should be 16 in. (41 cm) for 3/8 in. (10 mm) plywood siding and 24 in. (61 cm) for 1/2 in. (13 mm) plywood siding.

Particleboard and other exterior panels are applied in a manner similar to plywood panels. Install in accordance with the manufacturer's recommendations and Table 9-5.

Stucco

Stucco (portland cement plaster) is widely used as an exterior finish in certain areas of the world. A minimum of three coats (scratch coat, brown coat, and finish coat) should be applied over wire or expanded metal lath. At least two coats (base coat and finish coat) should be applied over concrete or masonry. The minimum finished

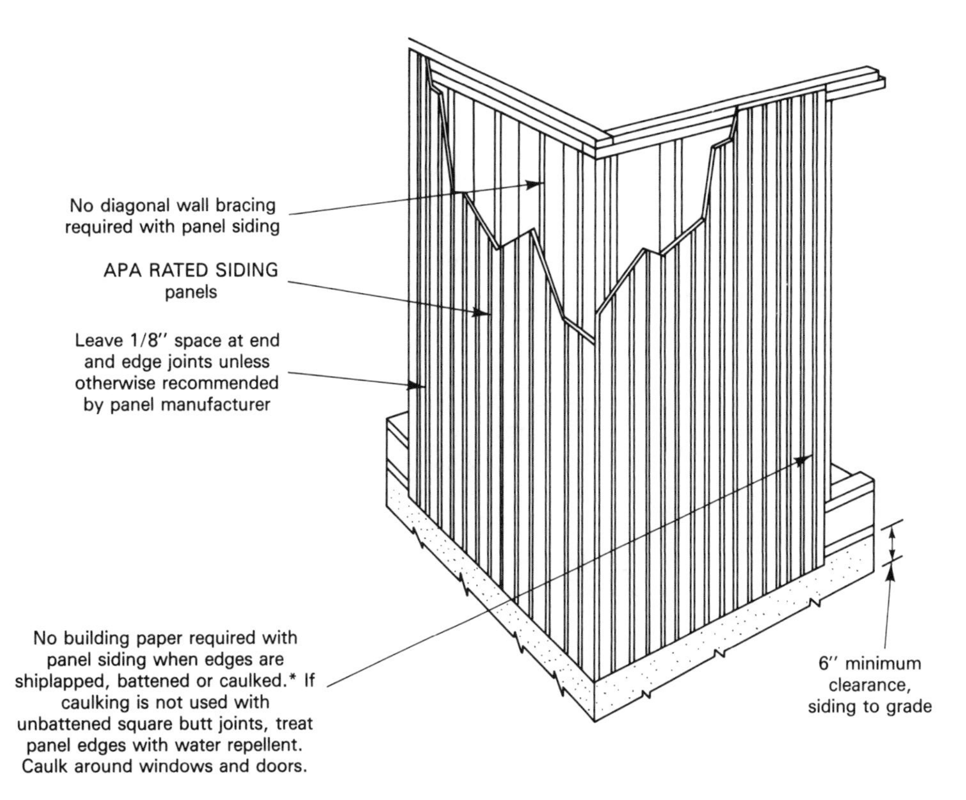

Figure 9-7 Plywood panel siding without sheathing (courtesy American Plywood Association).

thickness of stucco should be 1/2 in. (13 mm) over masonry, 3/4 in. (19 mm) over expanded metal lath, and 7/8 in. (22 mm) over wire lath.

The application of stucco over wood frame construction is illustrated in Figure 9-8. Sheathing may be plywood, particleboard, structural insulating board, or rigid foam insulating board. A weather-resistant membrane is normally installed between the sheathing and lath. However, an air infiltration barrier cloth or a rigid foam insulating board with taped joints may be sufficient. The procedures used for attachment of lath to the building frame should conform to the building code and the lath manufacturer's recommendations.

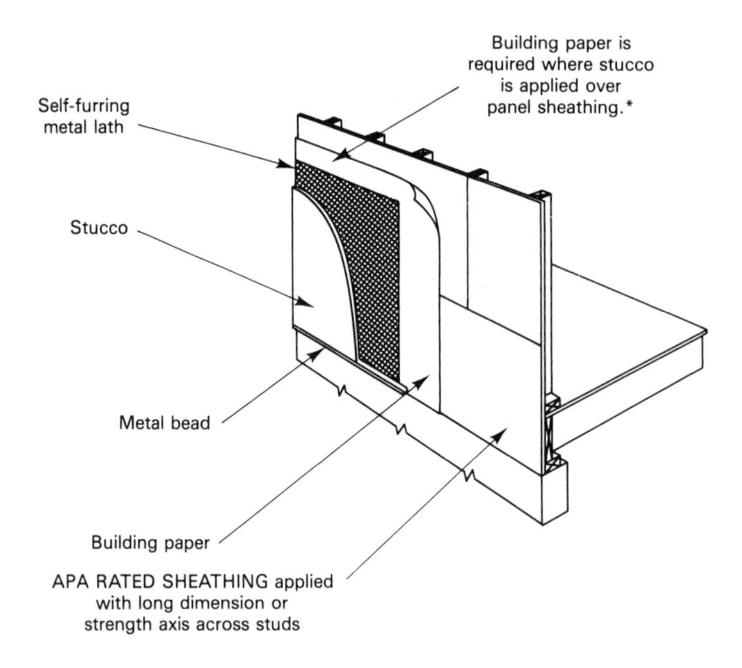

*Check local building code and applicator for specific requirements.
Note: Uniform Building Code requires two layers of grade D paper over wood-based sheathing.

Figure 9-8 Stucco over wood frame (courtesy American Plywood Association).

Masonry Veneers

Brick veneer is the most commonly used form of masonry veneer finish. However, stone, clay tile, concrete, and other masonry veneers are sometimes used. Masonry veneers are supported by the foundation wall and tied to the building frame by corrosion-resistant metal ties.

Brick veneer over wood frame construction is illustrated in Figure 9-9. Note the use of a 1 in. (25 mm) air space between the veneer and the sheathing. This air space plus flashing and weep holes should be used as shown to prevent the penetration of moisture into the building. *Weep holes* are constructed by omitting the end joint mortar of the bottom brick course at specified intervals. Requirements for the spacing of veneer ties are governed by the building code. Generally, it is recommended that each tie not support more than 3.25 sq. ft. (0.3 m^2) of veneer exterior surface area.

Brick veneer may also be applied over wood frame construction without sheathing. In this case, a weather-resistant membrane must be applied to the wood frame and a 1 in. (25 mm) air space maintained between the membrane and the veneer. Brick veneer may be applied over concrete block walls in a manner similar

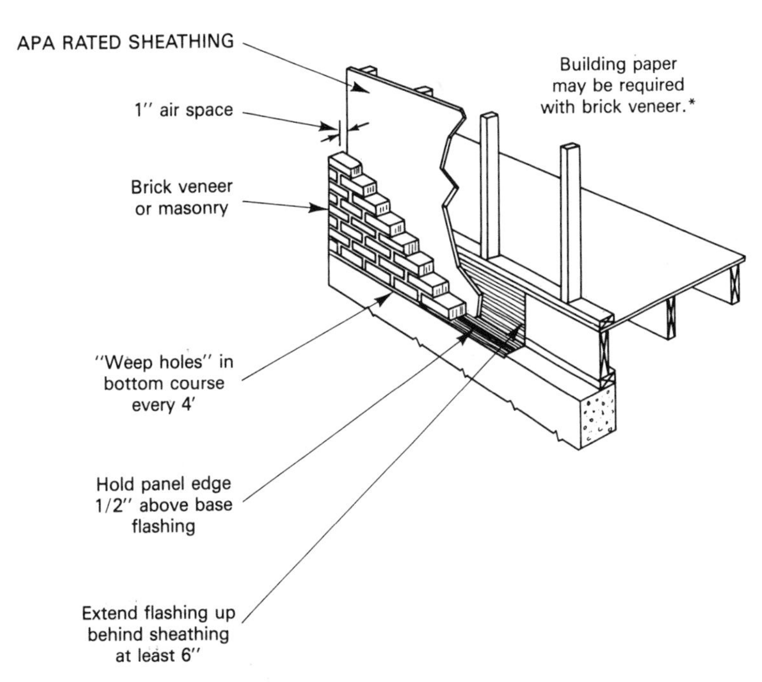

*Check local building code for requirements.

Figure 9-9 Brick veneer over wood frame (courtesy American Plywood Association).

to that used for frame construction. In addition, composite walls may be constructed using an exterior brick wythe bonded to an interior concrete block wythe.

Stone, clay tile, and concrete veneers may be used in a manner similar to brick veneer. When used, they should be installed in accordance with the manufacturer's recommendations and Table 9-5.

9-3 WATERPROOFING JOINTS

Types of Joints

In addition to joints that occur naturally at the intersection of different surfaces and materials, expansion and control joints are often placed in structures to control differential movement. Some causes of differential movement include: the shrinkage of concrete and wood components due to drying; temperature and moisture changes; and foundation settlement. In concrete and masonry structures, *expansion joints* are used to permit differential movement without damage to the structure. *Control joints* consist of grooves deliberately placed in concrete and masonry components to control the location of shrinkage cracking. In both types of joints, tongue and groove construction or flexible metal ties are used to provide structural integrity across the joint while permitting differential movement.

Some typical expansion joints in brick masonry walls are illustrated in Figure 9-10. Expansion joints should be placed in concrete and masonry walls where differential movement is likely to occur. Long, straight walls should be broken into sections separated by expansion joints. Expansion joints should also be placed at wall junctions, wall offsets, columns, and pilasters as illustrated in Figure 9-10. Expansion joints are also commonly required at window and door openings.

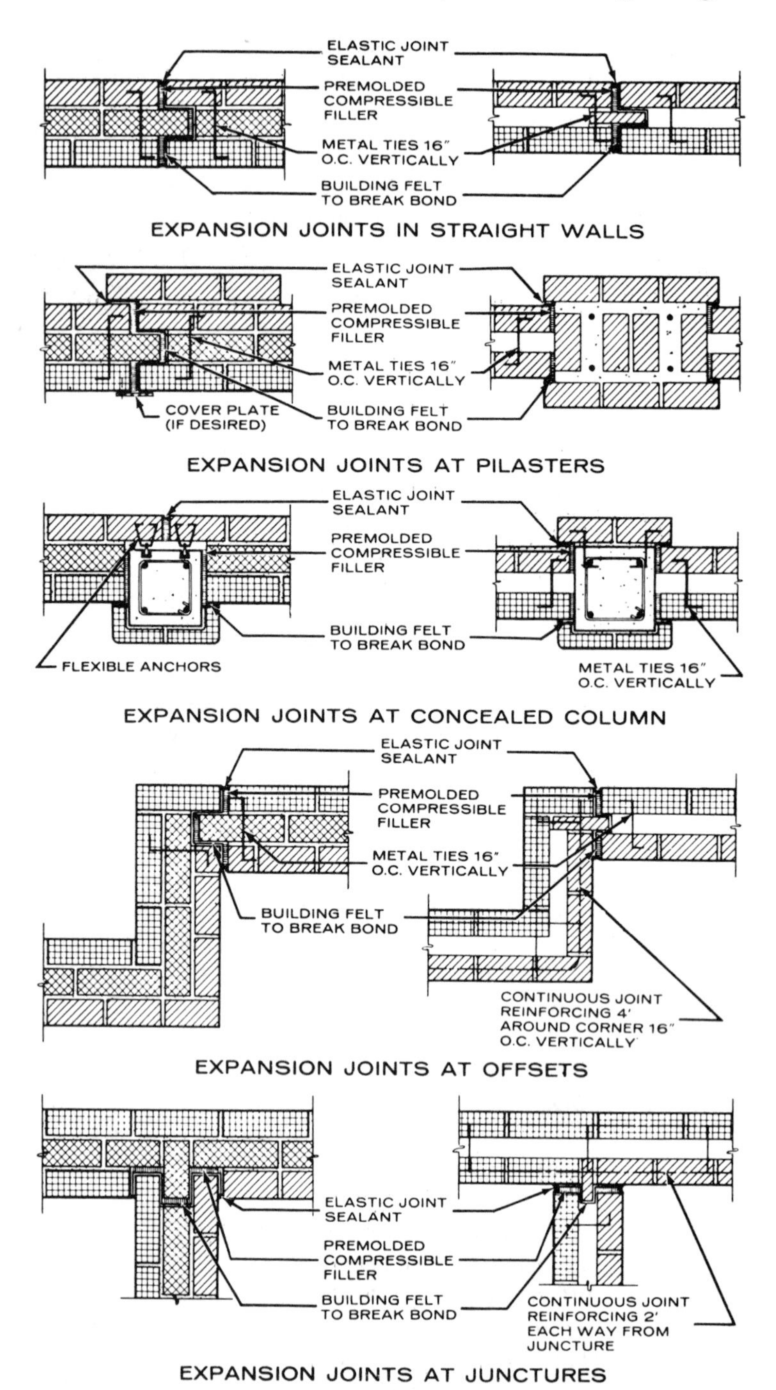

Figure 9-10 Expansion joints in brick walls (courtesy Brick Institute of America).

Joints must be sealed to prevent the penetration of moisture. Flashing and waterstops are used for this purpose.

Flashing and Waterstops

Flashing consists of thin layers of impervious material used to prevent the penetration of moisture through joints. Flashing is also designed to direct any moisture that might penetrate the joint back to the exterior of the building. When used in a concrete or masonry joint, such a waterproofing material is often called a *waterstop*. Materials commonly used for flashing include sheet metals (see Section 2-3), asphalt-impregnated felt, and plastics.

Some applications of flashing were illustrated in Figures 9-5 (window opening), and 9-9 (brick veneer). Other common applications of flashing include roof edges (Figure 9-11), roof valleys (Figure 9-12), intersections of walls and roofs (Fig-

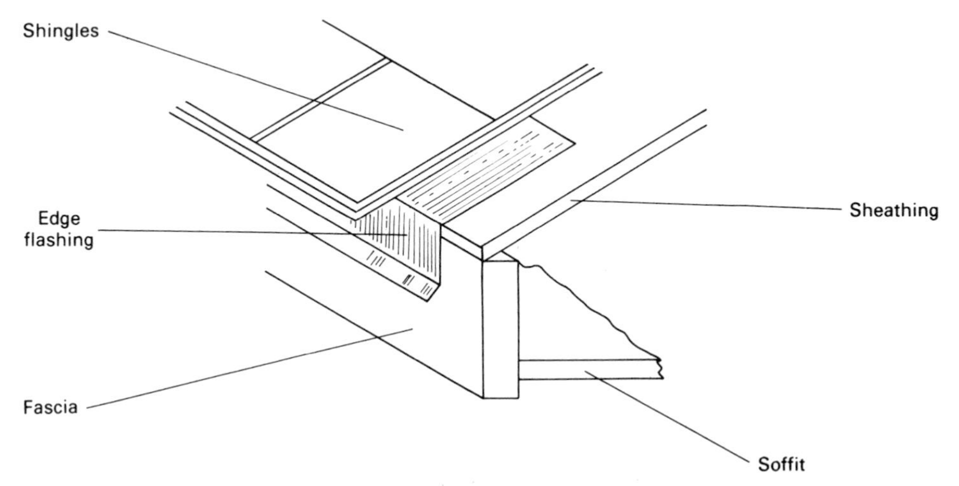

Figure 9-11 Flashing at roof edge.

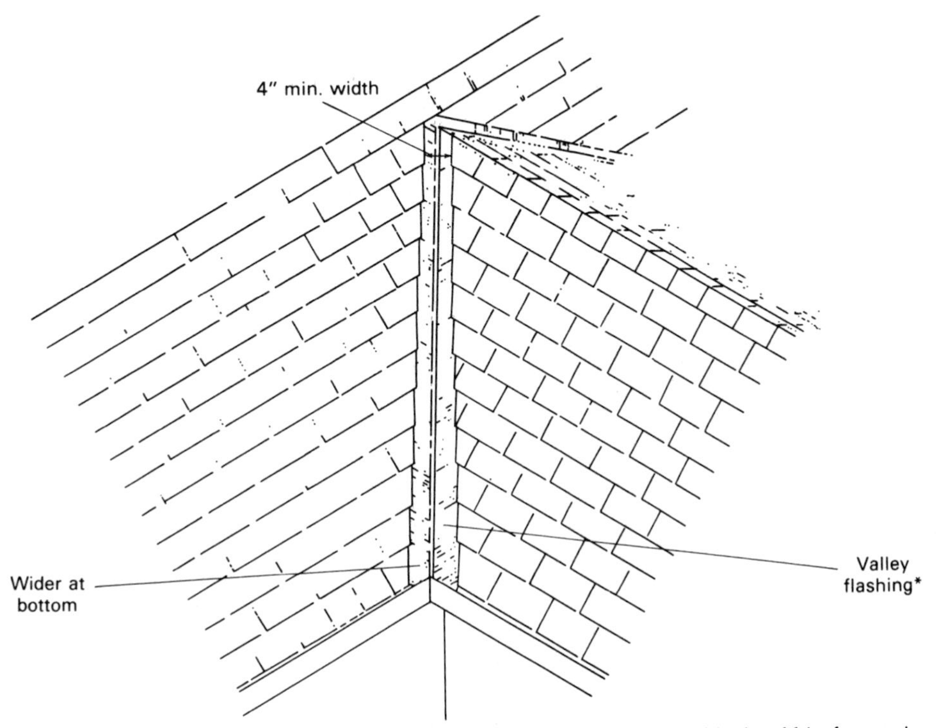

*Extend metal flashing each side of center line at least 8 in. for asphalt or wood shingles, 11 in. for wood shakes.

Figure 9-12 Valley flashing (adapted from U. S. Department of Agriculture).

ure 9-13), and around chimneys (Figure 9-14). Notice the use of *counterflashing* in Figure 9-14. Counterflashing is simply an upper layer of flashing that overlaps the lower layer of flashing. Counterflashing is commonly used where a sloping roof intersects a brick wall or chimney. The upper edge of the counterflashing is turned into the brick mortar joint and caulked to prevent the entry of moisture behind the counterflashing.

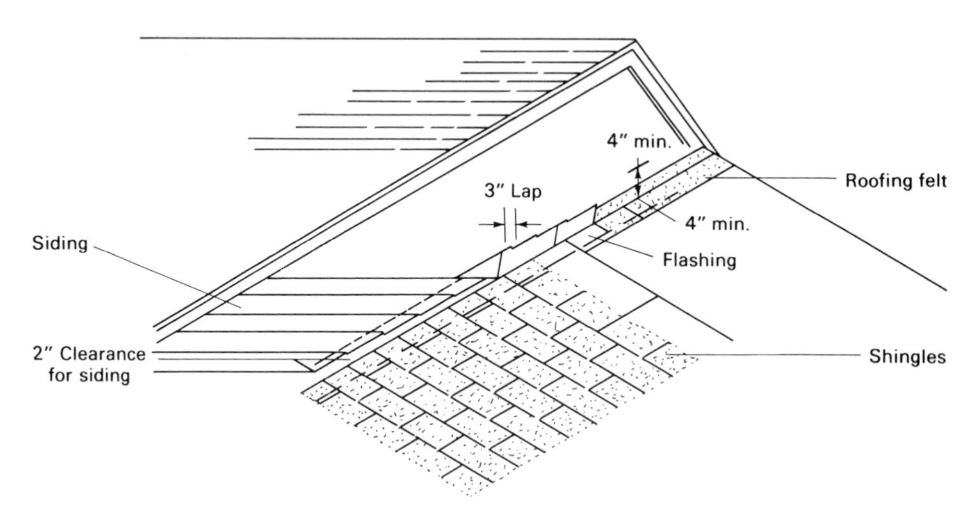

Figure 9-13 Flashing at roof and wall intersection (U. S. Department of Agriculture).

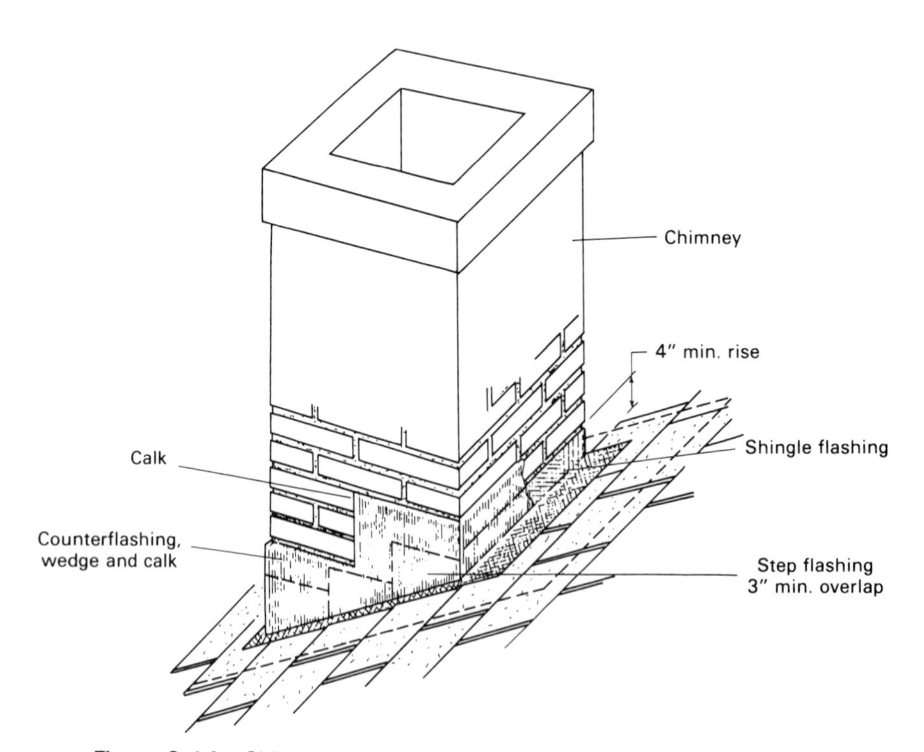

Figure 9-14 Chimney flashing (adapted from U. S. Department of Agriculture).

9-4 ROOFING

Types of Roof Coverings

Roofing, or roof covering, is applied over roof sheathing to provide a waterproof covering for the roof. Roofing must also have sufficient strength and durability to resist damage from the elements (sun, wind, and hail) as well as the traffic of maintenance workers. Some of the major types of roofing include: asphalt shingles; asphalt roll roofing; built-up roofing; asbestos-cement shingles; slate shingles; metal sheets and shingles; clay and concrete tile; wood shingles and shakes; and rubber and plastic membranes.

Installation

Roof sheathing must be clean and dry before roof underlayment and roof coverings are applied. Underlayment, when used, should be installed as described in Section 9-1. The requirements of the CABO building code for the fastening of roof coverings are shown in Table 9-6. Note that corrosion-resistant fasteners must be used in all cases. Fasteners should completely penetrate the roof sheathing or extend a minimum of 3/4 in. (1.9 cm) into the sheathing.

Asphalt Shingles

Asphalt shingles, the most widely used roof covering for residential structures, are manufactured of asphalt-impregnated felt covered with mineral granules. A newer type of shingle uses a glass fiber mat in place of a felt mat. Such shingles, often called *fiberglass shingles,* have greater fire and mildew resistance and longer life than do conventional asphalt shingles. A typical asphalt shingle measures 12 in. by 36 in. (30 by 91 cm) and has three tabs as shown in Figure 9-15. Other sizes and styles of shingle are also available. Shingle weights [pounds (kg) per square (100 sq. ft. or 9.3 m^2)] range from about 225 to 300 lbs. (102–136 kg) with the heavier shingles carrying a longer manufacturer's warranty of service life.

Asphalt shingles should not be used on roof slopes of less than 2 in 12. For roof slopes of 2 in 12, but less than 4 in 12, double coverage shingles should be installed over two layers of No. 15 felt underlayment installed as described in Section 9-1. For roof slopes of 4 in 12 or greater, only a single layer of No. 15 underlayment is required.

The usual procedure for installing asphalt shingles is illustrated in Figure 9-15. After the underlayment and flashings have been installed, a starter course of shingles is installed along the lower roof edge. This starter course may consist of shingles that have had their tabs trimmed off or of whole shingles placed with their tabs facing toward the ridge. The starter course should overhang the roof edge about 1/2 in. (1 cm) to provide a drip edge. A chalkline as shown in often used to help align the tab slots.

The first regular shingle course is then installed, with its outer edge flush with the outer edge of the starter course. Four nails or staples, placed as shown in Figure 9-15, are commonly used to fasten each shingle, although more fasteners may be required in areas of high wind. The second course of shingles is then installed with its outer edge flush with the top of the tab cuts in the first course. Note that the second course tabs are offset horizontally by one-half tab width from the first course. This results in the tab pattern illustrated. The bottom of each shingle tab is coated at the factory with a self-sealing compound which is activated after installation by the heat of the sun to seal the tab edges to the underlying shingle.

TABLE 9-6

Fastening Roof Coverings*

Roof covering material	*Fastener specifications*			*Spacing specifications[4]*
	Fastener style[2]	*Minimum O.D. crown*	*Minimum O.D. leg lengths[1]*	
Base ply and roofing plies	12 ga. roofing nail[6]		1[5]	Nails or staples driven through tin discs, spaced maximum 12″ on center.
	16 ga. staple	15/16	1[5]	
Asphalt shingles	12 ga. 3/8″ HD Roofing Nail		1 1/4[5]	4 nails or staples per each 36″-40″ section of shingle.
	16 ga. staple	15/16	1 1/4[5]	
Asphalt hip and ridge shingles	12 ga. 3/8″ HD Roofing Nail		1 1/4[5]	2 nails or staples are required for each hip and ridge shingle.
	16 ga. staple	15/16	1 1/4[5]	
Wood Shingles[3]	.076 shingle nail		1 1/4	16″ and 18″ shingle—2 fasteners per shingle. 24″ shingle—2 fasteners per shingle.
	.080 T-nail		1 1/4	
	16 ga. staple	7/16	1 1/4	
	.080 shingle nail		1 1/2	
	.080 T-nail		1 1/2	
	16 ga. staple	7/16	1 1/2	
Wood shakes[3]	.0915 shingle nail		2	2 nails or staples per each shake.
	.0915 to.099 T-nail		2	
	16 ga. staple	7/16	2	

*Reproduced from the 1986 edition of the *CABO One and Two Family Dwelling Code,* copyright 1986, and 1987 Amendments, copyright 1987, with permission of the publishers—Building Officials and Code Administrators International, International Conference of Building Officials and Southern Building Code Congress International, Inc.

[1] Shingles and shakes attached to roof sheathing having the underside of the sheathing exposed to visual view may be attached in these locations with nails or staples having shorter lengths than specified so as not to penetrate the exposed side of the sheathing.

[2] All nails and staples shall be corrosion resistant.

[3] Nails may have T-heads, clipped round heads, or standard heads.

[4] Roof coverings shall be fastened in an approved manner.

[5] Nails or staples shall be long enough to penetrate into the sheathing 3/4 inch or through the thickness of the sheathing, whichever is less.

[6] Annularly threaded nails with minimum 1-inch-diameter heads shall be used for plywood decks.

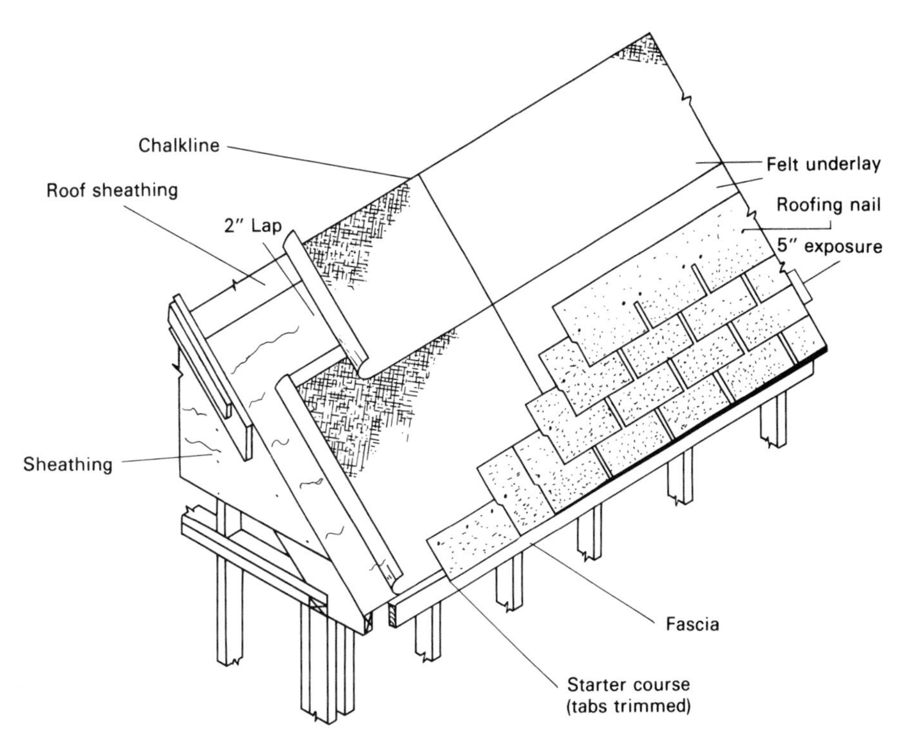

Figure 9-15 Installation of asphalt shingles (adapted from U. S. Department of Agriculture).

The roof ridge is closed by covering it with ridge shingles. Ridge and hip shingles may be supplied by the manufacturer or created by cutting apart standard shingles at the tab slots. Ridge and hip shingles should have the same exposure as that used on the roof itself. Fasten ridge shingles in accordance with Table 9-6.

Asphalt Roll Roofing

Asphalt roll roofing is a material similar to asphalt shingles but supplied in roll form, usually 36 in. (91 cm) wide. It is primarily used for utility and small industrial buildings. It should not be applied on roof slopes less than 2 in 12.

Application is similar to that used for roof underlayment. While roll roofing may be applied either horizontally (parallel to ridge) or vertically (perpendicular to ridge), horizontal application is most common and is described below. To minimize wrinkling and buckling, the roofing material should be cut into lengths not exceeding 18 ft. (5.5 m) and allowed to lay flat under warm conditions for 24 hrs. before installation. When installing roll roofing, follow the manufacturer's installation instructions, including the use of flashing and ridge caps. The installation procedures described below are often used.

For horizontal installation, work from the lower edge of the roof laying successive strips with side joints lapped 2 in. (5 cm). Allow the roofing to overlap roof edges at least 1/4 in. (6 mm) to form a drip edge. Cement laps and nail with roofing nails working from the center of the strip toward each end. For roof slopes less than 4 in 12, increase side lap to 3 in. (7.6 cm). Nails should be spaced 3 in. (7.6 cm) apart and 3/4 in. (1.9 cm) from the edge of the strip. Lap ends of strips at least 6 in. (15 cm), cement, and fasten with two rows of nails placed 3/4 in. and 1 3/4 in.

(1.9 and 4.4 cm) from the end. Again, nails should be spaced 3 in. (7.6 cm) apart. Fasten the roofing edges with a single row of nails spaced 2 in. (5 cm) apart and 1 in. (2.5 cm) from the edge. Coat all nail heads with roofing cement.

Built-up Roofing

Built-up roofing is often used on flat and low pitch roofs. It consists of three or more layers of roofing felt cemented together by hot asphalt or tar and covered with a layer of gravel embedded in hot asphalt or tar. The installation of a typical built-up roof is illustrated in Figure 9-16(a).

The base ply (30 lbs. felt in the figure) should be fastened to nailable decks in accordance with Table 9-6. On nonnailable decks, the base ply should be cemented to the roof deck in accordance with the deck manufacturer's recommendations. The succeeding felt layers (15-lbs. felt in the figure) are then applied by cementing them to the previous layer with a mop coat of hot asphalt or tar. A typical minimum requirement for such a mop coat is 20 lbs. (9.1 kg) of hot asphalt per square (100 sq. ft. or 9.3 m^2). Finally, the outer surface is formed by embedding a layer of gravel or slag in a coating of hot asphalt or tar. For this, a typical minimum requirement per square is 400 lbs. (182 kg) of gravel embedded in 60 lbs. (27 kg) of hot asphalt.

In Figure 9-16(b), note the use of a gravel stop to help retain loose gravel on the roof. Figure 9-16(c) illustrates the typical flashing used where the roof intersects a vertical wall.

Wood Shingles

The use of wood shingles for siding is described in Section 9-2. Wood shingles, usually No. 1 grade red cedar, are also used for roof coverings. When used for roof coverings, the maximum weather exposure permitted is less than that permitted for siding. The maximum exposure of No. 1 grade red cedar shingles and shakes (including hip and ridge shingles) should not exceed the following limits.

	Maximum exposure [in. (cm)]		
Length in. (cm)	*Shingles under 4/12*	*Shingles 4/12 & steeper*	*Shakes 4/12 & steeper*
16 (41)	3.75 (9.5)	5.00 (12.7)	—
18 (46)	4.25 (10.8)	5.50 (14.0)	7.50 (19.0)
24 (61)	5.75 (14.6)	7.50 (19.0)	10.00 (25.4)

The installation of a typical wood shingle roof is illustrated in Figure 9-17. While shingles may be applied over solid or spaced sheathing, solid sheathing is more commonly used. When used, spaced sheathing should not be smaller than nominal 1 × 4 in. (2.5 × 10.2 cm) boards installed at a center-to-center spacing equal to the weather exposure of the shingles. An underlayment of at least one ply of No. 15 roofing felt should be used when shingles are installed on roof slopes of less than 4 in 12. A single 36 in. (91 cm) wide strip of No. 15 or heavier roofing felt is usually installed flush with the eave for eave protection when an underlayment is not used.

The offset of joints (side lap) should not be less than 1 1/2 in. (3.8 cm) for adjacent courses and 1/2 in. (1.3 cm) for alternate courses. A spacing of 1/4 to 3/8 in. (6.4 - 9.5 cm) should be used between adjacent shingles to allow for expansion when wet.

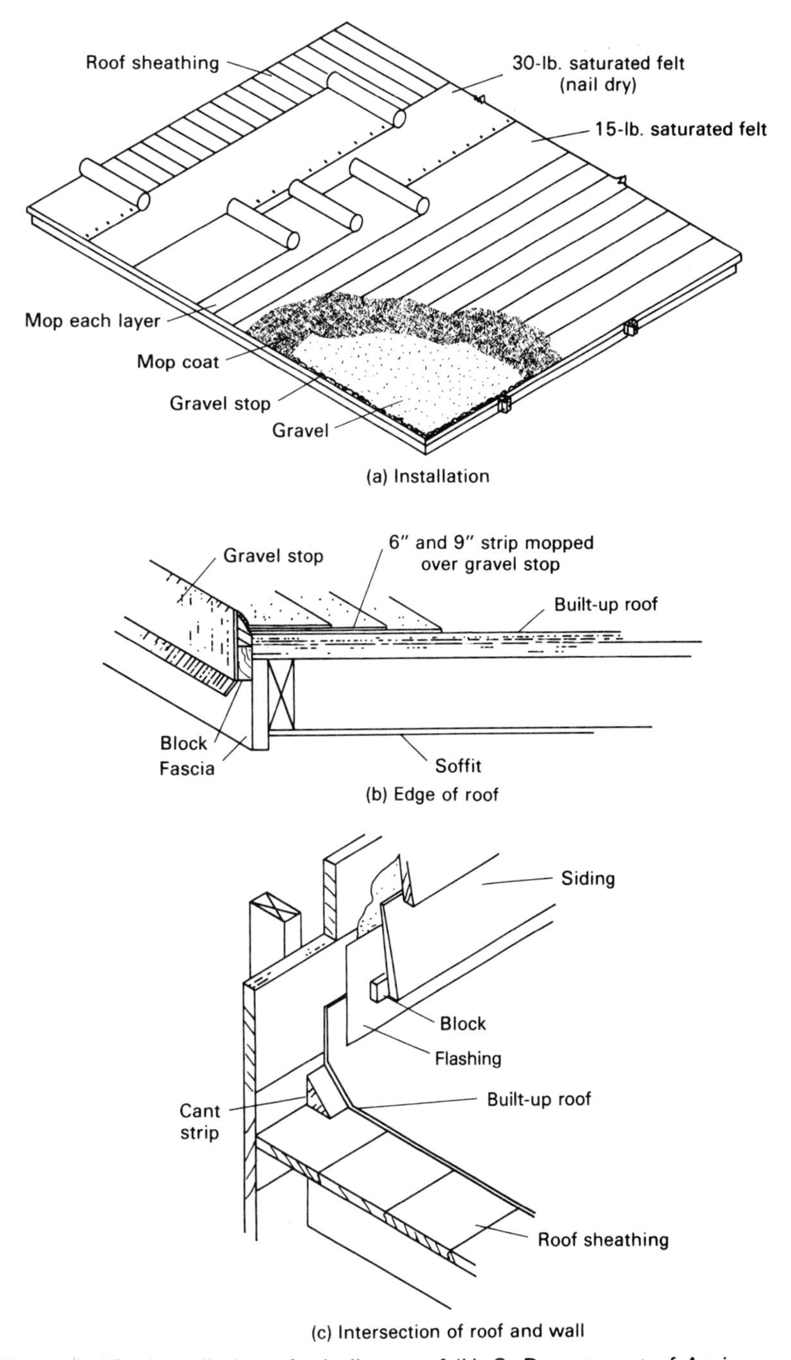

Figure 9-16 Installation of a builtup roof (U. S. Department of Agriculture).

The first course of shingles should be doubled as illustrated. The shingles should overhang the roof eave about 1 1/2 in. (3.8 cm) and the gable edge about 3/4 in. (1.9 cm) to form a drip edge. Shingles should be fastened as indicated in Table 9-6. Place nails 3/4 in. (1.9 cm) in from each side and 1 1/2 in. (3.8 cm) above the butt end of the following course (see Figure 9-17). Hip and ridge shingles may be supplied by the shingle manufacturer or may be fabricated on site. The weather

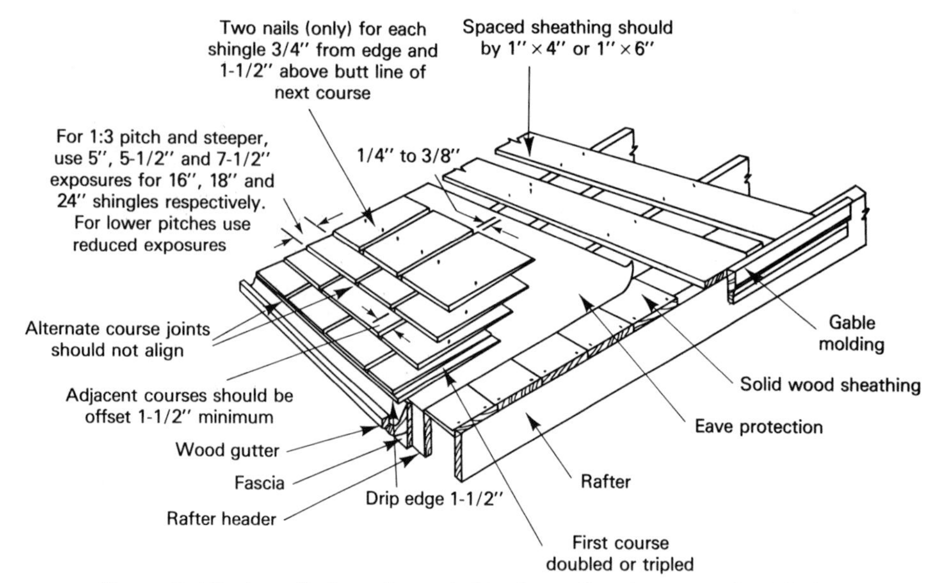

Figure 9-17 Installation of wood shingle roofing (courtesy Red Cedar Shingle and Handsplit Shake Bureau).

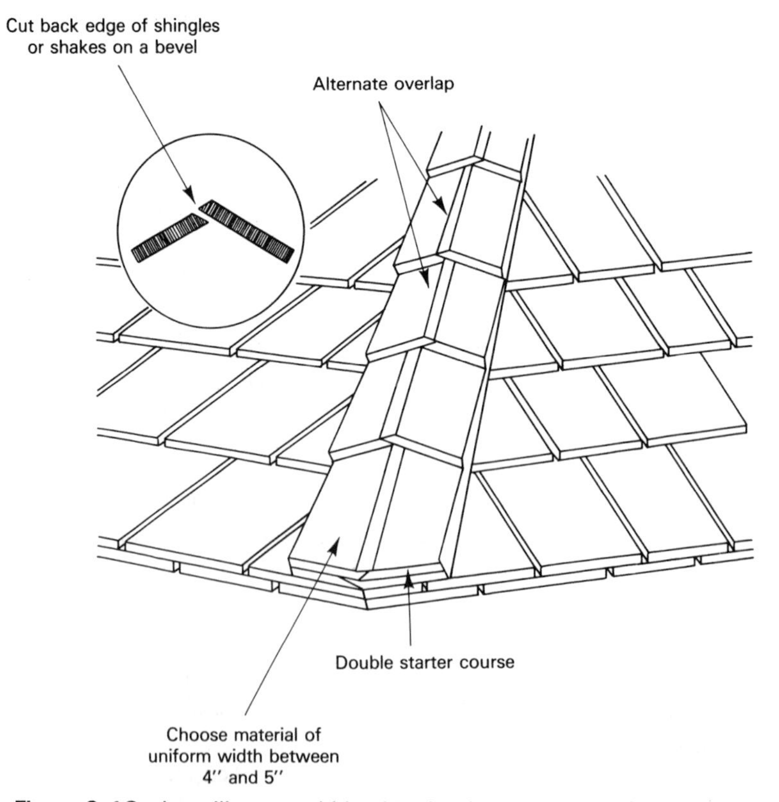

Figure 9-18 Installing wood hip shingles (courtesy Red Cedar Shingle and Handsplit Shake Bureau).

exposure of hip and ridge shingles should be the same as that used for roof shingles. A recommended procedure for installing hip shingles is illustrated in Figure 9-18.

Wood Shakes

Wood shakes are thick wood shingles that have been hand split and resawn. As a result, shakes are thicker than wood shingles and are rough on both faces. The installation of wood shakes as a roof covering is similar to the installation of wood shingles except as follows. When installed over spaced sheathing, the sheathing should be nominal 1 × 6 in. (2.5 × 15.2 cm) boards installed at a center-to-center spacing equal to the weather exposure of the shakes. Solid sheathing and an underlayment of No. 30 roofing felt should be used for roof slopes of 4 in 12 or less. Solid sheathing should also be used in areas subject to snow accumulation, with an underlayment of No. 15 or heavier roofing felt. When solid sheathing is used, the shakes may be fastened directly to the sheathing (see Figure 9-19) or they may be fastened to nailing strips applied over the sheathing at a center-to-center spacing equal to the weather exposure of the shakes.

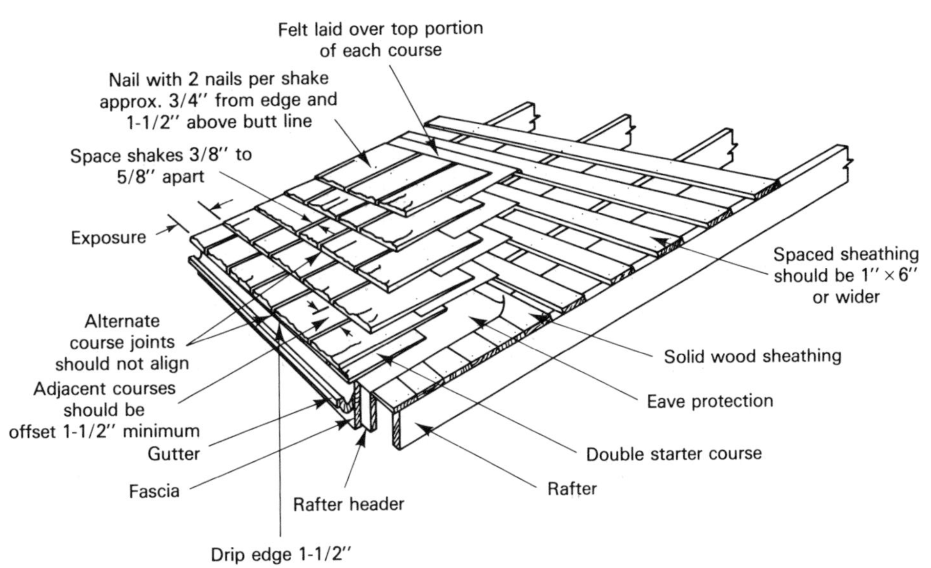

Figure 9-19 Installation of wood shake roofing (courtesy Red Cedar Shingle and Handsplit Shake Bureau).

The first course of shakes should be doubled, as illustrated in Figure 9-19. The spacing between adjacent shakes should be 3/8 to 5/8 in. (9.5 to 15.9 mm) unless otherwise specified by the building code. After each course of shakes is installed, an 18 in. (46 cm) wide strip of No. 30 roofing felt is applied to cover the upper end of the shake as shown in Figure 9-19. The lower edge of this strip should be located at a distance equal to twice the weather exposure above the butt end of the shake it covers.

Other Roofing

Some of the other available roof coverings include metal sheets and shingles, clay tile, concrete tile, slate shingles, asbestos-cement shingles, and elastomeric roof coatings described below. In all cases, roofing must be installed in accordance with the manufacturer's recommendations and the requirements of the local building code.

Metal shingles should not be installed on roofs having a slope of less than 3 in 12. They should be applied over an underlayment of No. 30 or heavier roofing felt. Some of the materials used for *sheet metal roofing* include aluminum, copper, galvanized steel, lead, stainless steel, and terneplate (sheet metal coated with an alloy of tin and lead). Many of these sheet metal roofings incorporate proprietary systems for fastening and waterproofing the joints.

Clay tile and *concrete tile* should not be installed on roofs having a slope of less than 3 in 12. An underlayment of No. 30 roofing felt or two layers of No. 15 felt should be employed. Many tiles incorporate anchor projections, which rest on nominal 1 × 2 in. (2.5 × 5.1 cm) wood strips during installation. Tiles should be secured by corrosion-resistant nails or anchors.

Slate shingles are split from natural slate, shaped into rectangular shingles, and drilled for fasteners. They are installed in a manner similar to wood shingles. An underlayment of No. 30 roofing felt or two layers of No. 15 felt should be employed.

Asbestos-cement shingles should not be installed on roof slopes of less than 3 in 12. Use an underlayment of No. 15 roofing felt. Installation is similar to that of asphalt shingles. Preformed ridge and hip shingles are used to cover these joints.

Elastomeric roofing consists of a coating of a rubberlike synthetic polymer (such as silicone rubber) applied to a sound roof deck. Such roofing may be quickly applied by spraying or brushing. Since it can be completely seamless, the opportunity for leakage is minimized. However, many of the available materials are relatively new and have only a limited performance history. To date, such roofing has primarily been used on commercial and industrial buildings, but its use on residential structures is expected to increase in the future.

PROBLEMS

1. Explain the difference between structural sheathing and nonstructural sheathing.
2. What is No. 15 roofing felt? How is it used?
3. Briefly explain how bevel wood siding should be installed on a wood frame structure.
4. Describe how a brick veneer is installed on a wood frame building sheathed in plywood.
5. Explain the purpose and construction of expansion joints and control joints in concrete and masonry construction.
6. What is "flashing"?
7. Briefly describe the installation of a double-coverage asphalt shingle roof.
8. What is a roofing underlayment? When is it used?
9. Briefly describe the installation of a 3-ply built-up roof.
10. How do wood shakes differ from wood shingles?

REFERENCES

1. American Institute of Timber Construction. *Timber Construction Manual.* Third ed. New York: Wiley, 1985.
2. Anderson, L. O. *Wood Frame House Construction.* Agriculture Handbook No. 73. U.S. Department of Agriculture, Washington, DC, 1975.
3. American Plywood Association. *APA Design/Constrution Guide: Residential and Commercial.* Tacoma, WA, 1988.
4. Council of American Building Officials. *CABO One and Two Family Dwelling Code.* Falls Church, VA.
5. Red Cedar Shingle and Handsplit Shake Bureau. *Red Cedar Shingle and Shake Design and Application Manual for New Roof Construction.* Bellevue, WA, undated.

chapter 10

Doors and Windows

10-1 FRAMING OPENINGS

Typical framing for exterior door and window openings is illustrated in Figure 10-1. Notice that double studs (stud plus jack stud) are used on each side of openings wider than 3 ft. (91 cm) to carry the vertical loads transmitted by the header and wall plate(s). The framing for openings in nonloadbearing interior walls and partitions is similar to that shown in Figure 10-1 except that only a single stud is required on each side of the opening, and the header may be a single 2 × 4 in. (5 × 10 cm) member placed with its 4 in. (10 cm) dimension horizontal. The cost of framing can be minimized by the use of modular frame dimensioning and by locating door and

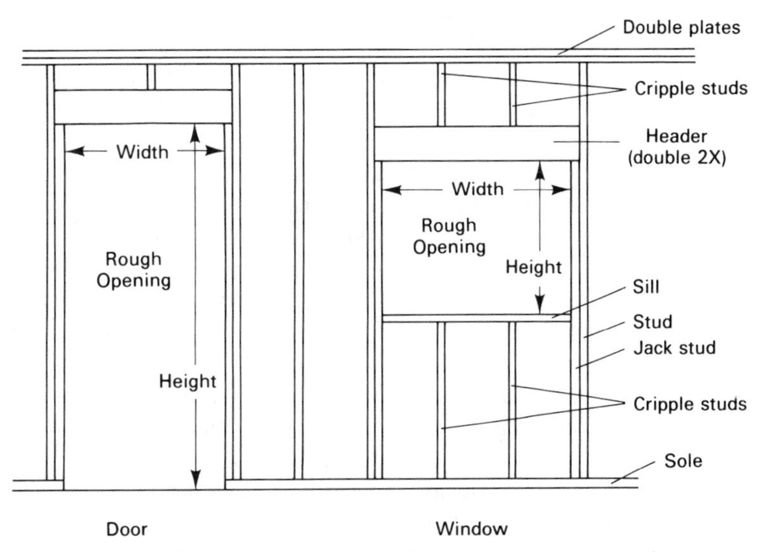

Figure 10-1 Typical framing for exterior door and window.

window openings on the module. Techniques for minimizing framing cost are described in Chapter 7.

The interior dimensions of the rough opening for doors and windows should be 3/4–1 1/2 in. (1.9–3.8 cm) greater in width and 3/4–1 in. (1.9–2.5 cm) greater in height than the door or window frame to be inserted. This provides space for plumbing and leveling of the door or window frame. For doors, a rough opening 2 1/2–3 1/2 in. (6.4–8.9 cm) wider and 3 in. (7.6 cm) higher than the nominal size of the door is often used.

The framing for a typical floor opening is described in Section 7-2 (see Figure 7-9). Framing for chimney and other roof openings is described in Chapter 8 (see Figure 8-14). Again, the size of the rough opening must provide the necessary clearance for the item to be inserted, which may be a chimney, a skylight, a stairway, or other item. If not specified on the construction drawings, the required size of the rough opening must be obtained from the manufacturer of the item to be inserted. A clearance of at least 2 in. (5 cm) must be provided between wood framing and the outside of a masonry chimney located partially or completely within the structure. A clearance of at least 3/4 in. (1.9 cm) should be provided between the chimney and floor, wall, or roof sheathing. A clearance of at least 1 in. (2.5 cm) should be provided between masonry chimneys located entirely outside the structure and any combustible material.

10-2 DOORS

Characteristics

Doors are available as exterior or interior types in a variety of materials, styles, and sizes. *Exterior doors* are intended for building entrance and exit and must be weather and vandal resistant. *Interior doors* are used for privacy and to subdivide areas within a building. The materials used to construct doors include wood, aluminum, steel, plastic, and glass. Hollow metal doors are often used in commercial buildings for strength and fire resistance. They are also increasingly being used as exterior doors of residential structures because of their weather and vandal resistance. When the interior of the door is filled with an insulating material, the door become quite energy efficient.

Doors are available in widths from 18 in. (46 cm) to 40 in. (102 cm) in 2 in. (5 cm) increments, heights of 72 in. (183 cm) to 84 in. (213 cm), and thicknesses of 1 3/8 and 1 3/4 in. (3.5 and 4.4 cm). The most common door height is 80 in. (203 cm), although entrance doors are sometimes 84 in. (213 cm) high. Main entrance doors are usually 36 in. (91 cm) wide and other exterior doors 32 in. (81 cm) wide. Common interior door widths include bedrooms 30 in. (76 cm), bathrooms 28 in. (71 cm), and closets 24 in. (61 cm). Exterior doors are normally 1 3/4 in. (4.4 cm) thick, and interior doors are 1 3/8 in. (3.5 cm) thick.

Some common styles of exterior doors are illustrated in Figure 10-2. The *panel door* is widely used for its decorative effect. Panel doors are often identified by the number of panels used. Thus, the door illustrated is a six-panel door. In panel doors, the vertical door members are *stiles,* the horizontal members are *rails,* and the inserts are *panels.* Panels can be solid or glazed. *Flush doors* have smooth exterior faces covering a frame (or edge strips) and a core. The core may consist of lightweight reinforcing such as strips, cells, or honeycomb fibers (*hollow core*), solid wood or particleboard (*solid core*), or plastic foam insulation (*foam core*). Wooden flush doors for exterior use should have solid cores to resist warping. Flush doors may contain glazed openings or have decorative designs applied to the exterior face.

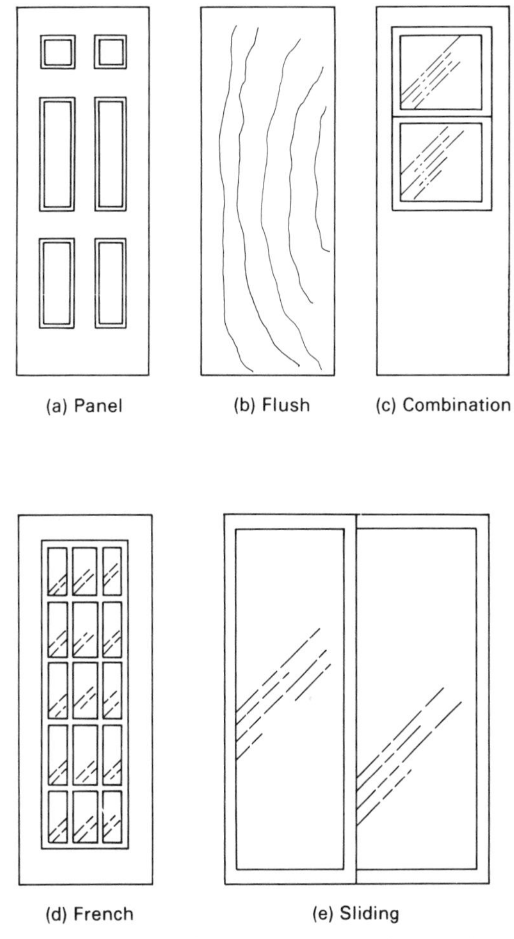

Figure 10-2 Common exterior door styles.

Combination doors combine a screened operable or removable glass section with a panel or flush door. *French doors* have a large number of glass panels to provide maximum visibility and are often used to provide access to porches or patios. *Sliding glass doors* consist of two or more sliding glass panels, framed in wood or aluminum, which move within a track. Such doors are used to provide maximum visibility and access to the exterior. However, in the usual case (two-panel), only one-half of the total opening is available for access or ventilation. However, it is possible to install the door as a *pocket door* so that all panels recess into the wall when the door is open. However, sliding glass doors typically offer poor resistance to air infiltration.

Typical interior door styles are illustrated in Figure 10-3. The interior panel door is a thinner version of the design used as an exterior door. The interior flush door is commonly a hollow core door, but a solid or foam core door may be used for reduced sound transmission. The *louvered door* is used for increased ventilation, primarily in closets and storage rooms. *Bifold doors* are assembled in pairs mounted and hinged as illustrated. As a result, each pair of panels folds together and projects into the room only the width of one panel. Bifold doors may consist of two panels or four panels (illustrated). Due to their design, bifold doors are able to cover a large opening and yet occupy only a minimum of space when open; they are often used as closet doors.

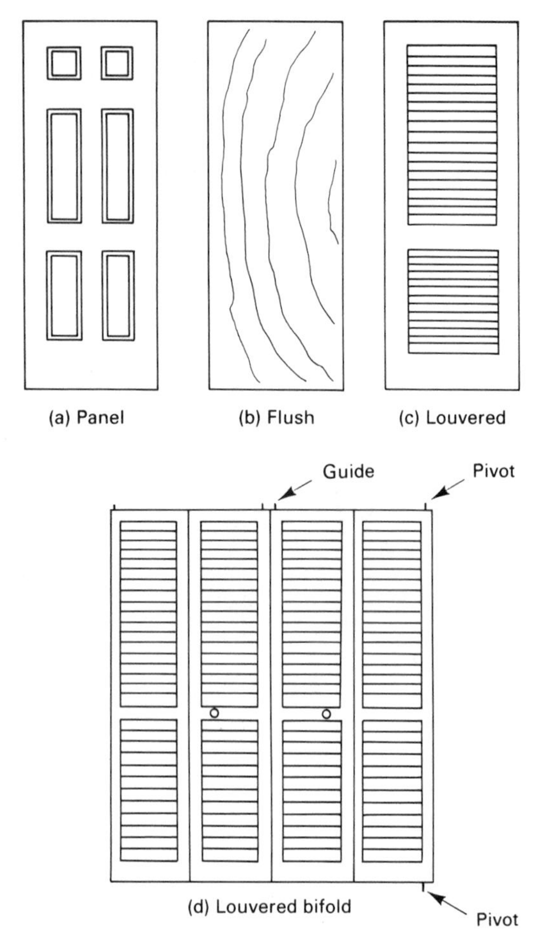

Figure 10-3 Common interior door styles.

Installation

Traditionally, a door frame is installed in a rough opening, hinges mounted, and the door hung within the frame. However, prehung doors are now available and widely used. A *prehung door* consists of a door frame (including stop), hinges, and premounted door assembled as a unit. The door is often also prebored for the installation of the door lock. Thus, it is only necessary to install the unit plumb and level within the rough opening and apply the casing to complete the door installation. The door lock and strike plate are installed after the door is hung in either case.

The major components of a door frame are illustrated in Figure 10-4. The sides of the frame are known as *jambs* and the top as the *head* (or head jamb). The *casing* covers the gap between the jambs and the rough opening. When fully closed, the door rests against the *stop*. A *threshold* is normally used for exterior doors to aid in preventing entry of water into the building. Wedges are used to position the door frame within the rough opening so that the frame (and door) is level and plumb. Four or five wedges are used on each side of the door frame as needed. The wedges are held in place by nailing through the frame and wedges into the wall studs. After installation, wedges are cut off flush with the door jamb, and the casing is installed by nailing it to the jamb, studs, and header.

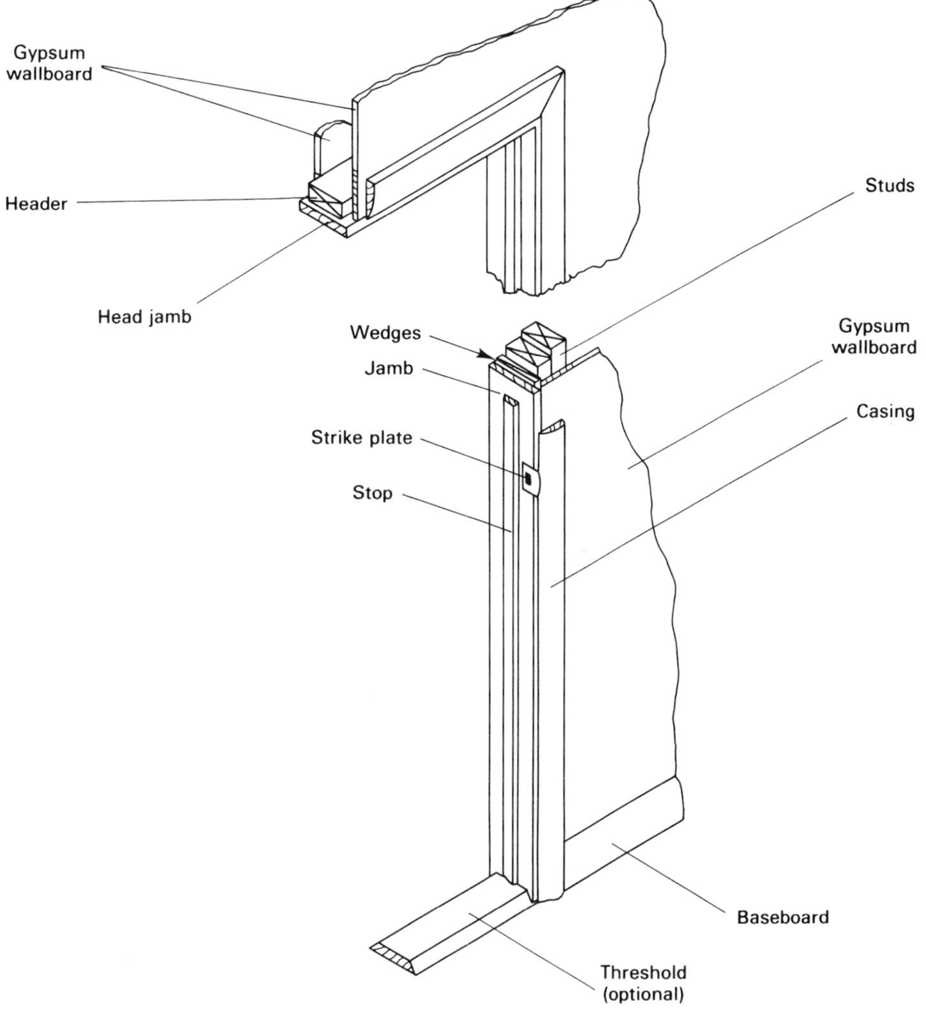

Figure 10-4 Typical door frame.

Hardware

Door hardware normally consists of hinges, door lock, and strike plate. The usual location for hinges and locks is illustrated in Figure 10-5. The loose-pin *butt hinge* illustrated in Figure 10-6 is employed for interior and exterior doors. A rounded-corner version of this hinge is also available and often used. The height and weight of the hinge needed is determined by the size and weight of the door and by its frequency of use. The required width of the hinge is determined by the door thickness, the backset distance, the inset distance, and the clearance needed. As shown in Figure 10-6, hinge clearance is the distance between the wall and the door surface when the door is fully open and parallel to the wall. The usual backset is 1/4 in. (6 mm), but a backset of 3/8 in. (10 mm) is sometimes used for heavy exterior doors. The inset distance is usually zero because the door surface is normally flush with the edge of the door frame.

For exterior doors 1 3/4 in. (4.4 cm) thick, three 4 × 4 in. (10.2 × 10.2 cm) hinges are commonly used. For interior doors, 1 3/8 in. (3.5 cm) thick, two 3 1/2 × 3 1/2 in. (8.9 × 8.9 cm) hinges are usually employed. Low-friction and ball bearing butt hinges are available for doors having a high frequency of use.

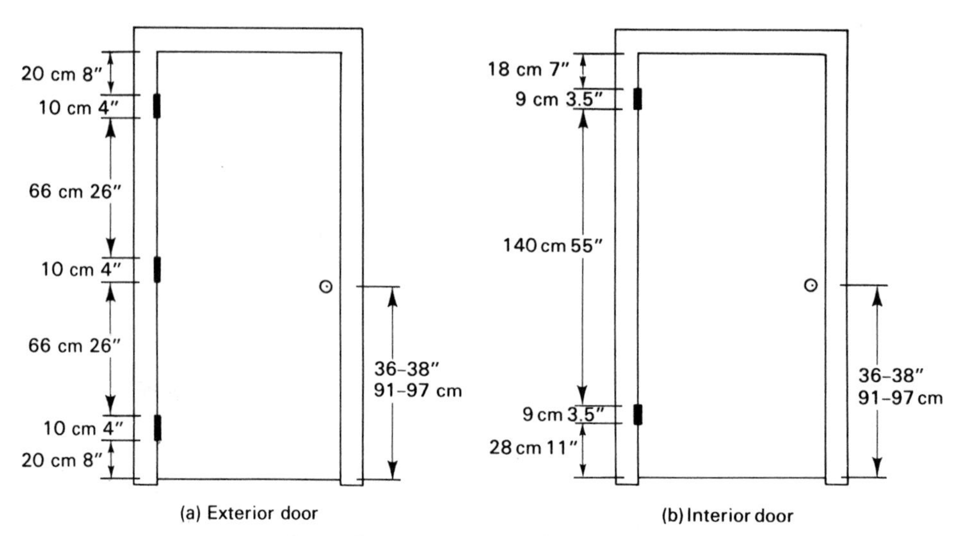

Figure 10-5 Typical location of door hardware.

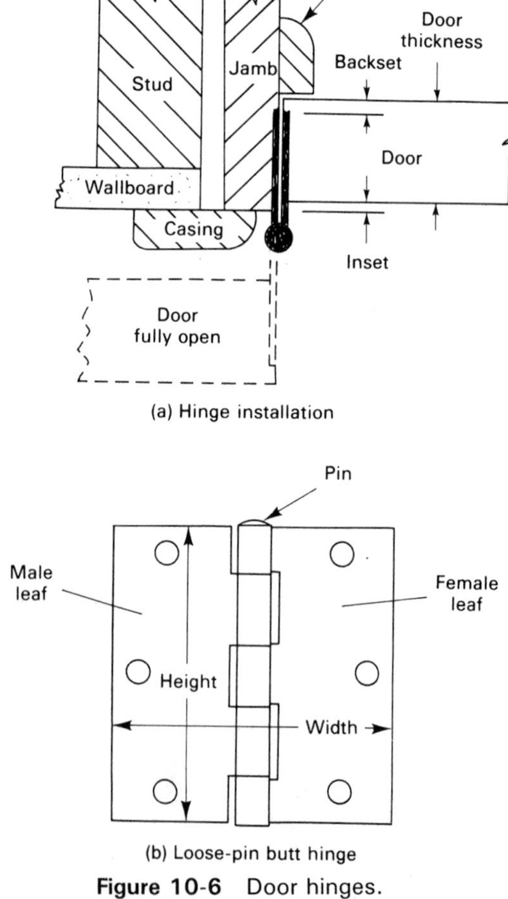

Figure 10-6 Door hinges.

Door locks consist of entry locks, privacy locks, and passage locks. *Entry locks* open by a key from the outside and contain a lock control on the inside. *Privacy locks,* commonly used for bedroom and bathroom doors, incorporate a lock control button on the inside together with a safety hole or slot that allows the lock to be opened by a screwdriver from the outside in an emergency. A *passage lock* is simply a door latch and handle without any locking device. Install the door lock in accordance with the lock manufacturer's instructions.

The *strike plate* provides a recess to hold the door latch in the closed position and prevents the door latch from damaging the door jamb. To install a strike plate, close the door (with the door lock installed) against the stop and mark on the jamb the position of the latch plunger. Place the strike plate over this mark so that the latch plunger falls into the latch recess. Mark the outline of the strike plate and latch recess. Drill or chisel out a recess in the jamb deep enough to accommodate the latch recess of the strike plate. Then, chisel out the remaining area within the outline of the strike plate deep enough so that the surface of the strike plate will be flush with the jamb surface. Finally, fasten the strike plate to the jamb with the screws provided.

10-3 WINDOWS

Characteristics

Windows are available in a variety of styles, materials, and sizes. Although fixed windows are sometimes used, most windows can be opened to allow ventilation and emergency egress (exit). (*Note:* Building codes have requirements for exits and emergency egress.) Windows also provide a view of the outside, admit light to the interior, and provide an architectural effect. The most common styles of window include double-hung, single-hung, casement, awning, and horizontal sliding units (see Figure 10-7). Windows are usually supplied by the manufacturer as complete ready-to-install units equipped with hardware, weatherstripping, and glazing (glass). They are also available without glazing, to be glazed after installation.

Double-hung windows consist of two sash (upper and lower) which may be raised or lowered within tracks to provide ventilation. Like doors, the sides and top of the window frame are *jambs.* The frame bottom is a *window sill* (or *sill*). When a sash is divided into smaller glass panels, the dividers are known as *mutins.* A system of springs or counterbalances is commonly used to offset the weight of the sash and enable it to be easily raised or lowered. However, lightweight sash may use a simple compression strip to hold the sash in place at any elevation. The maximum ventilation opening of a double- or single-hung window is about 45 percent of the window area.

Single-hung windows are similar to double-hung windows except that the upper sash is fixed so that only the lower sash can be opened. Because of this, single-hung windows do not provide as much rain protection when open as do double-hung windows.

Casement windows use sash that are pivoted vertically about one edge to open out. The major advantages of a casement window are that it provides a large ventilation area (about 90 percent of window area) and good emergency egress. However, screens must be placed inside the window, and an operating mechanism must be provided for opening and closing the sash.

Awning windows use sash that are pivoted horizontally near their top edge to open outward. Like casement windows, they provide a large ventilation area (about 75 percent of the window area). In addition, they provide better rain protection while open than do other window types. However, the small size of the opening

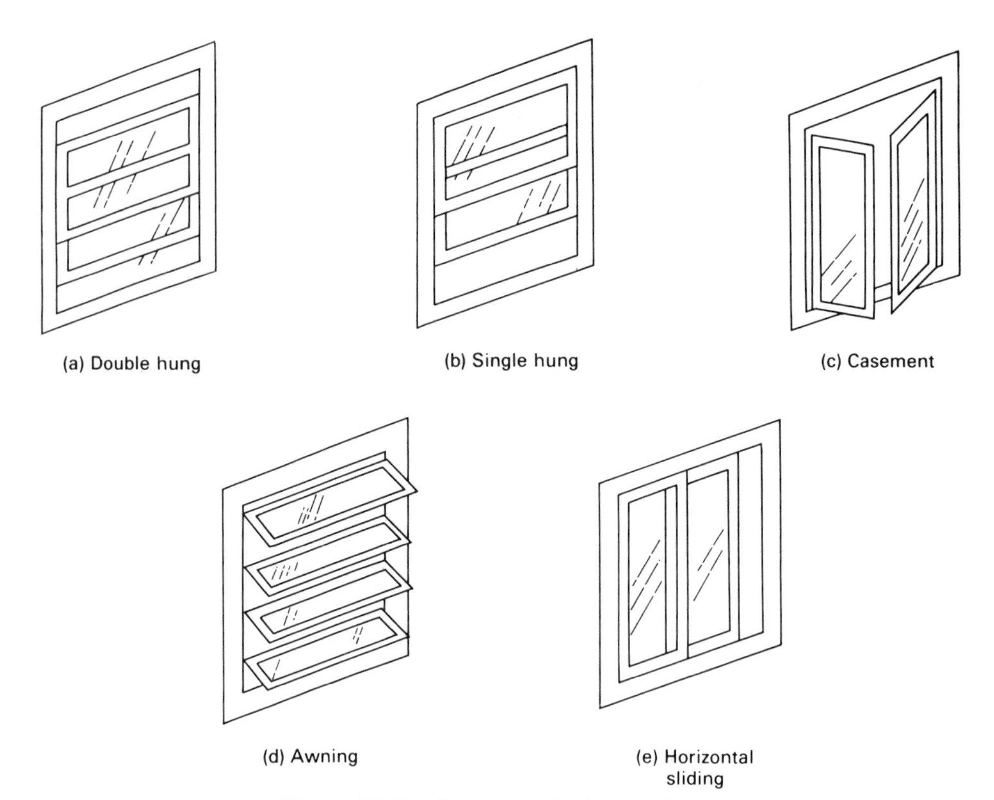

Figure 10-7 Common window styles.

between sash may limit emergency egress. Again, screens must be located inside the window.

Horizontal sliding windows consist of two or more sash that slide horizontally within tracks. Like sliding patio doors, the maximum opening for a two-panel unit is only one-half the total opening. Ventilation opening is about 45 percent of the window area, the same as single- and double-hung windows.

The principal materials used in constructing window frames and sash include wood, alumimum, vinyl, and steel. Wood units provide good energy efficiency (see Chapter 12) and are often used in residential construction. However, they are expensive and subject to damage from the weather and insects. To help resist such damage, wood components are treated with a preservative during manufacture. Vinyl and metal-clad wood frames and sash are also available. Aluminum windows are economical and corrosion resistant. However, they are not as energy efficient as wood windows. Vinyl windows do not corrode and have good insulating qualities. Steel windows are more rigid than aluminum units but require galvanizing or other treatment to resist corrosion. They are most commonly used in commercial and industrial buildings.

Glass is the principal glazing material; however, plastic is sometimes used to resist breakage. Insulated, tinted, reflective, and other special types of glass are available to reduce the heat passing through window panes (see Chapter 12).

Installation and Hardware

The installation of window frames is similar to the installation of door frames, as described in Section 10-2. Frequently, the exterior casing is attached to the window frame by the manufacturer. In this case, the window frame is inserted into the rough opening from the outside. The frame is then levelled and plumbed, taking care that the window frame remains square. Wedges are used under the sill and along each side jamb to level and plumb the frame. The frame is then fastened into place. Space between the rough opening and the window frame should be sealed to aid in preventing air infiltration into the building. Foam caulk is recommended for this purpose. The interior casing is now attached and the exterior casing sealed to the siding with caulk to further reduce air infiltration.

Window hardware is usually furnished and preinstalled by the window manufacturer. Hardware includes the operating mechanism for awning and casement windows, as well as locks or latches for securing sash in the closed position. After the window frames and sash are painted, insect screens are installed in accordance with the window manufacturer's instructions.

10-4 RESISTING AIR INFILTRATION

Windows and doors should meet industry standards for resisting air leakage when tested under the procedures of the American Society for Testing and Materials (ASTM) E-283. This test measures the volume of air leakage in cubic feet per minute (cfm) per foot of crack (l/s/m) at a wind pressure of 1.56 psf (75 Pa). The leakage of windows should not exceed 0.375 cfm/ft (0.58 l/s/m) and that of doors should not exceed 0.50 cfm/ft (0.77 l/s/m).

PROBLEMS

1. Sketch and label the components of a framed (rough) opening for a window in an exterior wall.
2. What is the minimum size of the rough opening that should be provided for the installation of a door or window frame?
3. How much clearance is required between wood framing and the outside of a masonry chimney located completely within the building?
4. State the usual height and thickness of exterior wood doors.
5. Why are foam core hollow metal doors finding increasing use as exterior doors for residential buildings?
6. Briefly describe the installation of a prehung wood door unit within a framed (rough) wall opening.
7. Identify the size and location of the hinges used for a typical exterior door.
8. Briefly describe the five major styles of windows described in this chapter.
9. Which two window styles provide the greatest ventilation area for a given window size?
10. When installing a window unit, how should it be sealed to the building frame to minimize air infiltration?

REFERENCES

1. American Institute of Architects. *Architectural Graphic Standards.* 7th ed. New York: Wiley, 1981.
2. Anderson, L. O. *Wood-Frame House Construction.* Agriculture Handbook No. 73. U.S. Department of Agriculture, Washington, DC, 1975.
3. National Forest Products Association. *Manual for House Framing.* Wood Construction Data No. 1. Washington, DC, 1970.
4. Vieira, Robin K., and Kenneth G. Sheinkopf. *Energy-Efficient Florida Home Building.* Florida Solar Energy Center, Cape Canaveral, FL, 1988.

chapter 11

Special Framing

11-1 ROOF EAVE

The portion of the roof that projects beyond the exterior wall is the roof *eave*. The underside of the eave is the *soffit*. The method of construction used to connect the exterior wall and the underside of the eave is commonly identified by the type of soffit employed. However, the portion of the building connecting the exterior wall with the underside of the eave is also known as a *cornice*. Thus, the terms "cornice" and "soffit" are sometimes used interchangeably.

The three principal forms of soffit construction are the *open soffit* (or *open cornice*), the *horizontal closed soffit* (or *box soffit with lookout*), and the *sloped closed soffit* (or *box cornice without lookout*). These are illustrated in Figure 11-1. The term *close cornice* is applied to the form of construction used to join the roof and the wall when the roof does not project beyond the exterior wall. In this case, there is no eave or soffit.

Some construction details for a horizontal closed soffit are shown in Figure 11-2. Note that the horizontal member supporting the soffit is known as a *lookout*. Thus, the name "box cornice with lookout" is also applied to this method of construction. The soffit normally contains louvered vents or a continuous screened slot to provide attic ventilation (see Chapter 12). The outside edge of the eave is known as the *fascia*. The *frieze board*, when used, provides a decorative effect and connects the top of the siding with the soffit. Although soffits have traditionally been constructed of plywood or composition board, the use of perforated aluminum or plastic soffit is becoming increasingly popular. Such materials resist weather damage, and the perforated panels eliminate the need for screening vent openings.

The soffit panels for sloped closed soffits are fastened to the underside of the rafters or truss extensions. Thus, the soffit surface is parallel to the roof surface. Open soffits and sloped closed soffits are often used for buildings with a wide roof overhang to minimize interference with window and door placement. Open soffits are used primarily for economy or for architectural effect.

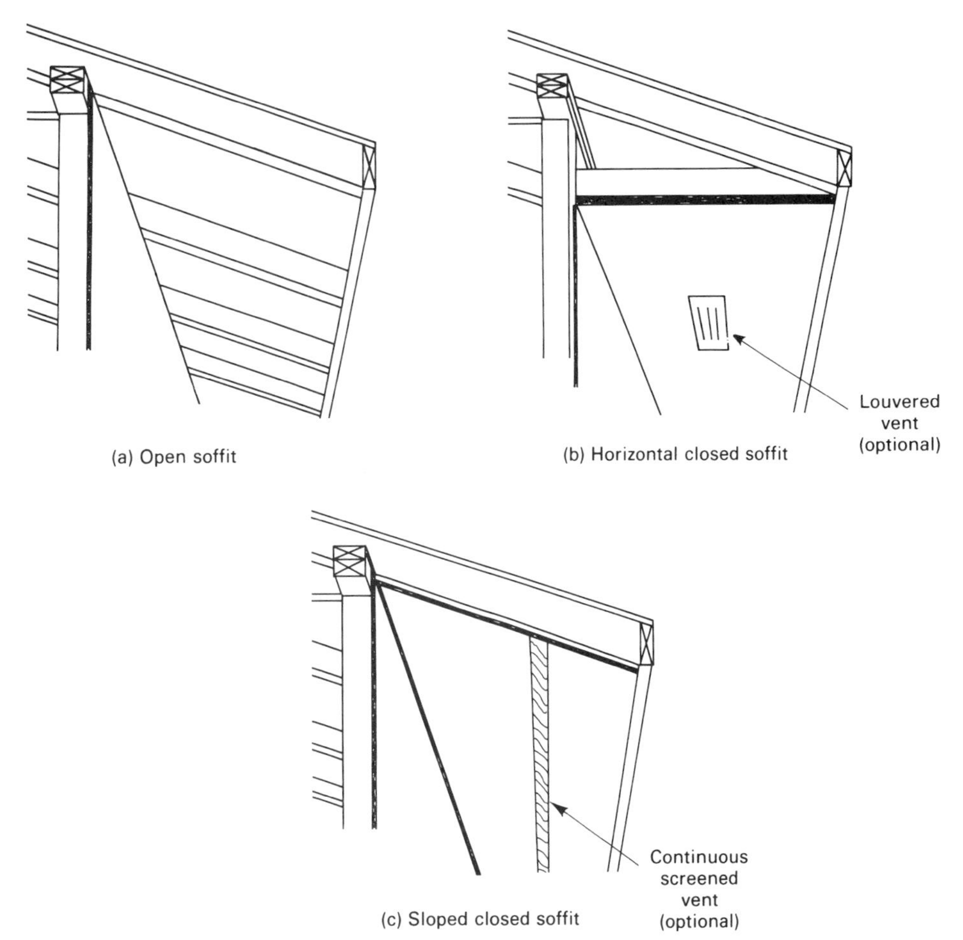

Figure 11-1 Types of roof soffit.

11-2 FRAMING FOR UTILITY SYSTEMS

The principal requirements that must be met in framing for utility systems are that the framing provide space for the passage of the necessary duct, pipe, and conduit, and that the framing be sufficiently strong to support all design loads. Some precautions to be observed when boring holes, cutting, or notching wood framing members are described in Chapter 7. These precautions are summarized below for your convenience.

To provide additional space within walls for vertical ducts, as well as waste and vent pipe, it may be necessary to use 2 × 6 or 2 × 8 in. (5 × 15 or 5 × 20 cm) soleplates and top plates in certain walls. When the sole or plate must be cut or notched more than one-half its width, reinforce the cut section using a metal tie at least 18 gauge by 1 1/2 in. (3.8 cm) wide across the cut section. Fasten the tie with four 16d nails at each end.

Wall studs should not be cut or notched to a depth greater than 25 percent of their width for bearing walls, or 40 percent of their width for nonbearing walls (see Figure 11-3). The maximum diameter of holes bored in studs should not exceed 40 percent of the stud width for bearing walls or 60 percent of stud width for nonbearing walls. However, a hole diameter as great as 60 percent of the stud width may be used for bearing walls when studs are doubled and not more than two successive doubled studs are involved. Bored holes should not be located closer than 5/8 in. (16 mm) from the edge of the stud.

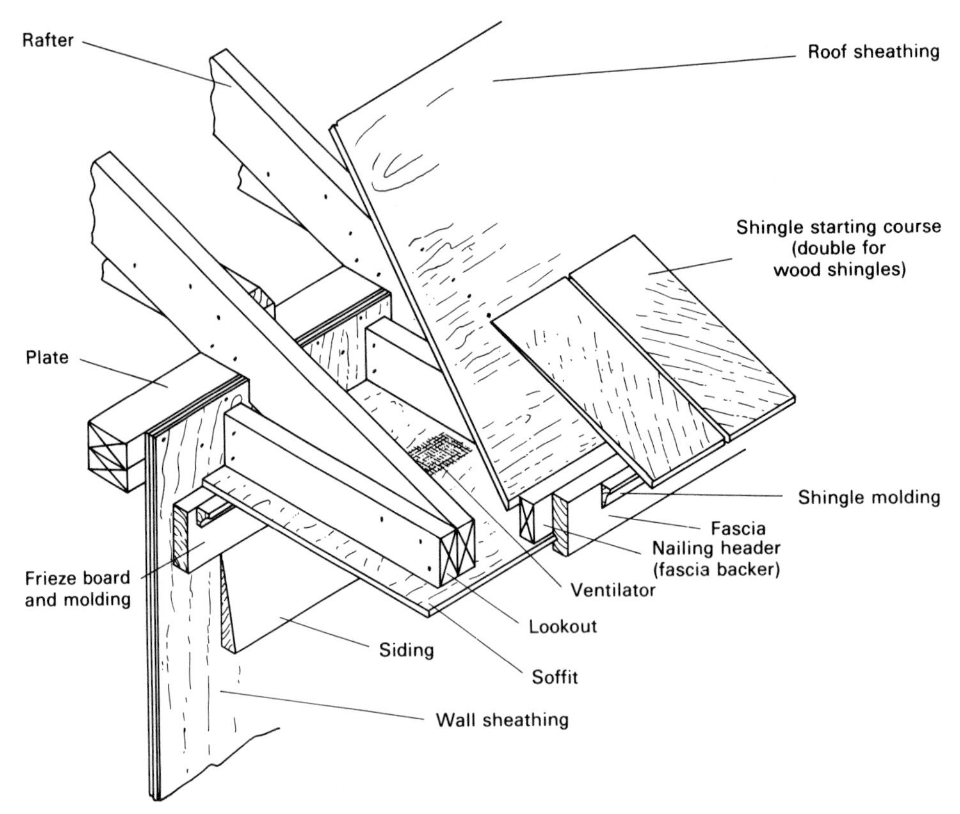

Figure 11-2 Horizontal closed soffit (adapted from U. S. Department of Agriculture).

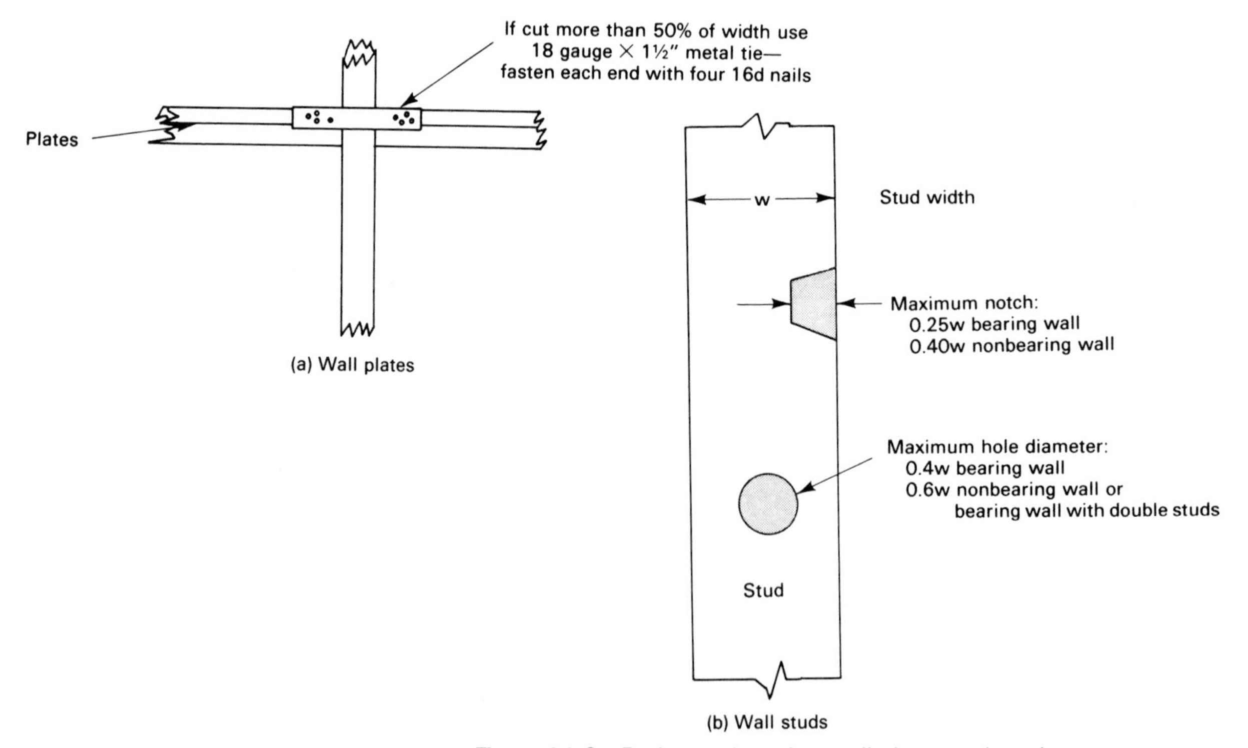

Figure 11-3 Boring and cutting wall plates and studs.

As discussed in Chapter 7, floor joists running parallel to a bearing wall should be doubled under the bearing wall. In this situation, when the wall contains vertical duct or pipe, separate the doubled joists supporting the wall enough to permit the duct or pipe to pass between the two joists. Floor joists should also be doubled under the outside edge of bathtubs in order to support the weight of the filled bathtub and its occupant.

Notches on the end of joists should not exceed one-fourth of the joist depth (see Figure 11-4). Other notches on the top or bottom edge of joists should not be deeper than one-third of the joist depth and should not be located within the center one-third of the span. The diameter of holes bored in joists should not exceed one-third of the joist depth and should not be located closer than 2 in. (5 cm) from the joist edge.

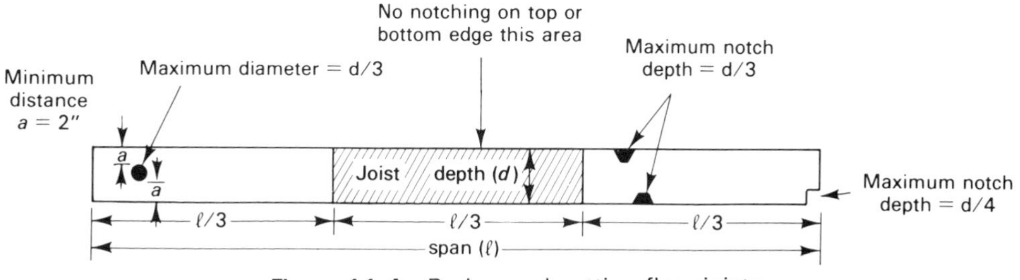

Figure 11-4 Boring and cutting floor joists.

11-3 STAIRWAYS

Basic Principles

Stairways provide access between floors of a building and are a means of emergency egress (escape). Because of their importance to occupant safety, building codes have a number of requirements relating to stairways, including minimum stairway width and headroom, installation of handrails and guardrails, maximum riser height, and minimum tread width.

The total horizontal distance occupied by a stairway is the *stairway run* (or total run), and the vertical distance from top to bottom (floor-to-floor) is the *stairway rise* (or total rise). These terms are illustrated in Figure 11-5.

Some common building code requirements for stairs include the following. The minimum stairway width should be 3 ft. (91 cm). Handrails may project into this space not more than 3 1/2 in. (9 cm) on each side. The headroom should not be less than 6 ft. 8 in. (2 m). Handrails should be provided on at least one side of stairs consisting of three or more risers. The height of handrails should be 30–34 in. (76–86 cm) above the front edge of the stair treads. Handrails should extend at least 6 in. (15 cm) beyond the top and bottom risers. Open side(s) of stairs with a total rise of more than 30 in. (76 cm) should be protected by guardrails extending 36 in. (91 cm) above the front of stair treads. Openings between rails or decorative closures of guardrails should not allow passage of an object larger than 6 in. (15 cm) in diameter.

Some terms and dimensions applying to stair components are illustrated in Figure 11-6. The *tread* is the horizontal portion of a step, and the *riser* (when used) is the vertical board closing the space between treads. When risers are used, the stairs are referred to as "closed riser" stairs. Although both closed risers and open risers are illustrated in Figure 11–6, they should not be mixed within a single set of stairs. *Riser height* or *tread rise* is the vertical distance between successive stair treads. *Tread width* is the horizontal distance between the front edge of successive

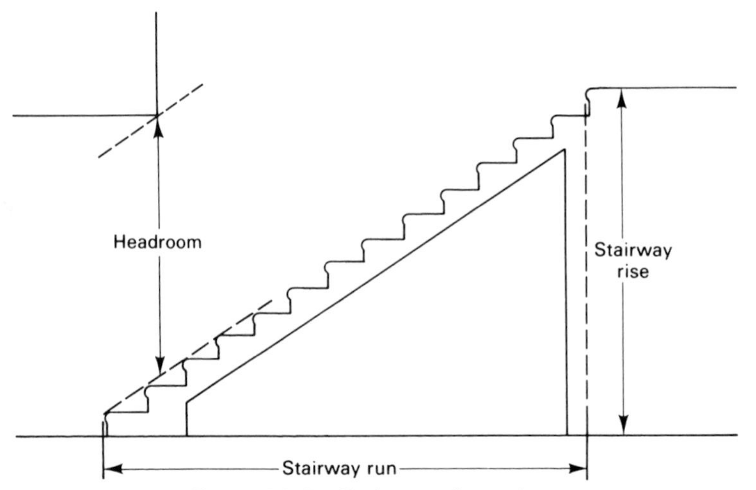

Figure 11-5 Stairway dimensions.

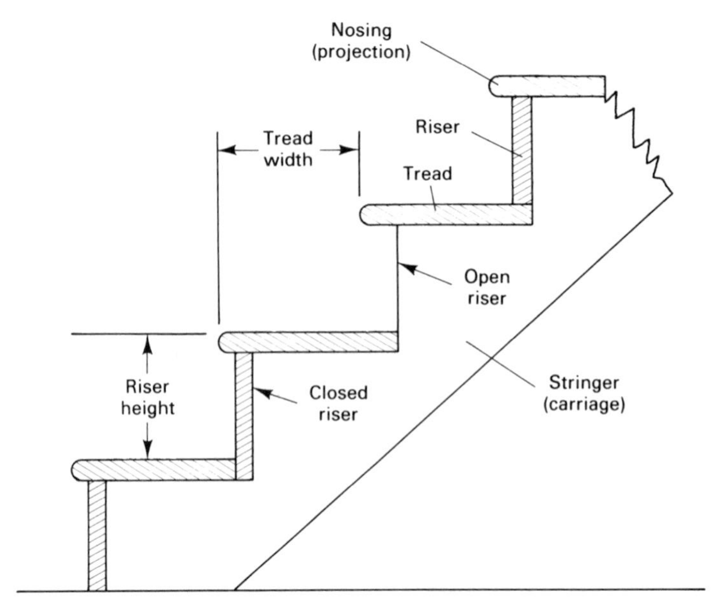

Figure 11-6 Stair components.

stair treads. Stairs with closed risers should have a *nosing* (or projection) of at least 1 in. (25 mm). *For safety, it is important that the rise and width of all tread in a stairway be uniform.* The maximum variation in these dimensions is usually specified by code but should not exceed 3/16 in. (5 mm).

The relationship between riser height and tread width should fall within certain limits for user comfort and safety. Several rule-of-thumb equations have been developed to define this relationship. A suggested relationship is as follows:

$$\text{Tread width} + (2 \times \text{riser height}) = 24 - 26 \text{ in. } (61 - 66 \text{ cm}) \qquad (11\text{-}1)$$

As always, code limits on riser height and tread width govern. Unless otherwise specified, the maximum riser height should not exceed 8 1/4 in. (21 cm), and tread width should not be less than 9 in. (23 cm). A riser height of 7 1/2 to 8 in. (19–20 cm) with corresponding tread widths of 10 to 9 in. (23–25 cm) are often used for interior stairs. For safety, exterior stairs usually have a lower rise and wider tread

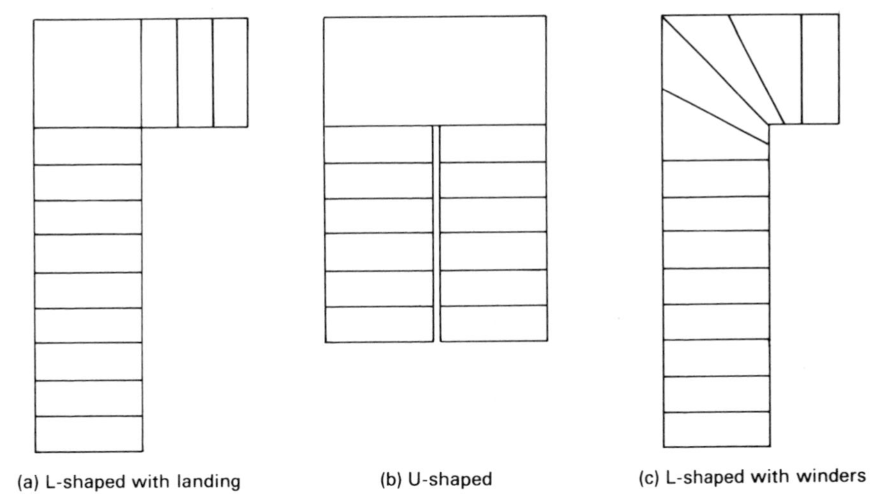

Figure 11-7 Stairways for limited space.

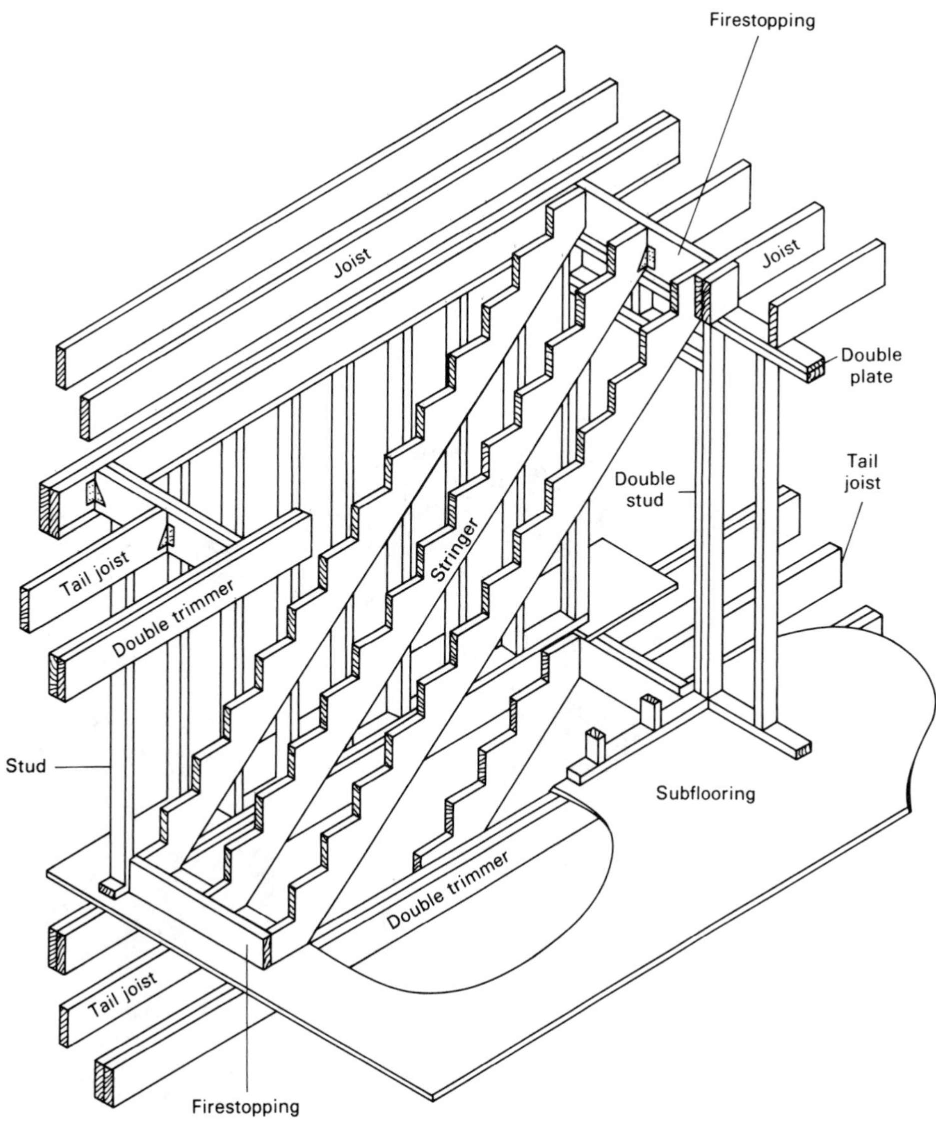

Figure 11-8 Framing for interior stairs (courtesy National Forest Products Association, Washington, DC).

than interior stairs. A rise of 6 in. (15 cm) and a tread width of 10 in. (25 cm) are often used.

When laying out stairs, the total horizontal distance (stairway run) occupied by the stairs and the floor-to-floor distance (stairway rise) may be calculated as follows:

$$\text{Stairway run} = n \times \text{tread width} \tag{11-2}$$

$$\text{Stairway rise} = (n + 1) \times \text{riser height} \tag{11-3}$$

where n = number of tread.

EXAMPLE 11-1

Problem: Find the stairway rise and run for a stairway consisting of 14 risers (13 tread). Tread width is 10 in. (25 cm), and riser height is 7 1/2 in. (19 cm).

Solution:

$$\text{Run} = 13 \times 10 = 130 \text{ in. (330 cm)} \qquad \text{[Eq. 11-2]}$$
$$\text{Rise} = (13 + 1) \times 7.5 = 105 \text{ in. (267 cm)} \qquad \text{[Eq. 11-3]}$$

Figure 11-9 Stairway framing.

When the space available for a stairway is limited, it may be necessary to break the stairway into sections separated by one or more *landings* (see Figure 11-7). The width and length of landings should be equal to or greater than the stairway length. *Winders* are pie-shaped treads used to conserve space. When permitted by code, winders should have a tread width at the narrowest portion of at least 4 in. (10 cm) and an average tread width of at least 9 in. (23 cm). *Spiral stairs* may also be used when permitted by code. If used, they should have a tread width at the narrowest portion of at least 6 in. (15 cm) and a tread width of at least 9 in. (23 cm) at a distance of 12 in. (30 cm) from the narrow end.

Construction

The rough framing for a typical wood interior residential stairway is illustrated in Figures 11-8 and 11-9. The notched beams that provide the principal support for stair treads are known as stair *stringers* or *carriages*. As illustrated, three stringers are commonly used.

The construction of plywood stair treads and risers is illustrated in Figure 11-10. Minimum plywood thickness should be 23/32 in. (18 mm) for treads and 15/32 in. (12 mm) for risers.

For a neater appearance, main stairways often use a *housed stringer* placed between the interior stair stringer and the wall, and a *finish stringer* (or *face stringer*) placed on the outside of an exposed outer stair stringer. Housed stringers are notched to receive stair treads and risers. When the treads and risers of main stairways are to be left uncovered, treads and risers are usually constructed of an appearance grade hardwood.

11-4 METAL FRAMING SYSTEMS

A number of proprietary metal framing systems are in existence. Like wood framing, most of these systems utilize studs, beams, and joists to construct the building

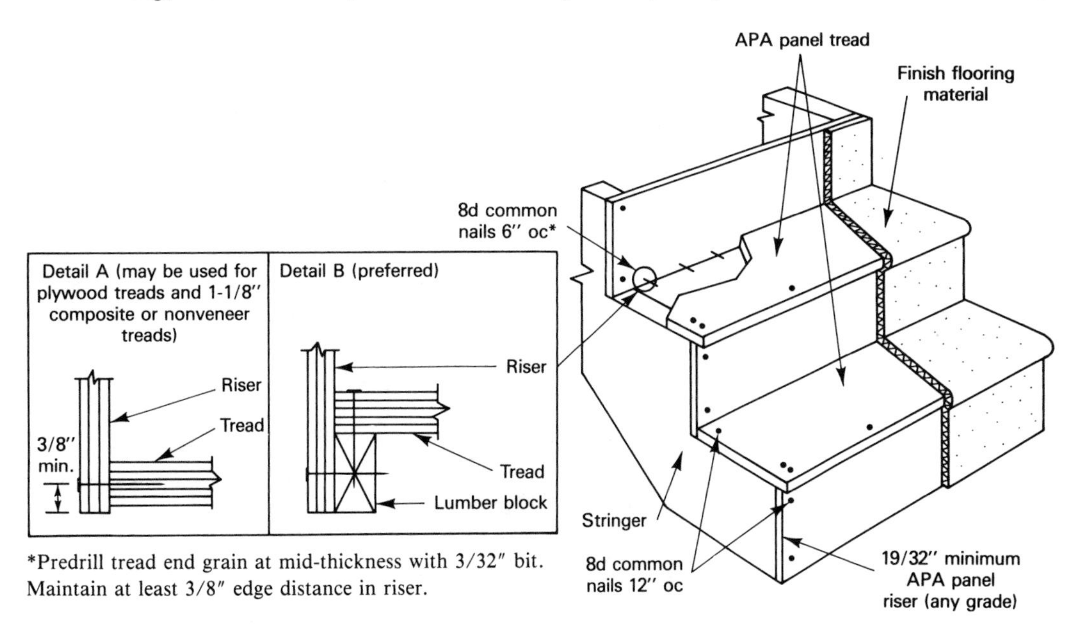

Figure 11-10 Stair treads and risers of plywood (courtesy American Plywood Association).

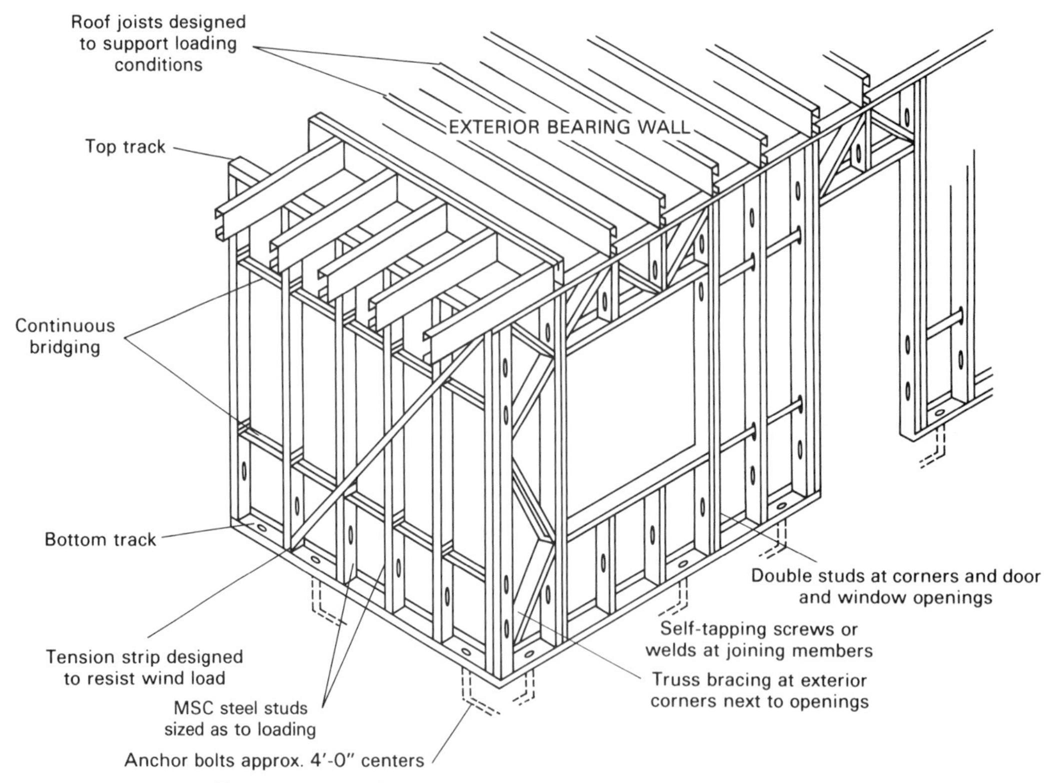

Figure 11-11 Metal framing system (courtesy Metal Studs Corp.).

Figure 11-12 Elements of metal framing system.

Figure 11-13 Construction using metal wall frame and wood roof trusses.

frame. The materials used are principally lightweight galvanized steel and aluminum, which resist rust, decay, insects, shrinkage, and warping. Because of their high strength, metal framing systems typically weigh less and take up less space than do equivalent wood framing systems. One such system is illustrated in Figure 11-11. Typical metal studs and track are shown in Figure 11-12.

The building under construction in Figure 11-13 combines metal wall framing with wood roof trusses.

PROBLEMS

1. Briefly explain the meaning of the following building construction terms:
   (a) Cornice
   (b) Fascia
   (c) Lookout
   (d) Soffit
2. Name and briefly describe the three principal forms of soffit construction.
3. Explain how you would construct a wood frame wall that must contain a 4-in. (10 cm) diameter waste pipe.

4. State the limitations that should be observed when boring holes in or cutting wood wall studs for the passage of utility lines.
5. When should floor joists be doubled?
6. What major factors relating to safety should be considered when constructing a stairway? Which of these is most important?
7. Explain the relationship between tread width and riser height within a flight of stairs.
8. When the space available for the installation of a stairway is limited, what alternatives to a straight stairway are available?
9. Calculate the floor-to-floor distance and stairway run for a straight stairway consisting of 13 risers having a riser height of 8 in. (20 cm) and a tread width of 9 in. (23 cm).
10. Briefly describe the characteristics of a metal framing system that might be used to construct a small one-story commercial building.

REFERENCES

1. American Institute of Architects. *Architectural Graphic Standards.* 7th ed. New York: Wiley, 1981.
2. Anderson, L. O. *Wood-Frame House Construction.* Agriculture Handbook No. 73. U.S. Department of Agriculture, Washington, DC, 1975.
3. American Plywood Association. *APA Design/Construction Guide: Residential and Commercial.* Tacoma, WA. 1988.
4. Council of American Building Officials. *CABO One and Two Family Dwelling Code.* Falls Church, VA.
5. National Forest Products Association. *Manual for House Framing.* Wood Construction Data No. 1. Washington, DC, 1970.

chapter 12

Insulation, Ventilation and Sound Control

12-1 COMFORT AND ENERGY EFFICIENCY

General

Through the ages, occupant comfort has been a major consideration in the design and construction of buildings. Since World War II, the comfort of building occupants has been increased largely through the use of mechanical equipment to produce a comfortable climate-controlled environment. That is, the air within living and working spaces is heated or cooled as necessary to produce a combination of temperature and humidity which is comfortable to the occupants. That phase of building construction concerned with the mechanical systems used to effect climate control is commonly referred to as *HVAC* (heating, ventilating, and air conditioning). The installation of HVAC systems is a specialized task that is normally performed by a specialty contractor (see Chapter 15). Other factors influencing occupant comfort include noise level and the supply of fresh air.

The rising cost of energy in recent years has forced designers and builders to pay increased attention to improving the energy efficiency of buildings without degrading occupant comfort. Although provisions relating to building insulation and energy efficiency have been incorporated into a number of local building codes and regulations, insulation alone is not the solution to the problem of minimizing the energy consumption of buildings while maintaining a comfortable environment. Many other factors, such as building design, building location and orientation, shading, roof color, and air tightness influence the heating and cooling load that must be satisfied for occupant comfort.

Many energy-conserving features have been incorporated in the "Energy Efficient Home" (Figure 12-1) suggested by the Owens-Corning Fiberglas Corporation. Considerations involved in the design and construction of safe, comfortable, energy-efficient buildings are discussed in this chapter. The end-of-chapter references provide additional information on many of these topics.

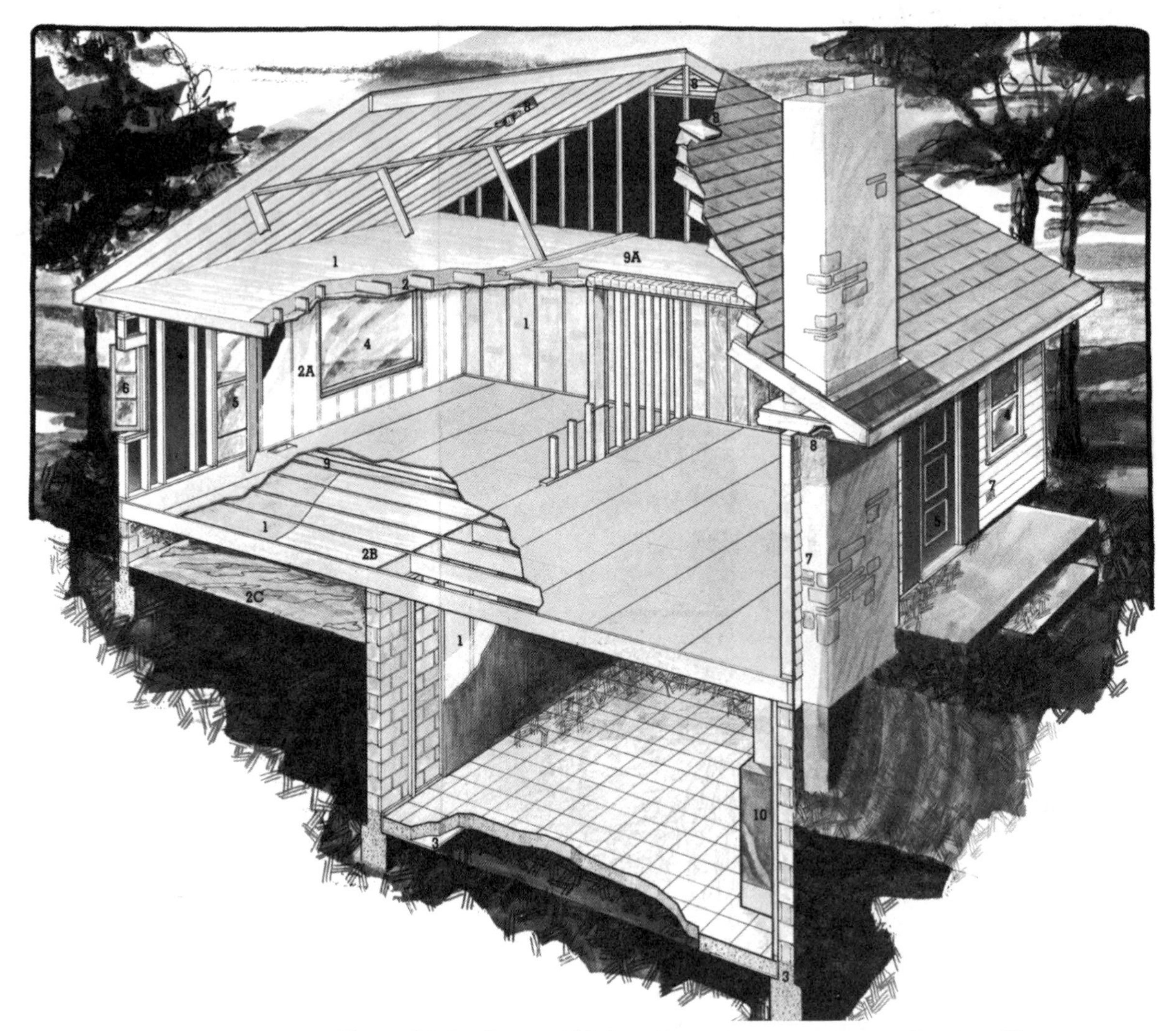

Figure 12-1 Energy Efficient Home (courtesy Owens-Corning Fiberglas Corporation).

Sources of Heat Gain and Loss

The sources of heating and cooling loads for a typical single-family residence located in a warm humid climate are illustrated in Figure 12-2. Notice that 62 percent of the total air conditioning load comes from heat and humidity from air infiltration and occupant-generated sources. Likewise, 39 percent of the total heating load comes from air infiltration. Although the distribution of heating and cooling loads varies with geographical location, the principles and techniques discussed in this chapter apply to all locations.

Methods of Heat Transfer

Before considering specific construction measures to improve building energy efficiency, it is helpful to consider the ways by which heat enters or leaves the conditioned space within a building. As indicated by Figure 12-2, the two principal mechanisms are air infiltration and heat flow through the building envelope. *Air*

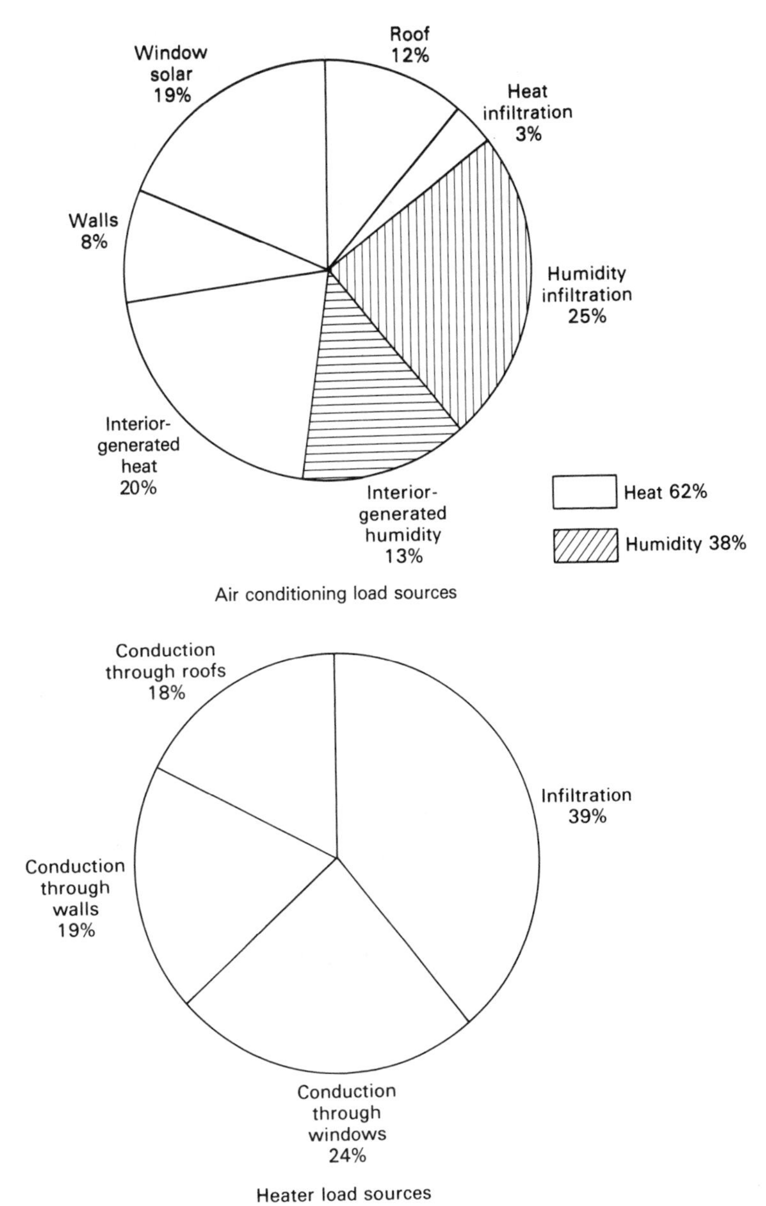

Figure 12-2 Sources of heating and cooling load for typical Florida residence (Florida Solar Energy Center).

infiltration is the leakage of air through the building envelope. Heat flow through the building envelope can take place in three ways: conduction, convection, and radiation.

Conduction is the flow of heat through a material by molecular contact. Heat flows from a hot spot to a cooler spot. The principal method for reducing heat conduction through a building envelope is to insert insulation into the heat path. *Insulation* is a material having a high resistance to heat flow.

Convection is the transfer of heat by air flow, with warm air rising and cool air falling. Convection is not a major source of heat loss or gain through the building envelope. It can be minimized by keeping the volume of air spaces within the building envelope small.

Radiation is the transfer of heat through air or space as electromagnetic waves or particles. Although less important than conduction, heat gain or loss through radiation can be reduced by the use of radiant barriers, as described in Section 12-3.

Building Design and Orientation

The heating and cooling loads imposed on a building are strongly influenced by the building design, location, and orientation. Following are some suggestions for increasing the energy efficiency of buildings through good building design, site selection, and orientation. Careful use of these techniques can cut total heating and cooling loads as much as 50 percent.

- Minimize the wall-to-floor area ratio. Other factors being equal, a compact building gains and loses less heat than does an elongated building.
- In hot climates, minimize window area on the east and west sides. In cold climates, minimize window area on the north side.
- Use a light color for roof and walls in hot climates and a dark color in cold climates.
- Locate HVAC equipment and ducts within conditioned space.
- In hot climates, locate water heaters and laundry rooms outside the conditioned air space.
- Incorporate into the building design the insulation, ventilation, air infiltration, and other measures described in this chapter.
- Locate the building to take advantage of existing vegetation (for summer shading and winter wind buffering) and geological features (for winter wind buffering and catching summer breezes).
- Orient the building so that major glass areas face south. Use roof overhang, mechanical shading, and window location to block summer sun but permit winter solar gain. Plant deciduous (leaf-shedding) trees on the south side and evergreen trees on the east and west.
- Orient the building so that seldom-occupied portions of the building (such as garages) buffer living areas. Locate buffers on the north side in cold climates and on the west side in hot climates.
- Locate operable windows to receive spring, summer, and fall breezes.
- In hot climates, use plants and trees to shade patios, windows, and air conditioners.

12-2 INSULATION

Materials

The material used to reduce heat conduction through the walls, ceiling, and floor of conditioned space is known as *insulation* (or thermal insulation). The characteristics of some common insulation materials are given in Table 12-1.

As indicated, these materials may be available in the form of batts or blankets, boards, or loose fill. Their insulating value varies with material density and type of facing employed. Cellulose, plastic foams, and wood fiber are combustible. The other materials are not.

TABLE 12-1
Characteristics of Common Insulation Materials

Material	*Form*	*R */in.*
Cellulose	Loose	3.2–3.7
Expanded polystyrene foam	Board	3.9–5.0
Fiber glass	Batt/board/loose	2.2–4.2
Perlite	Loose	2.5–3.7
Phenolic foam	Board	7.4–8.3
Polyurethane foam	Board	6.2–7.7
Polyisocyanurate foam	Board	6.2–7.7
Rock wool	Batt/loose	2.9–3.7
Vermiculite	Loose	2.4–3.0
Wood fiber	Board	2.1–2.4

*Coefficient of thermal resistance.

Measuring Resistance to Heat Flow

The *coefficient of thermal transmission,* or *U*-value, is a measure of the ability of a material to conduct heat. The *U*-value of a material represents the heat flow in BTU/hr./sq. ft. of surface area/°F temperature differential between the two sides of the material [$U = 5.678$ Watt/($m^2 \times$ °K)]. Unfortunately, *U*-values cannot be added. Since most building components are composites of several materials, it is desirable to have a unit of measure that *can* be added. Thus, the *R-value*, or *coefficient of thermal resistance,* is commonly used to measure insulation values. Insulation requirements of specifications and codes are commonly expressed as *R*-values. The *R*-value of an assembly is simply the sum of the *R*-values of the assembly components. The *U*-value of any element or assembly may be easily calculated as the reciprocal of its *R*-value. Calculation of *R*-value and *U*-value for an exterior wall is illustrated in Example 12-1.

EXAMPLE 12-1

Problem: Find the *R*-value and *U*-value of the exterior wall of a wood frame building composed of the materials below.

Component	*R-value*
Plywood siding 1/2″ (13 mm)	0.62
Insulating sheathing 1/2″ (13 mm)	1.32
Fiber glass batt 3 1/2″ (89 mm)	11.00
Gypsum drywall 1/2″ (13 mm)	0.45

Solution:

$$R = 0.62 + 1.32 + 11.00 + 0.45 = 13.39$$
$$U = 1 / 13.39 = 0.075$$

In the portion of the wall where a stud is present, the *R*-value is somewhat less than indicated above because the *R*-value of 3 1/2 in. (89 mm) of wood ($R = 4.38$) is less than the *R*-value of 3 1/2 in. (89 mm) of fiberglass ($R = 11.00$). When making heat-transfer calculations, the average *R*-value of the wall should be used to take this effect into account.

Insulating Walls

Wood Frame Walls. The principal method used for insulating a wood frame wall is to place fiber glass batt insulation into the wall cavity created by the studs. The insulation for a typical wood frame wall is illustrated in Figure 12-3.

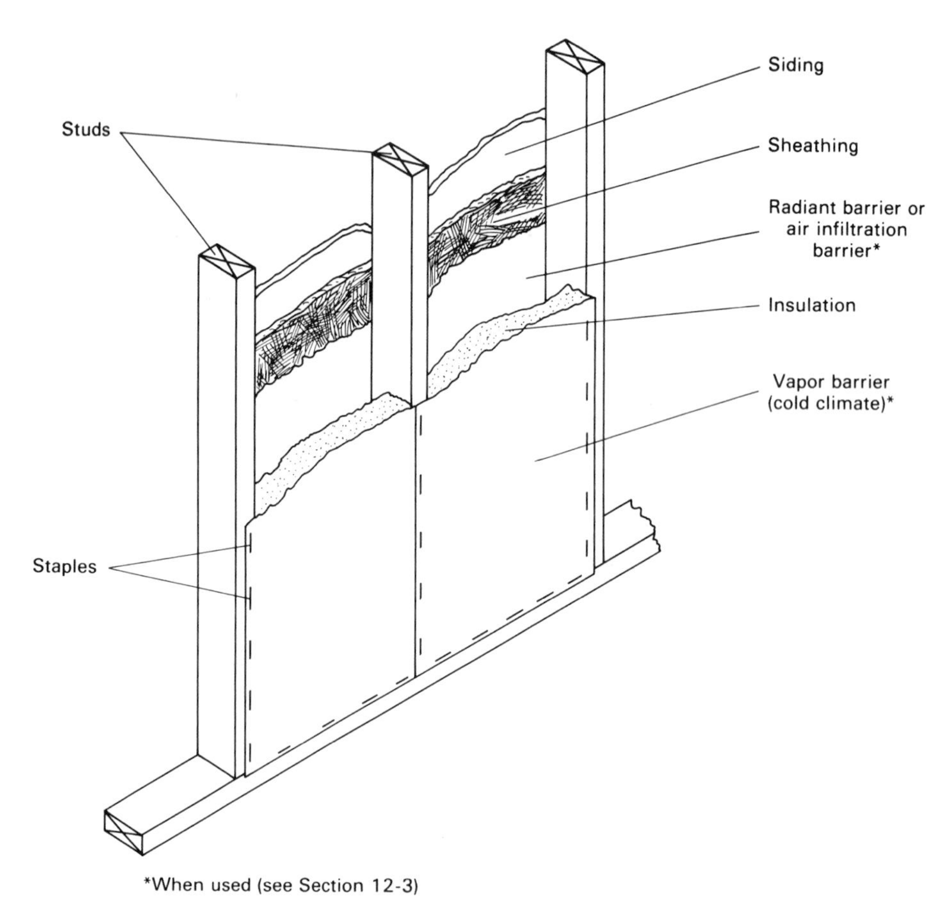

Figure 12-3 Insulation of wood frame wall.

A loose or blown-in insulating fill may be substituted for the fiberglass batt. In either case, the maximum amount of insulation that can be placed within the wall cavity is limited by the width of the studs used. To increase this value, 2 × 6 in. (5 × 15 cm) studs spaced 24 in. (61 cm) on center may be substituted for the usual 2 × 4 in. (5 × 10 cm) studs spaced 16 in. (41 cm) on center. The use of wide studs spaced further apart provides an added benefit in that the number of framing members is also reduced, which further increases the *R*-value of the wall as explained above. Another method for increasing the *R*-value of the wall is to add board insulation on the exterior of the wall frame.

Masonry Walls. Masonry walls typically have a lower *R*-value than do wood frame walls because there is less cavity in which to place insulation. Several forms of concrete block wall construction are illustrated in Figure 12-4. Notice that *R*-values range from 4.0 for an uninsulated wall to 11.6 for a wall using 3/4 in. (19 mm) insulating board between the block and furring strips. If the insulating board of Figure 12-4(c) were placed on the outside of the block, advantage could be taken of the thermal mass of the block to help stabilize interior temperatures. However, this design complicates the procedure for placing stucco on the exterior of the wall.

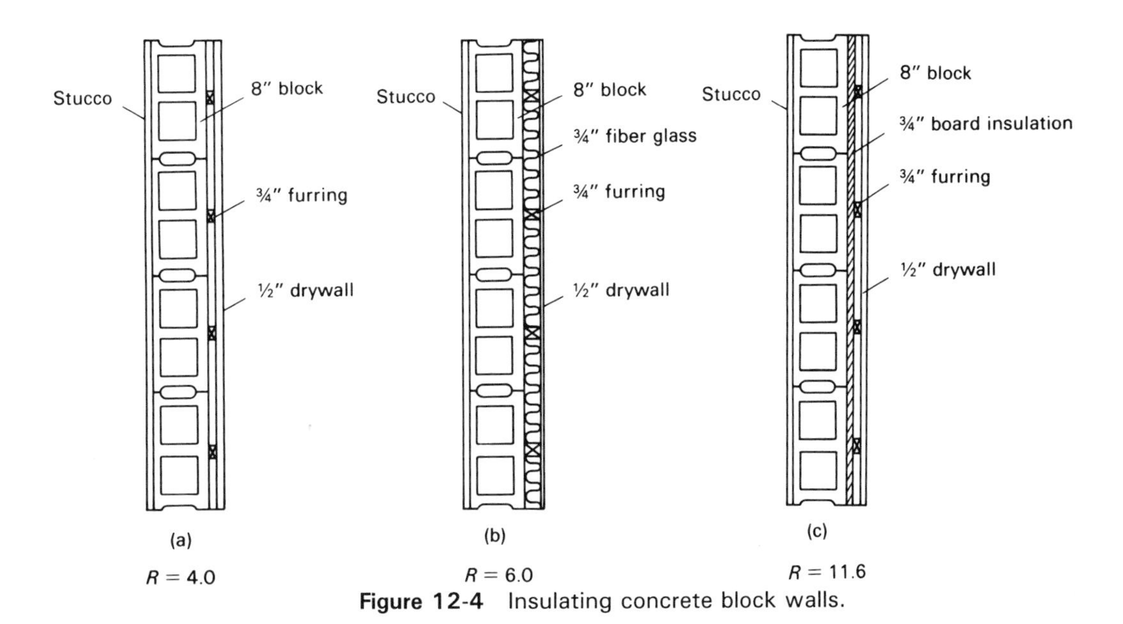

Figure 12-4 Insulating concrete block walls.

Construction Procedures. When installing insulation in wall cavities, it is important that no gaps be left in the insulation envelope. Areas requiring special attention include the corners of exterior walls, the intersection of exterior and interior walls, headers over exterior wall openings, and spaces around window and door frames.

Care must also be taken to minimize the compression of insulation around wiring or pipes located in the wall cavity. The compression of a 6 in. (15 cm) thick fiber glass insulation batt to a thickness of 3 1/2 in. (8.9 cm), for example, will reduce its insulating value from *R*-19 to *R*-13. In wood frame walls, wiring can be run along the bottom of the wall cavity by cutting notches into the wall studs. An alternative technique to minimize the compression of insulation by wiring is to split the insulation batt and place part of the insulation behind the wiring and part in front.

Insulating Attics and Ceilings

Principles. As shown in Figure 12-2, heating passing from the roof through the ceiling makes up the largest portion of the summer heat conduction load. In the winter, heat loss through the roof is comparable to that through the walls. Therefore, it is important to minimize the heat flow through the roof, attic, and ceiling. Minimizing heat flow through the roof requires a combination of proper roof materials, attic ventilation, insulation, airtight ceilings, and radiant barriers. In this section, we will consider only the use of insulation.

The *R*-value required for ceiling insulation depends on the climate and design requirements. However, a minimum *R*-value of *R*-19 to *R*-30 is often specified. Since ceilings usually provide a horizontal surface on which to place insulation, fiber glass batts or one of the loose forms of insulation shown in Table 12-1 are commonly used.

Construction Procedures. As usual, take care to ensure that no gaps are left in the insulation envelope and that all conditioned building space is covered. Avoid blocking attic ventilation openings with insulation. Cover attic access openings with removable insulated covers.

Avoid the use of recessed lights that penetrate the ceiling. Standard recessed lights must not be covered with insulation since this can create a fire hazard. If recessed lights are required, use lights manufactured with insulated covers.

Insulating Floors

Principles. The heat flow through a building floor may range from negative (that is, the heat flow reduces energy demand) in the case of a slab-on-grade floor in a hot climate, to a significant energy loss for a floor over a crawl space in a cold climate. Insulation requirements vary accordingly.

The two principal methods for insulating floors are illustrated in Figure 12-5. For slab-on-grade floors (Figure 12-5[a]), rigid foam insulation is placed around the slab perimeter using either of the locations shown as Alternate 1 and Alternate 2. If Alternate 1 is used, the insulation must be protected from damage by flashing, a shield, or other method. For floors above crawl space or other nonconditioned space [Figure 12-5(b)], batt insulation is commonly placed between floor joists and retained against the underside of the floor by wire rods or mesh as shown.

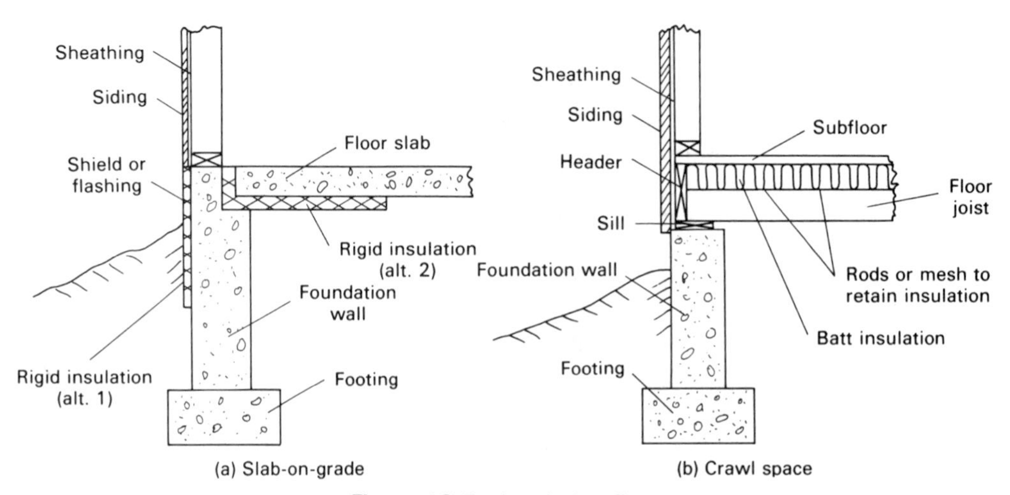

Figure 12-5 Insulating floors.

Construction Procedures. As always, ensure that no gaps are left in the insulation envelope. Floors that cantilever over exterior walls must be insulated and the insulation protected from damage.

12-3 INFILTRATION, VAPOR, AND RADIANT BARRIERS

Air Infiltration Barriers

As the name implies, *air infiltration barriers* are used to reduce air leakage through the building envelope surrounding conditioned space. As you recall, air infiltration may account for as much as 39 percent of a building's heating and cooling load (see Figure 12-2). Some air infiltration barriers permit the passage of moisture and some do not. Air infiltration barriers that do not pass moisture thus also serve as vapor barriers (see next section).

The techniques available to construct air infiltration barriers for building walls include: (1) applying stucco to the exterior wall, (2) sealing exterior wall sheathing, (3) wrapping the building in a barrier cloth ("house wrap"), and (4) sealing the interior drywall. Regardless of the technique employed for constructing an air infiltration barrier, all penetrations of the barrier, including window and door frames, must be carefully sealed to the barrier. Ceilings must also be sealed to the wall barrier to complete the air infiltration barrier envelope. The air leakage of windows and doors should not exceed the limits described in Section 10-4.

Stucco, widely used as the exterior finish for concrete block buildings, forms an effective air infiltration barrier for building walls. Alternatively, the exterior wall sheathing may be sealed to form an air infiltration barrier. To accomplish this, the sheathing must be sealed with an approved tape or sealant at the following locations: sheathing joints, where the sheathing meets the foundation wall (or seal the sheathing to the soleplate and the soleplate to the foundation wall), and where the sheathing meets the top plate. As another alternative, the exterior walls may be wrapped in an air barrier cloth (house wrap) as shown in Figure 12-6. The cloth must be sealed to the foundation wall, to the wall top plate, and to window and door frames. Joints between sections of cloth should be lapped and sealed as recommended by the cloth manufacturer. Finally, the drywall applied to the inside surface of exterior wall frames can be used to form the wall's air infiltration barrier. When employing this technique, the wall's drywall must be sealed to the wall top plate and soleplate, and the soleplate sealed to the floor.

Figure 12-6 House wrapped in air infiltration barrier cloth.

Vapor Barriers

A *vapor barrier* is used in cool climates to prevent condensation of moisture in wall insulation during the heating season. If allowed to occur, such condensation reduces the effectiveness of the insulation and may lead to fungus or decay damage inside the wall. The function of a vapor barrier is to prevent the migration of moisture during the heating season from the conditioned space into the wall cavity where it may condense as the temperature drops. Therefore, the vapor barrier should be placed on the interior (or warm-in-winter) side of the wall frame, unde the drywall. Materials commonly used as vapor barriers include polyethylene sheets and aluminum foil.

Recent research has indicated that vapor barriers are not needed in warm, humid climates (see Reference 6). However, if used in such climates, vapor barriers should be placed against the outside surface of the exterior wall frame. Since vapor barriers also serve as air infiltration barriers, either a vapor barrier or an air infiltration barrier may be applied to the outside of a building's exterior wall frame in hot, humid climates. When using either a vapor barrier or an air infiltration barrier, care must be taken to seal the barrier properly against air leaks as previously described.

Radiant Barriers

A *radiant barrier* is a material used to reduce the flow of radiant heat through the building envelope. Radiant barriers reflect radiant heat from the outside away from a building during the summer and reflect radiant interior heat into the building during the winter.

The material most commonly used in building construction as a radiant barrier is aluminum foil. Aluminum foil is a good reflector of radiant heat but a poor heat radiator. Thus, it reflects radiant heat back to the source but radiates very little heat in the other direction. Aluminum foil radiant barriers are available as single- or double-sided foil (with facing or reinforcing material), foil-faced insulation, and multilayered foil systems.

Research tests (see Reference 6) in a hot climate [air temperature 87°F (31°C)] have shown that the installation of a radiant barrier under the roof sheathing can reduce attic air temperature from 94°F (34°C) to 89°F (32°C) and the top of insulation temperature from 100°F (38°C) to 89°F (32°C). The radiant barrier was found to reduce summer heat flow through the ceiling by as much as 40 percent.

Radiant barriers for roofs may be installed using any of the following methods:

- Glueing or stapling barrier foil to the underside of roof sheathing before sheathing installation.
- Draping and stapling barrier foil across rafters or trusses before sheathing installation. In this method, the air space created by draping the foil serves to provide ventilation to the underside of the sheathing and helps reduce the temperature of the roof covering and sheathing.
- Stapling the barrier foil across the bottom of rafters or truss top chords after sheathing installation.

In all cases, the foil's shiny side must face down (toward the ceiling). In addition, an air space must be located adjacent to the foil in order for the barrier to perform properly.

Radiant barriers may also be used in building walls. They are most suitable for unshaded walls of buildings located in a hot climate. When used, the foil's shiny surface should face the interior of the building and an air space must be located adjacent to the foil.

12-4 VENTILATION

Requirements

Ventilation of buildings is required for a number of reasons, including:

1. To reduce summer attic temperature and the building cooling load.
2. To prevent moisture and radon gas from accumulating in the crawl space and passing through the building floor into conditioned space. Radon gas

has been found to be a health hazard. However, its occurrence depends on geological conditions. Excess moisture in the conditioned space must be removed by the air conditioning system for occupant comfort. Moisture in the crawl space can also lead to decay and fungus growth in the floor frame.

3. To remove moisture and odors from baths, showers, laundry rooms, and similar sources.
4. To provide adequate fresh air for occupant health.
5. To ensure an adequate supply of combustion air to fuel-burning appliances.
6. To remove combustion gases produced by fuel-burning appliances.

Ventilation and air supply requirements are commonly specified by code. Some typical requirements for items 1 through 4 above and their construction are described below. Vents for the supply of combustion air for fuel-burning appliances and for the removal of combustion gases are discussed in Chapter 15.

Attic Ventilation. The minimum *net free ventilating area* (N.F.V.A.) required for attic ventilation is usually set at 1/150 of the ceiling area. However, this requirement may be reduced to 1/300 of the ceiling area when the ceiling has a vapor barrier installed, and one-half of the ventilating area is located under eaves or cornices and one-half located in the roof at least 3 ft. (91 cm) above eave or cornice vents.

The upper roof vents may consist of ridge vents, gable vents, wind turbines, or power ventilators. Ridge and gable vents provide the most effective, trouble-free ventilating action. However, the effectiveness of gable vents is influenced by wind direction. Power ventilators are the least energy efficient since they consume energy and are no more effective than other ventilators. A combination of ridge vents and soffit vents, each supplying a vent area equal to 1/300 of the ceiling area is recommended (see Figure 12-7). A form of ridge vent that can be covered by shingles is available to provide unobtrusive appearance.

Crawl Space Ventilation. The minimum net free ventilating area required for crawl space above uncovered ground is commonly set at 1/150 of the floor area above the crawl space. A vent should be located within 3 ft. (91 cm) of each building

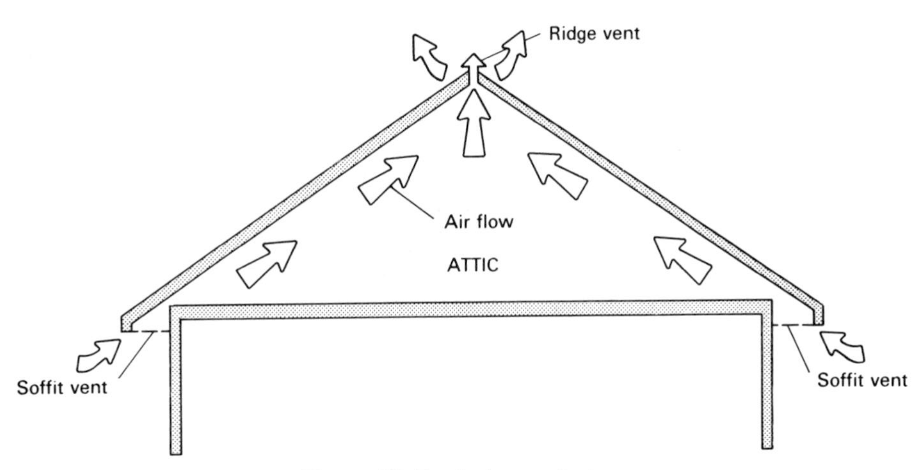

Figure 12-7 Attic ventilation.

corner. Vent openings should be covered with a corrosion resistant wire mesh having openings measuring 1/4 to 1/2 in. (6–13 mm) across.

It is recommended that the ground surface under crawl space be covered and sealed with an approved vapor barrier. This will help control moisture and radon gas migration into the crawl space. Some codes permit the net free ventilating area of crawl space to be reduced to 1/1500 of the floor area when the ground surface is covered with a vapor barrier and the crawl space is vented.

The under-floor area (crawl space) is sometimes used as a supply plenum for the distribution of conditioned air within the building. When such a system is used, the ground below the floor should be covered with a vapor barrier having a permeability rating not exceeding 1 perm and carefully sealed. The interior of foundation walls should be insulated with a fire resistant insulation. All floor openings other than conditioned air vents and ducts should be carefully sealed.

Exhaust Fans. For energy efficiency, exhaust fans should be provided for removal of heat and moisture originating in bathrooms, showers, laundry rooms, kitchens, and similar areas. Ventilation requirements for occupant health and safety are described in the section below.

Install exhaust fans so that they vent directly to the outside. Venting exhaust fans into attic or other concealed spaces may cause moisture buildup resulting in decay or fungus damage. Kitchen exhaust fans must not be vented into concealed spaces because of a possible fire hazard. Make exhaust ducts as short and straight as possible.

Occupant Health. To minimize the danger to occupants from indoor pollution, some experts recommend that the building air be completely changed at least three times per hour. When a building has been extremely well sealed against air infiltration, it is possible that infiltration will not supply sufficient fresh air. In this case, an outside air supply equipped with an air-to-air heat exchanger should be provided. The heat exchanger minimizes energy loss resulting from the outside air supply.

12-5 SOUND CONTROL

Importance

While sound control is not directly related to energy efficiency, sound control is important to occupant comfort. In many cases, walls, ceilings, and floors that are constructed with adequate insulation and air barriers also provide adequate sound control for noises originating outside the building envelope. However, it is also desirable to control the transmission of sounds originating within the building. Sleeping areas, for example, should be isolated from the noise of living and service areas. It is also desirable to isolate bathrooms, kitchens, family rooms, and utility rooms from other living spaces.

Sound control is particularly important in multifamily dwelling units. Codes may limit the sound transmission between adjacent dwelling units as well as between dwelling units and adjacent public areas, service areas, and commercial areas.

Measuring Sound Transmission

The most common method of rating the resistance of walls, ceilings, and floors to airborne sound transmission is by a *Sound Transmission Class* (STC) rating. In the test used for determining the STC of a building component, the sound transmission loss that occurs while passing through the component is measured at 16 frequencies between 125 and 4000 Hz. Results are then plotted against standard STC curves to determine the STC rating. The relationship between airborne sound control performance and STC rating is indicated below.

STC rating	*Sound control effectiveness*
Under 45	Poor
45–52	Fair to good
Over 53	Excellent

The impact sound created by walking or dropping an object on the floor is important to the comfort of persons occupying the space below the floor. The resistance of floor/ceiling structures to the transmission of impact sound is measured by an *Impact Insulation Class* (IIC) rating. The test procedure for the IIC rating test is similar to that used for the STC rating test except that the noise measured consists of the impact produced by a standard tapping machine. The relationship between impact sound control performance and IIC ratings is indicated below.

IIC rating	*Impact sound control effectiveness*
Under 50	Poor
50–60	Fair to good
Over 60	Excellent

Wall Construction

Methods for reducing the transmission of airborne sound through a wall include sealing all joints and openings in the wall, increasing the wall mass, placing insulation in the wall cavity, and breaking possible sound paths. Wall mass may be increased by using masonry materials or by using double layers of drywall on one or both sides of the wall. The sound path may be broken by using staggered studs placed on upper and lower plates that are wider than the studs, by constructing a double wall with a space separating the two halves of the wall, or by mounting drywall on resilient supports.

Some typical methods of wall construction used for sound control and their STC ratings are illustrated in Figure 12-8. Notice that STC ratings range from 35 for the usual stud wall consisting of 2 × 4 in. (5 × 20 cm) studs placed 16 in. (41 cm) on center, 1/2 in. (13 mm) drywall, and no insulation, to 63 for a double wall consisting of 2 × 4 in. (5 × 20 cm) studs placed 16 in. (41 cm) on center, one thickness of *R*-11 insulation, and double layers of 1/2 in. (13 mm) drywall on each side.

Floor/Ceiling Construction

Some typical forms of floor/ceiling construction and their impact sound ratings (IIC) are illustrated in Figure 12-9. Notice that IIC ratings range from 34 for a vinyl tile floor with a 5/8 in. (16 mm) plywood subfloor, 3/8 in. (10 cm) underlayment, 1/2 in. (13 mm) drywall ceiling attached directly to the joists, and no insulation, to 73 for a carpeted floor with the same plywood subfloor and underlayment, one thickness of R-11 insulation, and 1/2 in. (13 mm) drywall ceiling attached to resilient supports.

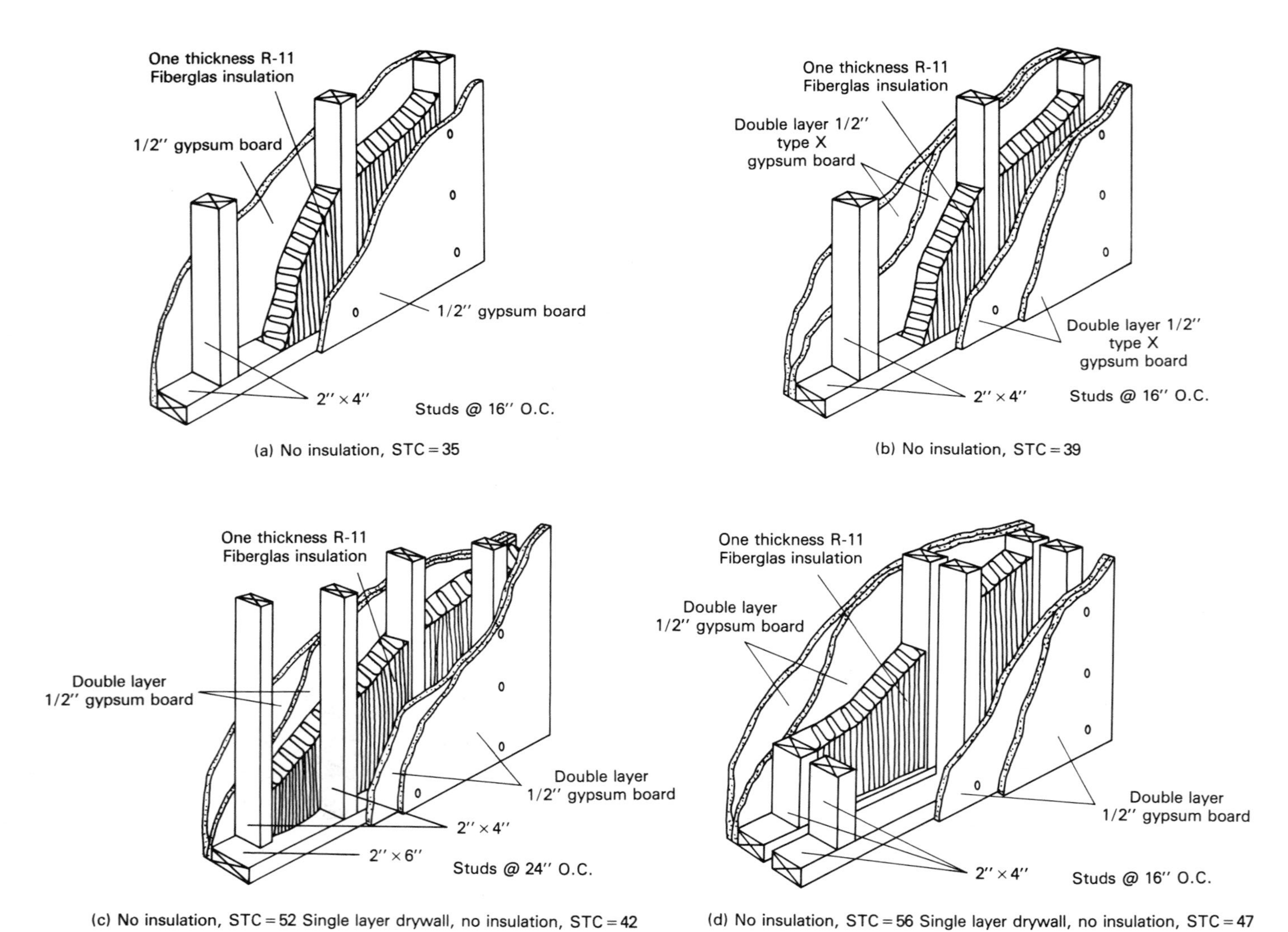

Figure 12-8 Wall construction for noise control (courtesy Owens-Corning Fiberglas Corporation).

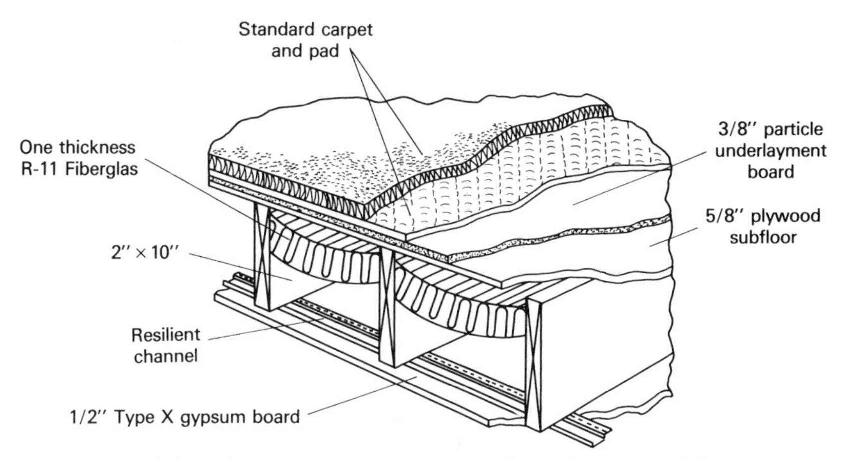

Figure 12-9 Floor/ceiling construction for noise control (courtesy Owens-Corning Fiberglas Corporation).

Construction Practices

A number of suggestions for improving the effectiveness of structural sound control are provided in Reference 5. Some of these recommendations include the following:

- Carefully seal openings in ceilings, floors, and walls created by electrical fixtures, switches, and outlets, and by utility lines.
- Do not locate doorbells, telephones, intercoms, or other noise-making equipment on corridor walls or walls between dwelling units.
- Do stagger electrical fixtures, switches, and outlets, as well as medicine cabinets, so that they are not located back-to-back on the same wall.
- Seal noise control walls to the ceiling and floor. Seal subfloor joints or cover with an underlayment when the space below the floor is occupied.
- Stagger doorways separated by a hallway.
- Use solid wood core or insulating core doors when sound control is needed.

PROBLEMS

1. Describe the sources of air conditioning load in a warm, humid climate and give their distribution.
2. Name four methods by which heat can enter or leave a building's conditioned space.
3. List six ways of reducing a building's heating and air conditioning load by using techniques of design, location, and orientation.
4. Calculate the *R*-value and *U*-value of the wall of Example 12-1 if the 1/2 in. (13 mm) insulating sheathing is replaced by an insulation/radiant barrier board having an *R*-value of 6.0.
5. **(a)** Explain how a vapor barrier acts to prevent moisture condensation inside an exterior building wall in a cold climate.
   (b) Considering your explanation in (a) above, why might a vapor barrier not be needed in a warm, humid climate?
6. List four ways of constructing an air infiltration barrier system for a building.
7. What purposes does ventilation of a building serve?
8. Recommend an energy efficient method for providing adequate attic ventilation in a single-family residence.

9. A code requires a wall with an STC rating of at least 45. Recommend a method of construction for a wood frame wall that satisfies this code requirement.

10. Suggest at least four construction techniques for improving the effectiveness of sound control within a building.

REFERENCES

1. Anderson, L. O. *Wood-Frame House Construction.* Agriculture Handbook No. 73. U.S. Department of Agriculture, Washington, DC, 1975.
2. Council of American Building Officials. *CABO One and Two Family Dwelling Code.* Falls Church, VA.
3. Owens-Corning Fiberglas Corporation. *44 Ways to Build Energy Conservation into Your Homes.* Toledo, OH, 1975.
4. National Forest Products Association. *Insulation of Wood-Frame Structures.* Wood Construction Data No. 7. Washington, DC, 1964.
5. Owens-Corning Fiberglas Corporation. *Noise Control Design Guide.* Toledo, OH, 1987.
6. Vierra, Robin K., and Kenneth G. Sheinkopf. *Energy-Efficient Florida Home Building.* Florida Solar Energy Center, Cape Canaveral, FL, 1988.

chapter 13

Finishes

13-1 DRYWALL

Technically, the term *drywall* applies to a wall finished without the use of mortar. This includes walls covered with gypsum board, plywood, plywood paneling, hardboard paneling, and other sheet materials. However, the term is commonly used to denote gypsum board. Hence, the terms *drywall, gypsum board, wallboard,* and *plasterboard* are often used interchangeably. Drywall is the most common form of interior wall finish used in light building construction.

Materials

Gypsum board consists of calcined gypsum (a mineral) faced on each side with paper or other sheet material. Gypsum and gypsum board are noncombustible and fire resistant. Gypsum board is manufactured in thicknesses of 1/4 in. (6 mm) to 1 in. (25 mm), with 3/8 in. (10 mm) and 1/2 in. (13 mm) sheets most commonly used. Standard sheets are 4 ft. (1.2 m) wide and 8, 9, 10, 12, 14, or 16 ft. (2.4, 2.7, 3.0, 3.7, 4.3, or 4.9 m) long. The available forms of gypsum board include regular gypsum board, Type X gypsum board, and predecorated gypsum board. *Regular gypsum board* is faced on the back with gray paper, and on the front with a cream-colored paper suitable for paint or other finish. *Type X gypsum board* is more fire resistant than regular gypsum board and is used when specified for fire-resistive assemblies. *Predecorated gypsum board* is faced on the front with a decorative design on paper or vinyl so that no additional finish is required.

Several other forms of gypsum board are available, including gypsum backing board, gypsum coreboard, and water resistant gypsum board. *Backing board* is intended to serve as the base layer when several layers of wallboard are being used. *Coreboard,* usually 1 in. (25 mm) thick, is used for free standing or temporary walls. *Water resistant gypsum board* is more resistant to water penetration than regular gypsum board. It is commonly used as a base for the application of tile in baths, showers, kitchens, and laundry areas.

Installation

Supports. Vertical supports for gypsum board should consist of nominal 2 in. (5 cm) or larger framing. Framing must be accurately aligned to provide a plane surface on which to mount the wallboard. The maximum spacing of supports and the fastener requirements for wallboard specified by the CABO code are shown in Table 13-1. Gypsum board may be fastened directly to above-grade masonry or concrete using adhesives. When applied to below-grade exterior walls or when the distance between supports exceeds the values given in Table 13-1, the gypsum board should be fastened to furring strips. Wood furring strips applied to solid masonry should be nominal 1 × 2 in. (2.5 × 5 cm) or larger. Wood furring strips applied to rafters or trusses should be nominal 2 × 2 in. (5 × 5 cm) or larger for nail attachment and nominal 1 × 3 in. (2.5 × 7.5 cm) or larger for screw attachment.

Fasteners. Either nails or screws may be used to attach gypsum board to framing. Ringed drywall nails (see Figure 8-18) should penetrate wood framing at least 3/4 in. (19 mm) unless otherwise specified by code. Drywall screws driven by power screwdrivers are rapidly replacing nails for fastening wallboard to framing. Drywall screws have Phillips heads and are designed to be self-penetrating. Type W screws are intended for wood framing, Type S screws for metal framing, and Type G screws for fastening wallboard to gypsum backing board. Unless otherwise specified, drywall screws should penetrate at least 5/8 in. (16 mm) into wood framing, 3/8 in. (10 mm) into metal framing, and 3/4 in. (19 mm) into gypsum backing board.

Adhesives may be used in combination with mechanical fasteners (nails or screws) to fasten gypsum board to framing. The use of adhesives in combination with mechanical fasteners reduces the number of mechanical fasteners required and produces a stiffer, stronger assembly.

Construction Procedures. Gypsum board may be applied with its long edges either parallel (*horizontal application*) or perpendicular (*vertical application*) to framing. Typical methods for attaching gypsym board to framing without adhesives are illustrated in Figure 13-1. Either nails or screws may be used as fasteners. When nails are used, either *single-nailing* or *double-nailing* may be employed. In double-nailing, two nails spaced 2–2 1/2 in. (50–63 mm) apart are driven at each location. Double-nailing produces tighter wallboard-to-framing contact than single-nailing, but increases nailing requirements. The additional holding power of screws permits fastener spacing to be increased for screw attachment.

Nail heads should be driven slightly below the surface ("dimpled") so that they can later be covered with compound. Don't fracture the face paper by overdriving nails or by driving nails at an angle.

Finishing Joints and Fasteners. To hide joints and fasteners, as well as create a smooth surface for paint or other final finish, joints and fastener locations should be finished as described below and illustrated in Figure 13-2.

1. Cover the joint with a layer of joint compound slightly wider than the joint tape to be used.
2. Apply joint tape centering it on the joint. Press the tape into the joint compound. Smooth the compound over the tape and feather the edges. Allow to dry (usually 12 to 24 hours).

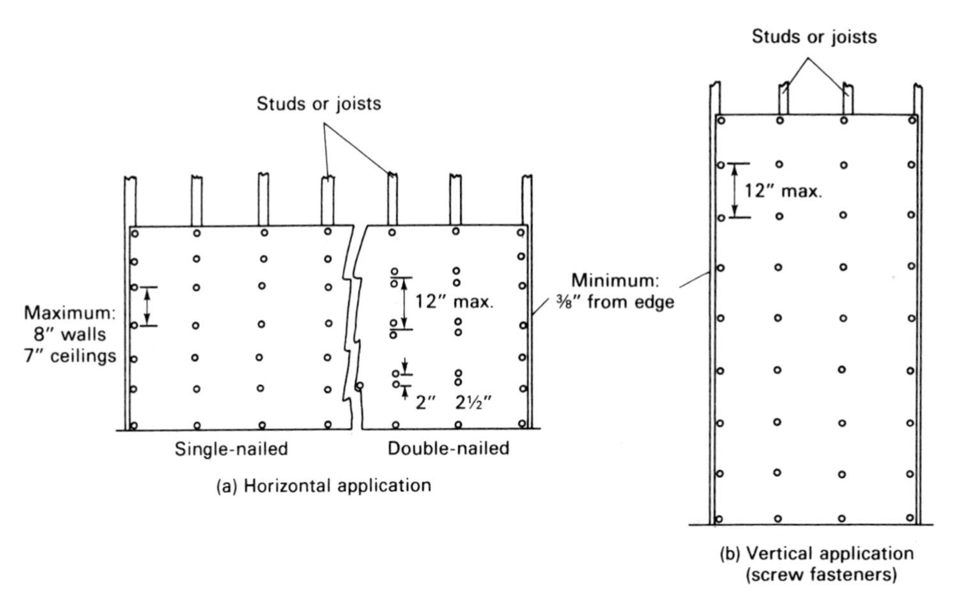

Figure 13-1 Installation of gypsum wallboard.

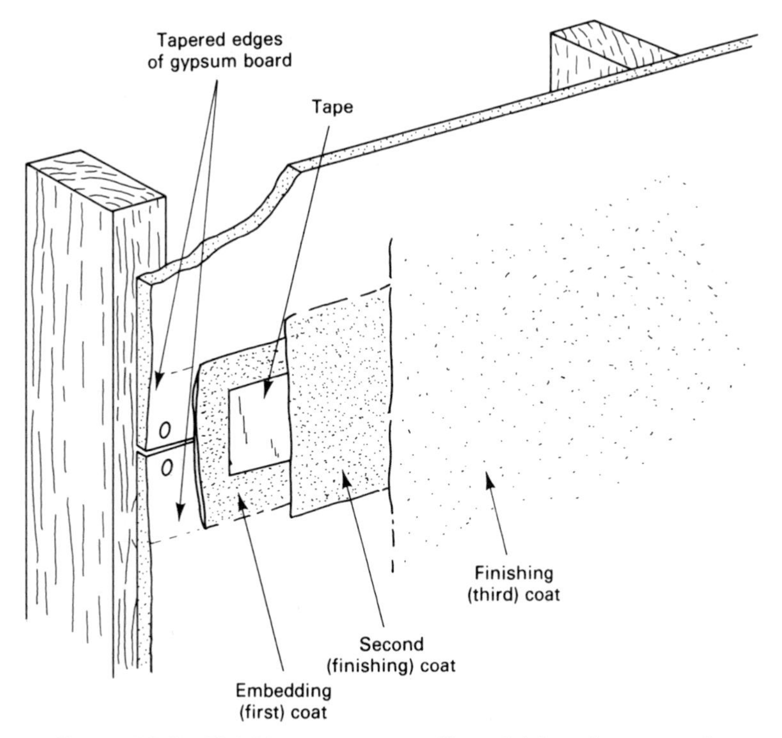

Figure 13-2 Finishing gypsum wallboard joints (courtesy Gypsum Association).

3. Apply a second coat of compound over the joint and feather the edges to a width of 2 to 4 in. (5 to 10 cm) beyond the first coat. Allow to dry.
4. Apply a thin finishing coat of compound and feather to a width of at least 6 in. (15 cm) on each side of the joint. Allow to dry.
5. Smooth the joint with fine sandpaper.

TABLE 13-1
Application of Gypsum Wallboard*

Thickness of gypsum wallboard (in.)	*Plane of framing surface*	*Long dimension of gypsum wallboard sheets in relation to direction of framing members*	*Maximum spacing of framing members (center-to-center, in inches)*	*Maximum spacing of fasteners (center-to-center, in inches)*		Nails[1]—to wood
				Nails [1,2]	Screws[3]	
Fastening required without adhesive application						
3/8	Horizontal[4]	Perpendicular	16	7	12	No 13 gauge, 1 1/2″ long, 19/64″ head, .098″ diameter, 1 1/2″ long, annular-ringed 4d cooler nail
	Vertical	Either direction	16	8	12	
1/2	Horizontal	Either direction	16	7	12	No. 13 gauge, 1 3/8″ long, 19/64″ head, .098″ diameter, 1 1/4″ long, annular-ringed 5d cooler nail
	Horizontal	Perpendicular	24	7	12	
	Vertical	Either direction	24	8	12	
5/8	Horizontal	Either direction	16	7	12	No. 13 gauge, 1 5/8″ long, 19/64″ head, .098″ diameter, 1 3/8″ long, annular-ringed 6d cooler nail
	Horizontal	Perpendicular	24	7	12	
	Vertical	Either direction	24	8	12	

With adhesive application						
3/8	Horizontal[4]	Perpendicular	16	16	16	Same as above for 3/8″
	Vertical	Either direction	16	16	24	
1/2 or 5/8	Horizontal	Either direction	16	16	16	As required for 1/2″ and 5/8″ gypsum wallboard, see above
		Perpendicular	24	12	16	
	Vertical	Either direction	24	24	24	
Two 3/8 layers	Horizontal	Perpendicular	24	16	16	Base ply nailed as required for 1/2″ gypsum wallboard and face ply placed with adhesive.
	Vertical	Either direction	24	24	24	

*Reproduced from the 1986 edition of the *CABO One and Two Family Dwelling Code*, copyright 1986, and 1987 Amendments, copyright 1987, with permission of the publishers—Building Officials and Code Administrators International, International Conference of Building Officials and Southern Building Code Congress International, Inc.

[1]Where the metal framing has a clinching design formed to receive the nails by two edges of metal, the nails shall be not less than 5/8 inch longer than the wallboard thickness and shall have ringed shanks. Where the metal framing has a nailing groove formed to receive the nails, the nails have barbed shanks or be 5d, No. 13 1/2 gauge, 1 5/8 inches long, 15/64 inch head for 1/2 inch gypsum wallboard; 6d, No. 13 gauge, 1 7/8 inches long, 15/64 inch head for 5/8 inch gypsum wallboard.

[2]Two nails spaced not less than 2 inches apart, nor more than 2 1/2 inches apart, and pairs of nails spaced not more than 12 inches center to center may be used.

[3]Screw shall be Type S or W per ASTM C 1002 and long enough to penetrate wood framing not less than 5/8 inch and metal framing not less than 1/4 inch.

[4]Three-eighths inch single-ply gypsum board shall not be installed if water-based spray-textured finish is applied nor to support insulation above a ceiling.

The procedure for taping interior corner joints is similar to that described above. Exterior corners should be covered with a metal *corner bead* (Figure 13-3) fastened to the framing. Joint compound is then applied to each side of the corner bead and feathered into the wallboard. Three coats of compound are usually required.

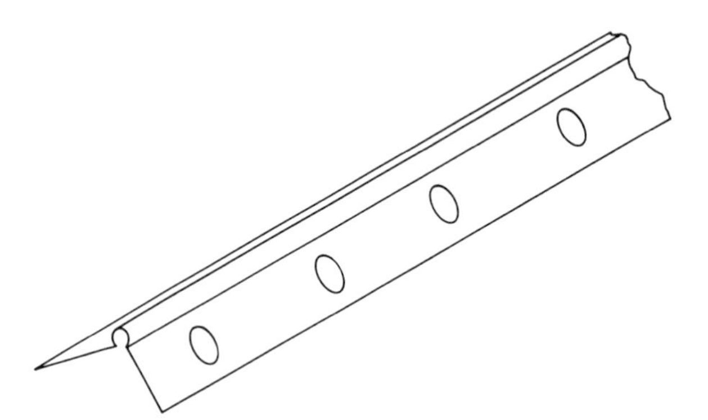

Figure 13-3 Corner bead.

The depression at each fastener location is filled with compound at the same time that the second joint coat is applied. Again, feather the edges as described above.

13-2 PLASTER

Although its use as an interior finish has been largely replaced by drywall construction, plaster provides an excellent wall finish. The materials and procedures used for gypsum plastering are described below.

Lath

The material that provides a base for the application of plaster is *lath*. While a number of types of lath are available, the most common types include gypsum lath, expanded metal lath, and wire lath. *Gypsum lath* (also called *rock lath*) is manufactured in 16 × 48 in. (41 × 122 cm) and 24 × 48 in. (61 × 122 cm) sheets, 3/8 or 1/2 in. (10 or 13 mm) thick, plain, or perforated. *Expanded metal lath* is available in weights of 2.5–4.0 lbs./sq. yd. (1.4–2.2 kg/m^2). *Wire lath* is available in weights of 1.4 and 1.95 lbs./sq. yd. (0.8 and 1.1 kg/m^2).

Lath should be applied with its long axis perpendicular to supports (horizontal application). The end joints of gypsum lath should be staggered in adjacent sheets and should fall on studs or other supports. However, end joints of gypsum lath above window and door openings should not be located on studs forming the side of the opening. Expanded metal and wire lath should be backed up by a water resistant building paper applied over framing. Some wire lath is manufactured with an integral paper backing.

The maximum spacing of lath supports and lath fastener requirements are usually specified by code. Typical maximum spacing of supports for gypsum lath is 16 in. (41 cm), increased to 24 in. (61 cm) for 1/2 in. (13 mm) plain lath. Maximum fastener spacing for gypsum lath is usually 5 in. (13 cm), reduced to 4 in. (10 cm) for 1/2 in. (13 mm) plain lath.

Plastering

Gypsum plaster is a mixture of gypsum, additives, aggregate, and water. The aggregate may be sand or a lightweight material such as perlite or vermiculite. Plaster made with lightweight aggregate provides additional insulating and sound-absorbing properties.

Plaster may be applied as a three-coat or a two-coat finish. For a three-coat finish, the first coat is a *scratch coat* or *base coat,* the second coat is a *brown coat* or *leveling coat,* and the third coat is a *finish coat.* In a two-coat finish, the scratch coat is eliminated. The minimum thickness (measured from the lath face) of the finished plaster should be 1/2 in. (13 mm) when using gypsum lath and 5/8 in. (16 mm) when using expanded metal or wire lath.

When applying plaster around window and door openings, *plaster grounds* are used to create an edge for the plaster. Three types of plaster grounds are illustrated in Figure 13-4: (a) the side door jambs serve as the plaster grounds: (b) temporary wood strips are used and removed after the plaster has dried; and (c) the narrow wood strips used are left in place and later covered by the door or window casing.

A single coat of lightweight plaster is sometimes sprayed on ceiling gypsum board to provide a textured, sound-absorbing finish. Another form of single-layer

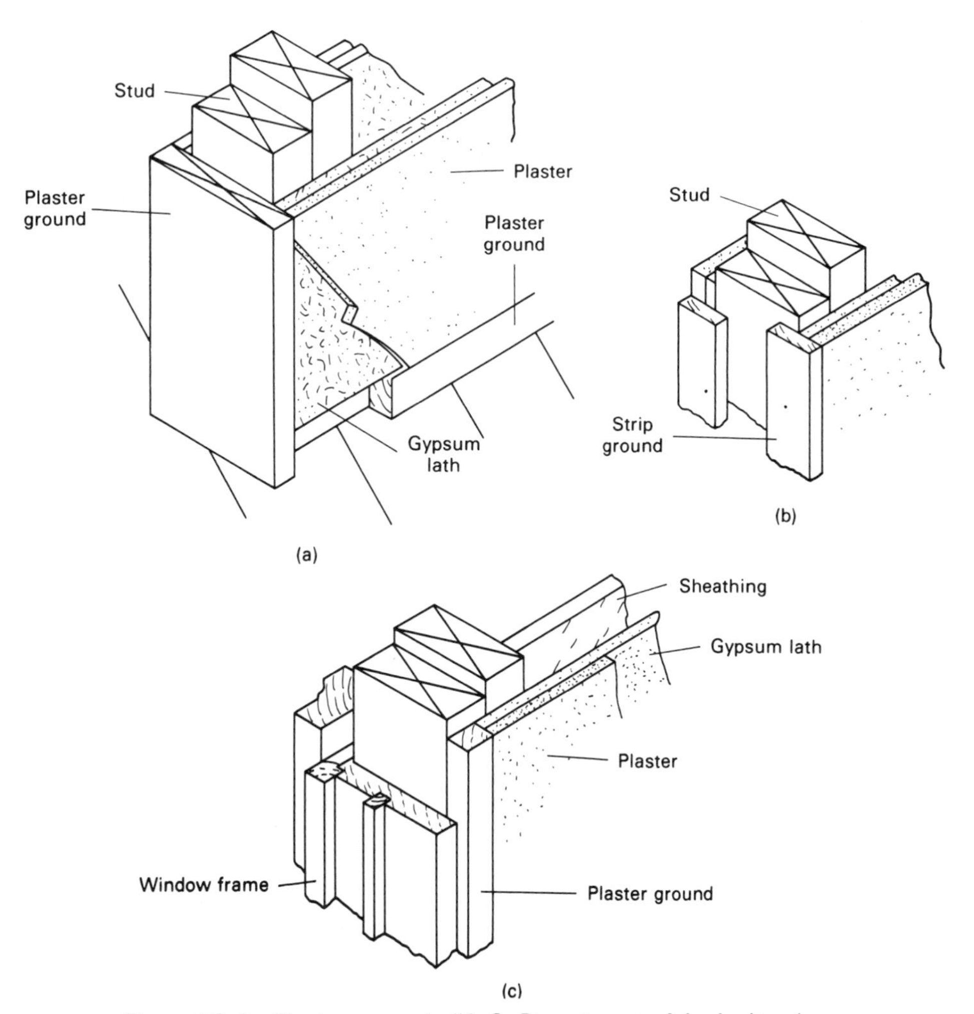

Figure 13-4 Plaster grounds (U. S. Department of Agriculture).

plaster is known as *veneer plaster.* This consists of a layer of plaster at least 1/16 in. (2 mm) thick applied over a *gypsum base for veneer plaster.*

13-3 OTHER FINISHES

Paneling

In addition to drywall and plaster, other common types of interior wall finish include prefinished sheet paneling of plywood, hardboard, and fiberboard, as well as wood strip paneling. Most wall paneling is produced in 4 × 8 ft. (1.2 × 2.4 m) sheets. For fire resistance, codes usually require that plywood and hardboard paneling less than 1/4 in. (6 mm) thick be applied over a base of gypsum board at least 3/8 in. (10 mm) thick.

Sheet paneling 1/4–5/16 in. (6–8 mm) thick should be supported by studs spaced 16 in. (41 cm) on center. For thicker panels, stud spacing may be increased to 24 in. (61 cm) on center. Space nails at 6 in. (15 cm) intervals along panel edges and 12 in. (30 cm) intervals within the panel. Minimum nail size should be 4d for 1/4 in. (6 mm) paneling, 6d for 5/16–1/2 in. (8–13 mm) paneling, and 8d for thicker paneling. Paneling may also be attached to framing or gypsum board by the use of adhesives.

Boards used for solid wood paneling should not be wider than 8 in. (20 cm) unless specially produced for such applications. Boards with tongue and groove edges are commonly used and may be installed either horizontally (perpendicular to studs), vertically (parallel to studs), or diagonally. When installed horizontally, stud spacing should not exceed 16 in. (41 cm) for 3/8 in. (10 mm) boards. Twenty-four inch (61 cm) stud spacing may be used for boards 5/8 in. (16 mm) and thicker. When installed vertically, provide blocking or other horizontal supports at an interval not exceeding 24 in. (61 cm). Nail paneling at each support with two or three 5d or 6d nails.

Trim

Interior. Common types of interior trim are illustrated in Figure 13-5. Primarily used for decorative effect, *ceiling molding* covers the joint between walls and ceiling. It also eliminates the need for taping and finishing this joint. *Baseboards*

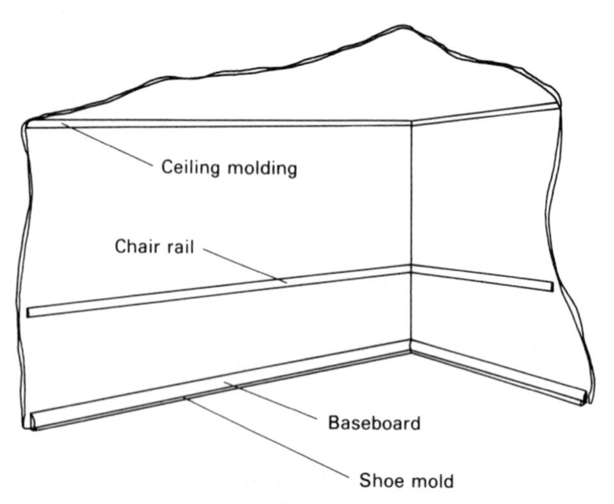

Figure 13-5 Interior trim.

are used to cover the joint between walls and floors. A *base cap* is sometimes used on top of the baseboard to form a two-piece baseboard. A *shoe mold* may be used in addition to the baseboard to hide the joint between the baseboard and the floor. A *chair rail* is a form of molding placed horizontally around the room several feet above the floor. Originally intended to protect the wall from scuffing by chair backs, chair rails are today used primarily for decorative effect. Since they divide the wall horizontally into two sections, they often serve as a divider for different wall finishes above and below the rail. For example, wood paneling might be used below the rail and wallpaper above the rail.

The window and door frames described in Chapter 10 also serve as interior trim.

Exterior. The major elements of exterior trim have been described in previous chapters. These include door and window frames (see Chapter 10), and cornices, fascia, soffits, and frieze boards (see Section 11-1).

13-4 PAINTING AND STAINING

Although prefinished materials are sometimes used, painting and staining usually provide the final exterior and interior surface treatment. For environmental reasons as well as convenience, most of today's paints and stains are manufactured with a water base. However, oil-based paints and stains are available for certain applications. The materials and procedures used for painting and staining are described below.

Painting

The two basic classifications of paint are *exterior* and *interior,* according to its intended use. Other classifications pertain to the type of paint surface produced. *Flat* paint produces a dull, nonreflective surface and is commonly used for interior walls. *Enamel,* available as gloss or semigloss, produces a hard, smooth surface. *Gloss* enamel provides a shiny surface and *semigloss* produces a surface with a slight luster. Water-base paints include latex, alkyd, and acrylic paint.

A minimum of two coats of paint are required for unfinished wood, and three coats are recommended. The first coat should be a stain resistant oil, alkyd, or latex primer. This is followed by one or two finish coats of an oil, latex, or acrylic latex paint that is suitable for the primer used. The paint should be applied in accordance with the manufacturer's recommendations. In general, paint may be applied by brush, roller, or spray gun. Interior wood trim should be sanded smooth and all dust removed before paint is applied. It may be necessary to apply a wood filler to porous wood trim before final sanding and painting.

Staining

Stain is often used as a preservative and waterproofing agent as well as a coloring agent. *Transparent* amd *semitransparent* stains (usually oil-based) allow the wood grain to show through the finish. *Solid color* or *opaque* stains (oil or water base) cover the wood grain but do not hide the wood's surface texture. Interior stains are usually transparent and are later covered with a sealer or varnish. Exterior stains are semitransparent or opaque and usually include preservative and waterproofing agents. Stains should be applied in accordance with the manufacturer's recommendations. Since it is important that transparent and semitransparent stains penetrate the wood surface, brush application is recommended. If spray gun application is used for exterior semitransparent stain, the stain should be worked into the wood by a brush or pad immediately after spraying.

For flooring and interior trim, it may be necessary to apply a wood filler to porous wood before staining. After the wood has been filled and sanded smooth, stain is applied as desired. Finally, the wood is coated with a transparent finish such as a sealer or varnish to provide the final surface.

13-5 FLOOR COVERINGS

There are a number of types of floor coverings (or finish flooring) available, including solid and laminated wood, carpet, resilient flooring, ceramic tile, and terrazzo. Some of the common materials and procedures used are described below.

Wood Flooring

A properly finished wood floor is beautiful and long lasting. Solid wood strip flooring is available in hardwood or softwood with tongue-and-groove or square edges, 1 1/2–5 1/8 in. (38–130 mm) wide, 3/8–1 1/2 in. (10–38 mm) thick, and up to 16 ft. (4.9 m) in length. A common size for tongue-and-groove hardwood strip flooring is 3/4 in. (19 mm) thick by 2 1/4 in. (57 mm) wide.

Before placing wood strip flooring, be sure the subfloor is clean and smooth. Drive all subfloor fasteners flush or slightly below the surface. It is recommended that a layer of 15 lbs. (6.8 kg) asphalt saturated felt be laid between the subfloor and flooring. The felt layer helps to deaden sound and reduce air infiltration through the flooring. Strips of tongue-and-groove flooring are nailed through the tongue edge to the subfloor using 7d or 8d flooring nails as illustrated in Figure 13-6. Nails should be driven at an angle of 45° to 50° to the floor. Best results are obtained when the flooring is laid perpendicular to the floor joists and nailed through the subfloor into the floor joists. One additional nail should be driven into the subfloor

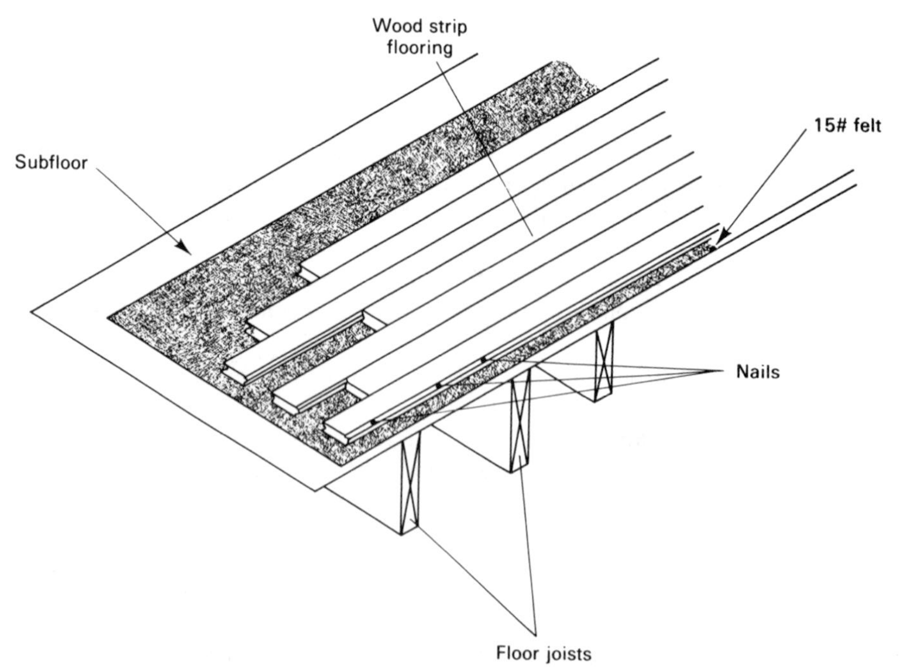

Figure 13-6 Installation of wood strip flooring.

between joists. Set nails below the tongue surface using a nail set to avoid damage to the edge of the flooring. When wood strip flooring is installed over a concrete slab, the flooring should be nailed to a 3/4 in. (19 mm) plywood underlayment or nominal 1 in. (25 mm) thick wood sleepers laid on top of the concrete slab.

Solid wood flooring is also available as wood block flooring. Blocks are available as squares ranging in size from 4 in. (10 cm) to 12 in. (30 cm). They are often made up of several narrow wood strips bonded together. When the block incorporates small square, rectangular, or diagonal patterns, the flooring is called *parquet*. Blocks are nailed or glued to the subfloor as recommended by the manufacturer.

There are also a number of prefinished laminated wood flooring products available. Laminated flooring may be glued in place or installed as a floating floor system. Flooring for glued installation is available in several styles and sizes including planks 3–7 in. (8–18 cm) wide and parquet blocks 6–19 in. (15–48 cm) square. They are installed using adhesives recommended by the manufacturer. Prefinished flooring used for floating floor systems is backed with a thin layer of foam. The flooring pieces are installed by glueing them together along their tongue and groove edges only. Thus, they ''float'' over the subfloor. Both glued and floating floor systems can be installed directly over concrete slabs.

Carpet

Carpet forms an attractive flooring covering, which also insulates and deadens sound. A carpet and pad (see Figure 12-9) may be installed over a subfloor and underlayment, a combined subfloor-underlayment, or a concrete slab. The carpet is held in place along its edges by a tack strip fastened to the floor. Special carpet installation tools are used to bond carpet seams and to stretch the carpet during installation. Although wool carpets are preferred by some, most carpets today are manufactured of synthetic fibers. Such carpets are relatively inexpensive, attractive, and durable.

Resilient Flooring

Resilient flooring includes vinyl sheet, linoleum, and tiles of vinyl, rubber, asphalt, and cork. Vinyl flooring with ''no wax'' finishes are becoming especially popular. Resilient flooring should be installed over a concrete slab, a combined subfloor-underlayment, or a subfloor and underlayment. Since the flooring is relatively thin and flexible, it is important that the underlying surface be as clean and smooth as possible. When plywood underlayment is used, install the underlayment immediately before placing the resilient flooring. Drive all underlayment fasteners flush or slightly below the surface. Do not fill nail holes when nails are set below the surface. Fasten the flooring to the underlying surface in accordance with the manufacturer's instructions.

Ceramic Tile

Ceramic tile is a fired clay tile, usually glazed, manufactured for use in covering floors, walls, and other surfaces. Due to its beauty and durability, it is often used to cover the floors of bathrooms, kitchens, and entrance foyers. The major types of floor tile include ceramic floor tile and quarry tile. *Ceramic floor tile* uses an unglazed or low gloss surface to make it less slippery when wet. It is available in squares of 1 in. (2.5 cm) to 6 in. (15 cm) as well as rectangles, hexagons, and other shapes. *Quarry tile* is an unglazed clay tile manufactured for use as a floor covering. It is available as squares, rectangle, hexagons, and other shapes in sizes from 3 × 3 in. (8 × 8 cm) to 8 × 8 in. (20 × 20 cm).

Floor tile may be set in a 1 1/4 in. (32 mm) mortar bed or bonded to the supporting surface using dry-set mortar, epoxy mortar, or an organic adhesive. The supporting surface should be either a concrete slab or plywood at least 15/32 in. (12 mm) thick. The specific subfloor and underlayment required depends on the flooring system and adhesive used (see Reference 2). After tile is set, tile joints are filled with mortar grout of the desired color. Any mortar film remaining on the tile after the grout has hardened is removed with a damp cloth or weak acid solution.

Terrazzo

Terrazzo is a floor covering made up of stone chips set in a portland cement matrix. After the matrix has hardened, the surface is ground to expose the inner portions of the stone chips. The surface is then polished and sealed to produce a smooth, durable, and attractive floor. While usually constructed in place on top of a concrete slab, terrazzo tile is also available.

PROBLEMS

1. What is Type X gypsum board?
2. How should gypsum board be fastened to a wood frame wall?
3. Describe the composition of a three-coat plaster finish and identify the purpose served by each element.
4. Briefly explain the meaning of the following terms.
   (a) Plaster grounds
   (b) Gypsum lath
   (c) Veneer plaster
5. Identify and briefly describe the major types of exterior wood stains.
6. What minimum paint finish should be applied to unfinished wood?
7. Explain how prefinished plywood paneling 1/8 in. (3 mm) thick should be applied to the interior of a frame wall having wall studs spaced 16 in. (41 cm) on center.
8. List the major types of interior wood trim.
9. Briefly describe the installation of solid wood strip flooring over a wood floor frame.
10. What is a floating floor system?

REFERENCES

1. Anderson, L. O. *Wood-Frame House Construction.* Agriculture Handbook No. 73. U.S. Department of Agriculture, Washington, DC, 1975.
2. American Plywood Association. *APA Design/Construction Guide: Residential and Commercial.* Tacoma, WA, 1988.
3. Council of American Building Officials. *CABO One and Two Family Dwelling Code.* Falls Church, VA.
4. Gypsum Association. *Gypsum Board Products Glossary of Terminology.* Evanston, IL, 1985.
5. Gypsum Association. *Using Gypsum Board for Walls and Ceilings.* Evanston, IL, 1985.

chapter 14

Electrical Systems

The construction of a building's electrical system is a specialized task normally performed by an electrical contractor. Since the electrical contractor usually works under a subcontract arrangement with the builder (prime contractor), the electrical contractor is commonly referred to as the *electrical sub.* Although the builder is seldom a qualified electrician, the builder is responsible for ensuring that all electrical work is carried out correctly and on schedule. As a result, all builders must understand the basic requirements for construction of a building's electrical system. This chapter provides a brief introduction to building electrical systems and their construction. Refer to the *National Electric Code* (see Section 14-2) and other end-of-chapter references for specific requirements and further information.

14-1 FUNDAMENTALS OF ELECTRICAL SYSTEMS

Basic Relationships

The fundamental relation governing the flow of electricity is expressed by Equation 14-1 (*Ohm's Law*). Voltage is expressed in volts (V), current in amperes or amps (A), and resistance in ohms (Ω). When any two values are known, the other may easily be calculated.

$$\text{Voltage} = \text{current} \times \text{resistance} \qquad (14\text{-}1)$$

Electrical power requirements are commonly expressed in watts (W) or kilowatts (kW = 1000 W). The relationship between power, voltage, and current is expressed by Equation 14-2.

$$\text{Watts} = \text{voltage} \times \text{current} \qquad (14\text{-}2)$$

Although 1 horsepower is equivalent to 746 watts, do not use this relationship to determine the current required by a motor. Use the motor's nameplate current rating instead.

The use of Equations 14-1 and 14-2 is illustrated by Example 14-1.

EXAMPLE 14-1

Problem: A water heater is rated at 4500 watts, 240 volts. Find:
a. The current required by the water heater.
b. The resistance of the heating element in the heater.

Solution:

a. Watts = voltage × current (Eq. 14-2)
4500 = 240 × current
Current = 4500/240 = *18.75 amps*

b. Voltage = current × resistance (Eq. 14-1)
240 = 18.75 × resistance
Resistance = 240/18.75 = *12.8 ohms*

Electrical Terms

Ampacity The current-carrying capacity of a conductor expressed in amperes.

Branch Circuit A circuit running from a power panel to one or more appliances, fixtures, or receptacles.

Circuit Breaker A device that automatically shuts off current flow in a circuit on overload. It can also be operated manually and reset after opening.

Conductor A wire or cable used to supply electricity.

Conduit A tube or pipe used to enclose and protect electrical conductors.

Feeder A cable connecting the service entrance to the service panel (main distribution panel), or the service panel to other power panels.

Fixture A light assembly attached to a wall or ceiling.

Fuse A device that interrupts current flow by melting when subjected to an overload. The fuse or its fusible element must be replaced to restore current flow.

Ground A rod, pipe, or plate buried in the ground, or a wire attached to it.

Ground Fault Interrupter (GFI) A device that shuts off current flow when it senses a small flow of electricity from a hot wire to ground. It is used to reduce the danger of electrocution in damp locations and near water.

Hot Wire The electrically live wire of a circuit, usually black or red in color.

Neutral Wire The neutral or grounded wire of a circuit, usually white in color.

Overcurrent Protection Device A device that protects against excessive current flow in a circuit. This includes circuit breakers, fuses, and GFI devices.

Receptacle An electrical outlet designed to accept an electrical plug.

Service Entrance The main electrical supply into a building.

Service Load The maximum expected (design) current load for a building.

Service Panel The main power panel for a building.

Switch A device for making or interrupting the flow of electricity in a circuit.

Electric Power Supply

The two basic types of electric power supply systems are alternating current (AC) systems and direct current (DC) systems. In *alternating current* systems, the direction of current flow reverses twice during each electrical cycle. The frequency of a system is expressed in *hertz* (Hz), which indicates cycles per second. U.S. utility systems provide 60 Hz AC power; many other parts of the world utilize 50 Hz AC power. In *direct current* (DC) systems, current always flows in one direction. Direct current power is seldom available from utility systems for building use. If direct current is required for specialized equipment, it is generated locally from the building's AC power system.

Most U.S. dwellings and other small buildings use the single-phase 3-wire AC system illustrated in Figure 14-1(a). In a *single-phase* system, the voltage cycle is identical in all hot wires. Notice that the voltage between either hot wire (Line 1 or Line 2) and the neutral wire is 120 V, and the voltage between the two hot wires is 240 V. The use of this three-wire system makes it easy to provide either 120 V or 240 V branch circuits.

The three-phase AC system illustrated in Figure 14-1(b) is used for larger buildings, and, when needed, for three-phase motors or equipment. *Three-phase* simply means that the voltage cycles in the three hot wires are slightly out of step with each other. Notice that the voltage between any hot wire and the neutral is 120 V, and the voltage between any two hot wires is 208 V.

The electric power supply system used by most U.S. small buildings is single-phase AC, three-wire, 120/240 V. This is commonly abbreviated 1 ø 3W 120/240 V on plans and specifications.

Electrical Diagrams

Some common electrical symbols used on building electrical diagrams are shown in Figure 14-2. A portion of the electrical diagram for a small residence is shown in Figure 14-3. Notice that the electrical diagram has simply been superimposed on the building's floor plan. This is commonly done for dwellings and other small buildings. The plans for larger buildings commonly include separate sheets of electrical diagrams and listings of power panel circuits.

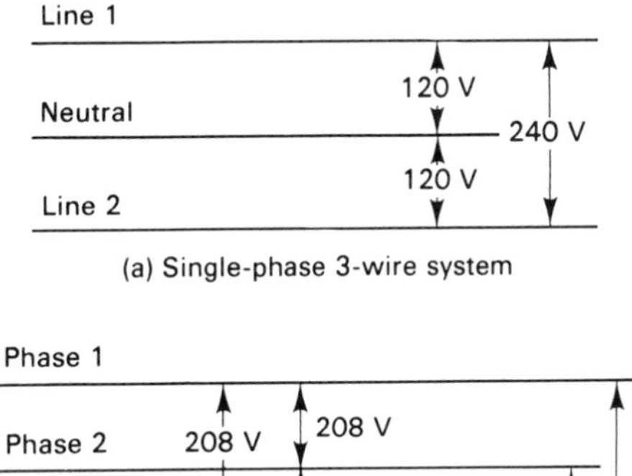

(a) Single-phase 3-wire system

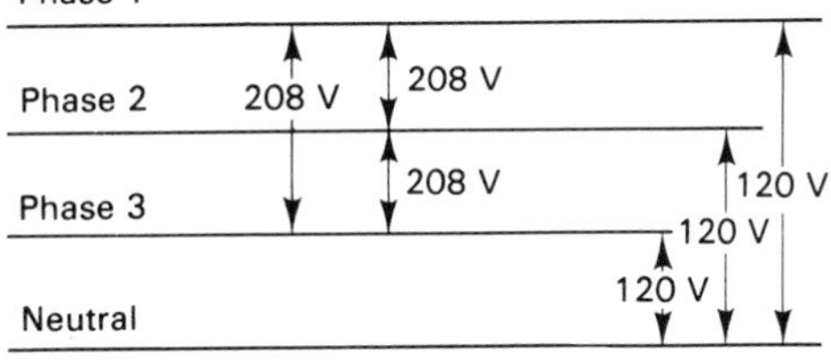

(b) Three-phase 4-wire system

Figure 14-1 Single-phase and three-phase systems.

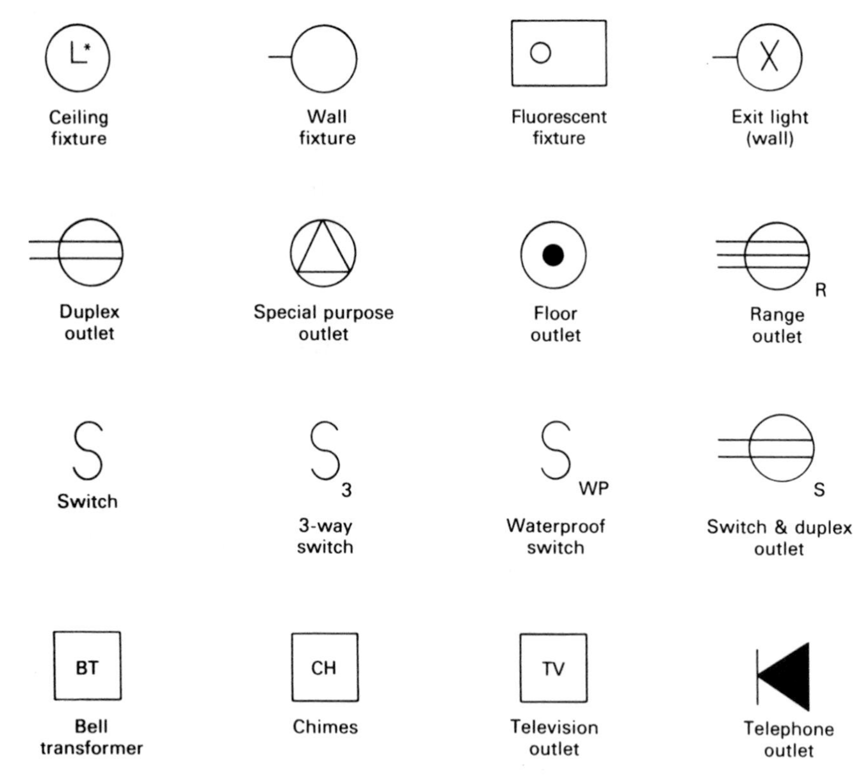

*Letter indicates type of fixture

Figure 14-2 Common electrical symbols.

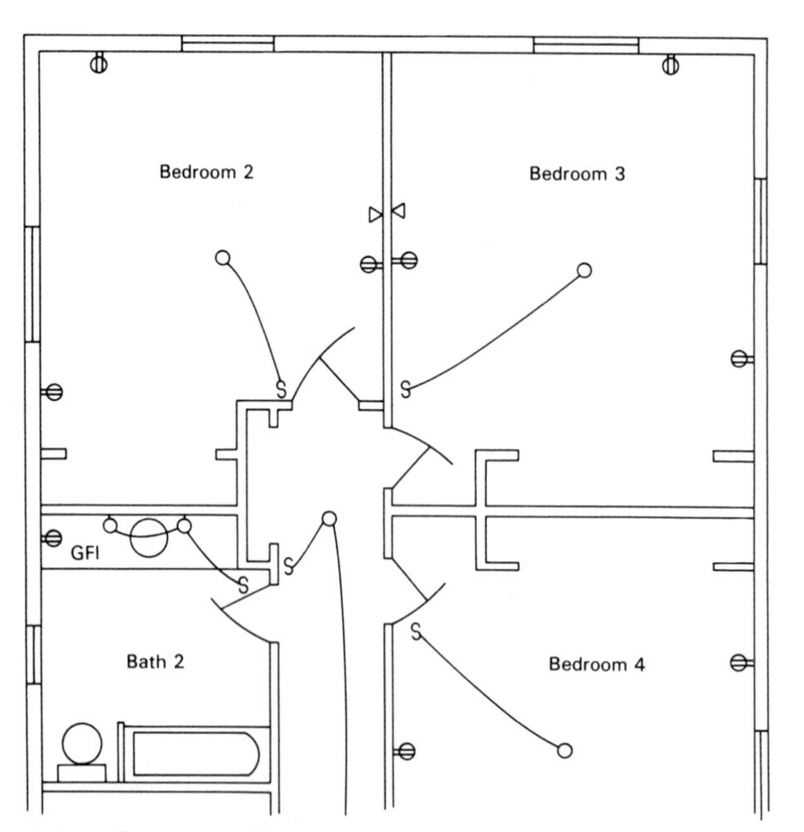

Figure 14-3 Portion of electrical diagram for house.

14-2 GENERAL REQUIREMENTS

Codes

In contrast to other construction codes, the *National Electrical Code* (NEC), published by the National Fire Protection Association, is the only U.S. national (or model) electrical code. As a result, it is widely used throughout the United States, and the number of local modifications is limited. The code requirements discussed in this chapter are based on the 1987 NEC. Always consult the current edition of the code for up-to-date requirements.

Electrical System Design

The design of the electrical system for a building should be done by a qualified professional. Some of the items that must be determined include the total current demand for all lights, appliances, and equipment, and the location of all fixtures, receptacles, and special equipment. Based on this information, the building's service load is calculated, branch circuits are laid out, conductors and circuit breakers are sized, and service panels, switches, conduit, and other electrical components are selected. The design of electrical systems is beyond the scope of this chapter, but design procedures and example calculations are provided in References 2, 3, and 6.

Conductors

The size of U.S. electrical wire and cable is designated by either its American Wire Gauge (AWG) number or its cross sectional area in thousand circular mils (MCM). [NOTE: One circular mil represents the area of a circle 1/1000 in. in diameter.] Single wire conductors (Figure 14-4) commonly consist of insulated copper or alumi-

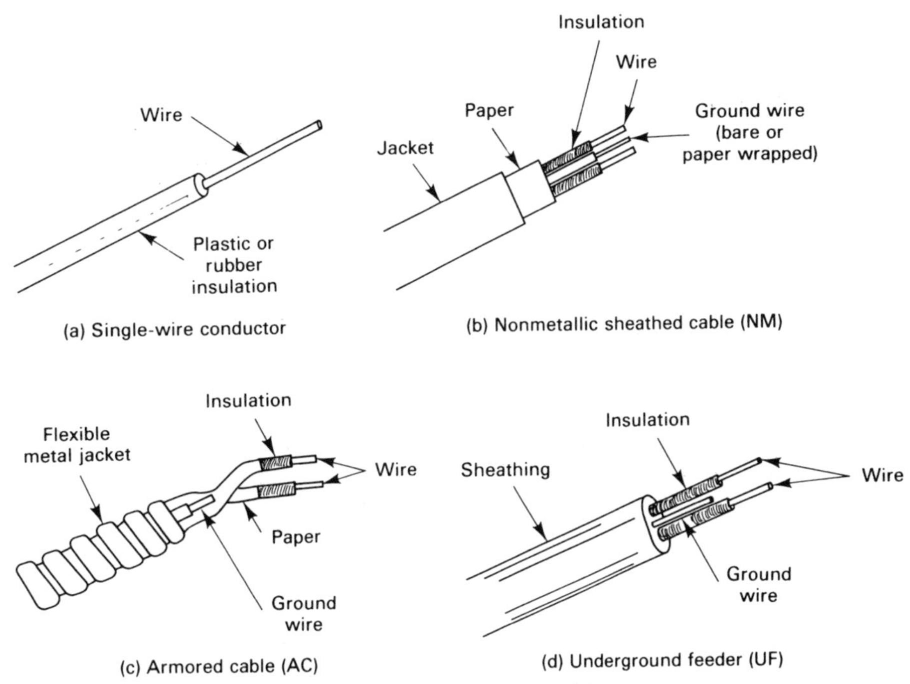

Figure 14-4 Wire and cable.

num wire ranging in size from AWG 18 to AWG 0000 (also written 4/0) and from MCM 250 to MCM 2000. Notice that wire size increases with decreasing AWG numbers. That is, AWG 18 is *smaller* than AWG 16. However, since the MCM number represents cross sectional area, wire size increases with increasing MCM number. That is, MCM 300 is larger than MCM 250. Conductor sizes larger than AWG 8 normally consist of stranded wire, and smaller sizes use a solid conductor. The minimum wire size required for a conductor is determined by the current to be carried, the allowable voltage drop in the circuit, the wire type, and the temperature.

Wire type is designated by an alphabetic symbol indicating the type of insulation used and its suitability for various uses. Although a number of wire types are available, some common types are shown in Table 14-1.

When several insulated conductors are assembled and covered by a single outer protective cover, the assembly is called a *cable*. The most common types of cable used in building construction are *nonmetallic sheathed cable* (Type NM and NMC), *armored cable* (Type AC and ACL), and *underground and branch-circuit cable* (Type UF), as illustrated in Figure 14-4. Type NMC is nonmetallic sheathed cable incorporating a corrosion resistant outer covering. Type ACL is armored cable containing waterproof lead-covered conductors. Nonmetallic sheathed cable is commonly called *Romex*, and armored cable is commonly called *BX*.

Conduit and Accessories

When single-wire insulated conductors are used, they are placed inside a protective tube called *conduit*. Conduit is available in a number of forms including rigid and flexible, metallic and nonmetallic, conduit and tubing. *Electrical metallic tubing* (EMT), also called *thin-wall conduit,* is widely used. *Rigid metal conduit* is often used for the service entrance and for maximum physical protection. Increasing use is being made of nonmetallic forms of conduit including *rigid nonmetallic conduit* and *electrical nonmetallic tubing*. Flexible forms of metallic conduit include *flexible metallic tubing* and *flexible metal conduit*.

Some typical outlet boxes, fittings, and accessories used with conduit are illustrated in Figure 14-5. The fittings used for rigid conduit are similar to pipe fittings. Special fittings are used with thin-wall conduit (EMT) and flexible conduit. Conduit is fastened to the mounting surface with conduit straps. Outlet boxes are used for holding conduit or cable in position, making wire connections, and for mounting switches, outlets, or fixtures. Metal and plastic boxes are available in rectangular, square, octagonal, and round shapes.

TABLE 14–1
Wire Types

Wire type	*Description*	*Maximum temperature (°C)*
RH	Heat-resistant rubber	75
RHH	Heat-resistant rubber	90
RHW	Moisture- and heat-resistant rubber	75
THHN	Heat-resistant thermoplastic	90
THW	Moisture- and heat-resistant thermoplastic	75
TW	Moisture-resistant thermoplastic	60
UF	Underground feeder—single conductor	60 or 75

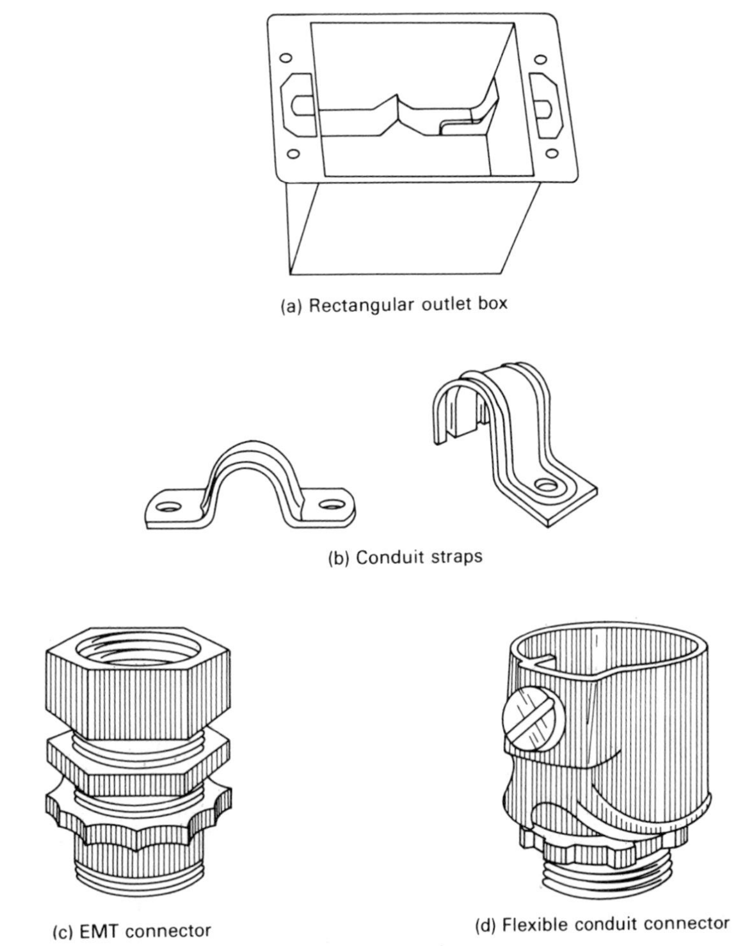
(a) Rectangular outlet box

(b) Conduit straps

(c) EMT connector

(d) Flexible conduit connector

Figure 14-5 Outlet box and conduit fittings.

Sheathed cable is used without conduit except when protection from physical damage is required. Nonmetallic sheathed cable (NM) is widely used for wiring in dry locations in dwellings and other buildings not higher than three stories above ground. Armored cable (AC) may be used in dry locations where more physical protection is desired. Underground feeder and branch-circuit cable (UF) may be used in wet or corrosive locations and for direct burial in earth. Some typical fittings and accessories for nonmetallic and armored cable are illustrated in Figure 14-6. Cables are fastened to the mounting surface by staples or clamps.

Receptacles

A *receptacle* is an electrical outlet designed to accept an electrical plug. The receptacle is mounted in an outlet box and connected to the circuit wires or cable. The most common type of receptacle is the double 2-pole, 3-wire, grounding receptacle shown in Figure 14-7, often called a *duplex outlet.* Notice that it accepts a plug consisting of two parallel blades connected to the hot and neutral wires and a round blade connected to circuit ground. This receptacle is rated 125 volts, 15 or 20 amps.

A number of receptacle shapes have been standardized by the National Electrical Manufacturers Association (NEMA) to designate the voltage and ampacity of the receptacle. Some common types of receptacles, their NEMA designations, and their electrical ratings, are illustrated in Figure 14-8.

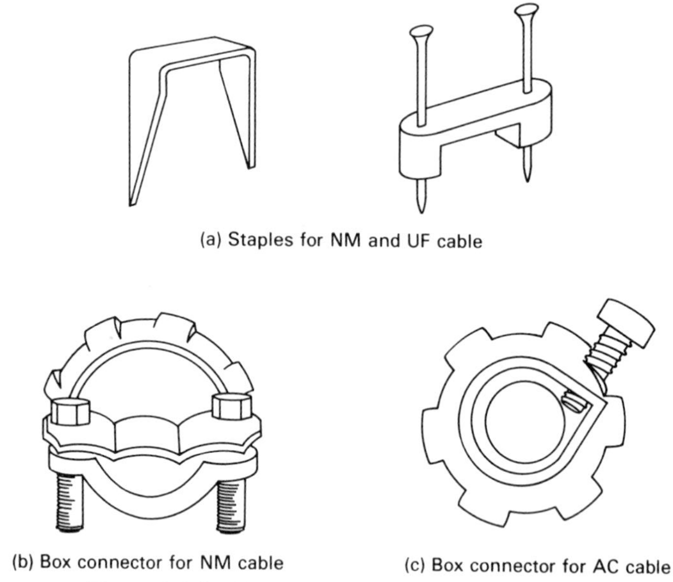

(a) Staples for NM and UF cable

(b) Box connector for NM cable (c) Box connector for AC cable

Figure 14-6 Staples and connectors for cable.

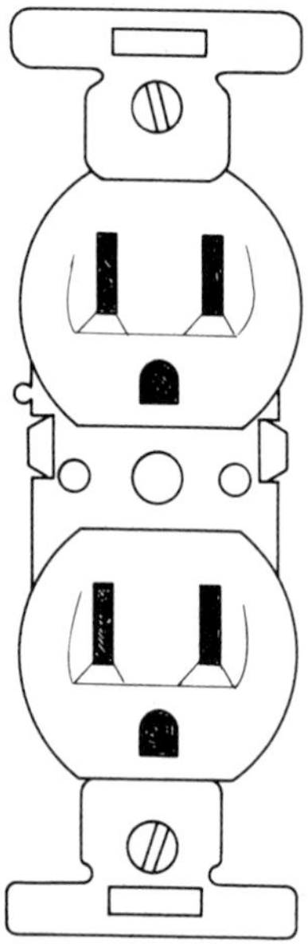

Figure 14-7 Duplex receptacle.

Electrical Equipment

Specifications or local codes often require that electrical equipment and components be approved for safety by Underwriters Laboratories, Inc. Underwriters Laboratory (UL) publishes lists of equipment that have been inspected and approved. In addition, such equipment can be identified by a UL trademark seal affixed to the equipment.

Conductors, switches, receptacles, conduit, and other electrical components must be rated for a voltage and current equal to or greater than that which they are required to carry. Power panels must accommodate the required number of circuits with provision for future expansion. Circuit breakers and fuses must have ratings corresponding to the selected ampacity. However, the current rating of circuit breakers and fuses must not exceed the ampacity of their associated conductors. When using aluminum conductors, ensure that connectors and other electrical components are approved for use with aluminum conductors.

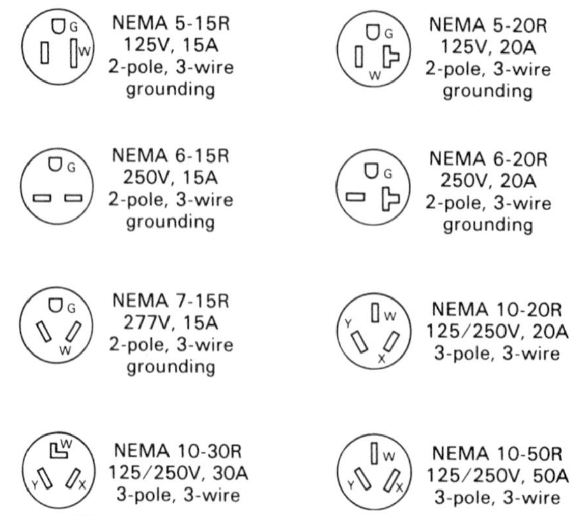

Figure 14-8 Electrical receptacles.

14-3 CIRCUITS

Branch Circuits

Branch circuits run from a power panel to one or more appliances, fixtures, or receptacles. The principal types of branch circuits and their use are described in the following paragraphs.

While the allowable current-carrying capacity (ampacity) of a circuit depends on wire size, wire type, and temperature, Table 14-2 gives the minimum size of copper wire permitted in branch circuits.

Lighting and General Purpose Circuits

The amount of light provided in a room is measured in *footcandles* (fc) (*lux* in metric units). One footcandle (10.76 lux) is equivalent to one *lumen* per square foot, and one lux equals one lumen per square meter. The level of illumination required in a room depends on the type of activity to be carried out in the room. This may vary from 10 fc to over 100 fc. For large buildings, the architect will determine the level of illumination required and specify the type, number, and location of lighting fixtures to be used. For dwellings and other small buildings, the lighting fixtures and their location are selected by the owner and/or builder. At least one wall switch-controlled light fixture should be provided in every habitable room, hallway, and garage, and at each outside entry. However, a wall switch-controlled receptacle may be substituted for the required light fixture in habitable rooms, except for bathrooms and kitchens.

Lighting fixtures may use either incandescent or fluorescent lamps. Incandescent lights are inexpensive but have a low lighting efficiency (footcandles per watt) and a relatively short life (750–1000 hours). Fluorescent lights are more expensive but provide more light per watt, last longer, and produce less heat than do incandescent lights.

For residences and small buildings, *general purpose circuits* are commonly used to serve lighting fixtures and general purpose wall receptacles. The code requires that a minimum electrical capacity of 3 watts per square foot of living area be provided for lighting and general purpose receptacles. The minimum number of 20 amp 120 volt general purpose circuits may be calculated using Equation 14-3. The use of this equation is illustrated in Example 14-2.

$$\text{Number of circuits} = \frac{\text{living area (sq. ft.)}}{800} \qquad (14\text{-}3)$$

TABLE 14–2
Minimum Wire Size for Branch Circuits

*Circuit rating** *(amps)*	*Minimum wire size*** *(AWG)*
15	14
20	12
30	10
40	8
50	6

*Use overcurrent protection device of the same rating.

**Includes copper wire types RH, RHH, RHW, THHN, THW, and TW.

EXAMPLE 14-2

Problem: Determine the minimum number of 20 amp general purpose circuits required for a house having a living area of 1800 sq. ft.

Solution:

$$\text{Number of circuits} = \frac{\text{living area}}{800} \qquad \text{(Eq. 14-3)}$$

$$= \frac{1800}{800} = 2.25$$

$$= 3 \text{ circuits}$$

General purpose receptacles should be spaced along interior walls so that no point on a room's baseboard is more than 6 ft. (1.8 m) horizontally from a receptacle. General purpose receptacles, or *convenience outlets,* usually consist of double grounding type receptacles, commonly called "duplex outlets."

At least one receptacle must be provided in each bathroom near the lavatory. Although the receptacle may be connected to a general purpose circuit, it must be equipped with ground fault interrupter (GFI) protection.

Appliance Circuits

A minimum of two 20-amp small appliance branch circuits are required in each dwelling unit to serve the kitchen, breakfast room, pantry, and dining room. Although the NEC code permits a refrigerator to be connected to these circuits, it is good practice to provide a separate circuit for the refrigerator.

The receptacles serving the kitchen countertop should be divided between the small appliance branch circuits. A receptacle should be located along each section of countertop wider than 12 in. (30 cm).

A separate 20-amp branch circuit with receptacle is required for the laundry area of each dwelling except where a common laundry area is provided for multifamily dwellings.

Equipment Circuits

For dwellings, additional branch circuits are required to serve air conditioners/heat pumps/furnaces, water heaters, clothes dryers, electric ranges/ovens/cooktops, dishwashers, garbage disposers, motors, and similar loads. The minimum circuit rating and the overcurrent protection device rating required for these circuits must be determined in accordance with the NEC code. A convenient summary of the design rules for dwellling branch circuits is provided in Reference 2.

For commercial and industrial buildings, power and circuit requirements are often more complex. The electrical plan for such buildings is commonly provided by the architect/engineer for the project.

Temporary Services

During construction of a building, a temporary power supply is required to serve the builder's tools and equipment. A typical temporary power supply for a small dwelling is illustrated in Figure 14-9. The builder's electrical sub usually installs the mounting pole, meter box, power panel, and receptacles. The electric utility com-

Figure 14-9 Temporary power supply for construction of a house.

pany installs the electric meter and connects the meter box to the power distribution system.

14-4 CONSTRUCTION PROCEDURES

Scheduling

One of the major responsibilities of the construction manager is the scheduling of work to minimize project delays and conflicts among elements of the construction work force. The coordination of the various subcontractors and trades is particularly important in building construction. Some common scheduling methods used in building construction include the *bar graph* or *bar chart* schedule (also called a *Gantt chart*) and the *Critical Path Method* (CPM) described in Reference 7.

Regardless of the scheduling method used, it is critical that conflicts between trades be minimized. It is desirable that all elements of the field construction force provide input into the development of a construction schedule. When some subcontractors or trades are not represented in the planning group developing the schedule, it is important that they review the proposed schedule before it is finalized to ensure that the project can be carried out as planned. The following are some areas in which conflicts are most likely to occur in building construction.

- Excavation and foundation work
- Under-the-slab utilities and concrete work

- Installation of electrical circuits during structural work
- Location of cutouts, sleeves, and other openings for electrical lines
- Location of hangers for electrical lines and supports for equipment

Construction Sequence

The principal steps in the installation of a building's electrical system include:

- Installation of temporary power
- Installation of under-the-slab conduit and cable
- Roughing in; that is, the installation of concealed conduit, cable, and accessories
- Pulling of wire in conduit (if applicable)
- Installation of fixtures, outlets, cover plates, circuit breakers, etc.
- Connection of electric meter and final testing

Some typical roughing in of electrical cable for a single-family residence is shown in Figure 14-10. After roughing in is completed and inspected, drywall is installed. Installation of receptacles, fixtures, and circuit breakers follows drywall finishing and painting.

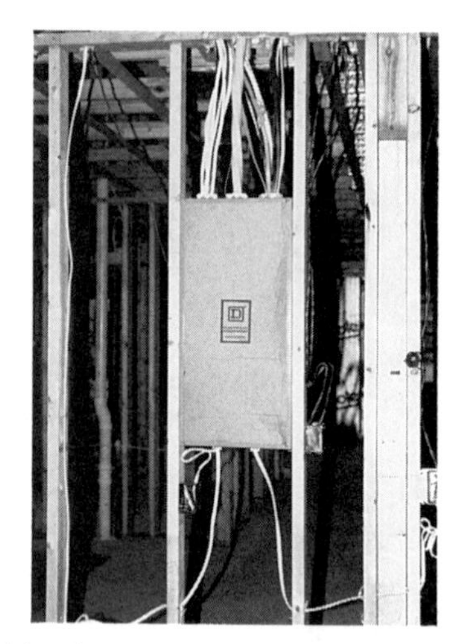

Figure 14-10 Electrical roughing in.

Quality Control

Although the installation of electrical systems is highly specialized work, the builder (prime contractor) remains responsible for ensuring that the electrical system is installed in accordance with the plans, specifications, codes, and the standards of good workmanship. If the builder is unable to judge whether a particular aspect of the electrical system meets these requirements, he or she should call on an independent expert for advice. On larger projects, the architect/engineer (A/E) often provides construction inspection services to the owner. In such cases, the A/E's advice may be sought to clarify the requirements of the contract.

Testing is an important element of quality control. When no test procedures are specified, the builder should verify with the electrical sub and an owner's representative that all components of the electrical system function as intended. Any test procedures detailed in the project plans and specifications must, of course, be carried out as specified.

PROBLEMS

1. Who is responsible to the owner for ensuring that all electrical work in a building meets the requirements of the plans and specifications?

2. An electric water heater is rated at 5000 watts, 208 volts. Find the current required by the heater.

3. An electric space heater requires a current of 25 amps at 240 volts. How many watts does the heater consume?

4. If electricity costs $0.08 per kilowatt-hour, how much does the electric space heater of Problem 3 cost to operate per hour?

5. Sketch the voltage relationships between the wires of a single phase 120/240 volt branch circuit having three wires plus ground.

6. Define or briefly explain the following electrical terms:
   (a) EMT
   (b) Romex
   (c) Ampacity
   (d) GFI

7. What is the minimum number of 20 amp general purpose branch circuits required for a 1500 sq. ft. dwelling?

8. How is temporary electrical power usually supplied to a dwelling or other small building during construction?

9. Briefly describe the major scheduling problems likely to be encountered in the construction of a small building.

10. How can the builder (prime contractor) ensure adequate quality in the construction of a building's electrical system?

REFERENCES

1. American Institute of Architects. *Architectural Graphic Standards.* 7th ed. New York: Wiley, 1981.
2. Calter, Paul. *Practical Math Handbook for the Building Trades.* Englewood Cliffs, NJ: Prentice-Hall, 1983.
3. Dagastino, Frank R. *Mechanical and Electrical Systems in Construction and Architecture.* Reston, VA: Reston, 1978.
4. Hettema, Robert M. *Mechanical and Electrical Building Construction.* Englewood Cliffs, NJ: Prentice-Hall, 1984.
5. McGuiness, William J., and Benjamin Stein. *Building Technology: Mechanical and Electrical Systems.* New York: Wiley, 1977.
6. National Fire Protection Association, *National Electric Code.* Quincy, MA.
7. Nunnally, S. W. *Construction Methods and Management.* 2nd ed. Englewood Cliffs, NJ: Prentice-Hall, 1987.

chapter 15

Plumbing and Mechanical Systems

15-1 GENERAL REQUIREMENTS

Scope

Plumbing systems, mechanical systems, and electrical systems make up the utility systems for a building. The construction of electrical systems is described in Chapter 14. As in electrical systems, the construction of plumbing and mechanical systems is specialized work that is normally performed by specialty contractors. Since these specialty contractors commonly work under a subcontract arrangement with the builder (prime contractor), they are often referred to as the *plumbing sub* and the *mechanical sub*. Although the builder is seldom a qualified plumber or mechanical specialist, the builder is responsible for ensuring that all plumbing and mechanical work is carried out correctly and on schedule. Thus, the builder must understand the basic requirements for construction of building plumbing and mechanical systems. This chapter provides a brief introduction to plumbing and mechanical systems and their construction. Refer to the end-of-chapter references for specific requirements and further information.

The plumbing system for a building consists of the hot and cold water supply lines, domestic water heater(s), waste and vent pipes, fixtures and traps, the building sewer line(s), and the storm drain line(s) within or adjoining the building. A local sewage disposal system (septic tank or package treatment plant) must also be provided when a public sewer system is not available. The plumbing drainage, waste, and vent system is often referred to as the building's DWV system. Construction requirements for a building's plumbing system are discussed in Section 15-2.

The mechanical system for a small building consists primarily of the building's heating, ventilating, and air conditioning (HVAC) equipment and its supply lines and ducts. Thus, the specialty contractor involved is often referred to as the *HVAC sub*. The construction requirements for a building's HVAC system are discussed in Section 15-3.

Codes

Unlike the electrical system, there is no single dominant code governing U.S. plumbing and mechanical work. However, the plumbing and mechanical requirements for one- and two-family dwellings have been compiled and standardized in the CABO Code (Reference 1). Other major U.S. codes governing plumbing and mechanical systems are as follows.

PLUMBING

1. Basic/National Plumbing Code
2. Standard Plumbing Code
3. Uniform Plumbing Code

MECHANICAL

1. Basic/National Mechanical Code
2. Code for the Installation of Heat Producing Appliances
3. Standard Gas Code
4. Standard Mechanical Code
5. Uniform Mechanical Code

Temporary Services

The temporary plumbing services required for the construction of a building consist primarily of domestic water supply and sanitary waste disposal. Temporary water supply is usually provided in a manner similar to that used for temporary electricity; that is, the local water company installs a supply line and water meter to the building site. The builder connects temporary lines and faucets for use during construction. When the building's plumbing system is complete, the temporary lines are disconnected, and the building's water supply line is connected to the water meter.

Sanitary waste disposal during construction is normally provided by placing portable toilets at the work site. These are commonly supplied and serviced by a company specializing in such services. Solid waste and hazardous liquid waste is collected by the builder and hauled to an approved waste disposal site by the builder or a waste disposal service.

Temporary mechanical services most often involve heating of the building during cold weather. When required, heat is normally provided by portable space heaters. Curtains of canvas or plastic may be used to temporarily enclose and weatherproof the building when necessary. When fuel-fired heaters are employed in enclosed spaces, care must be taken to provide adequate ventilation to avoid the buildup of carbon monoxide, a deadly gas. Temporary cooling equipment is rarely required during building construction. However, when required it may be supplied by portable refrigeration or air conditioning equipment used in a manner similar to that described for heating.

15-2 PLUMBING

Fundamentals

A building's water supply system must provide water at a satisfactory pressure in a quantity sufficient to meet the needs of the building's occupants and equipment. Likewise, the building's drainage, waste, and vent (DWV) system must be adequate to drain all fixtures, appliances, and equipment without overflow or clogging. Of particular concern is the prevention of backflow in the water supply system. *Backflow* is the flow of water or other liquids into the water supply system through a faucet or other outlet. Backflow may be prevented by maintaining at least a 1 in. (25 mm) vertical air gap between a water supply outlet and the maximum water level that can occur in the fixture or appliance. For outlets below the normal or maximum water level, backflow protection can be provided by inserting a vacuum breaker device into the water supply line near the outlet.

For hydraulic flow calculations, fluid pressure is commonly expressed in feet (m) of head. *Head* is the height of a column of water that will produce a pressure equal to that actually measured at a specific point. The calculation of head must include both the vertical elevation of the point (elevation head) and the pressure loss due to friction (friction head) which occurs in the pipe between the source and the specified point. Charts and procedures for calculating head and pipe flow friction losses are contained in Reference 6. Such calculations are required for independent water supply systems such as those used at isolated construction sites. However, building piping systems are commonly sized by using *water supply fixture units* (w.s.f.u.) as demand for water supply lines and *drainage fixture units* (d.f.u.) as demand for drain, waste, and vent (DWV) systems. Plumbing codes contain specific requirements for sizing building plumbing systems as discussed in the following paragraphs.

Other factors that must be considered in the installation of plumbing systems include pipe support, protection from breakage of pipes passing through or under walls and foundations, and the installation of buried pipes running parallel to and below the level of foundations. Requirements for these items are normally contained in plumbing codes, but the following general rules are suggested. The maximum spacing of supports for horizontal pipe should not exceed the following distances: rigid plastic water supply pipes, 3 ft. (0.9 m); plastic DWV pipe, 4 ft. (1.2 m); cast iron soil pipe, 5 ft. (1.5 m) for 5 ft. (1.5 m) lengths and 10 ft. (3 m) for 10 ft. (3 m) lengths; copper pipe and tubing, 6 ft. (1.8 m); threaded steel pipe, 10 ft. (3 m). The maximum spacing of supports for vertical pipe should not exceed the building story height. In addition, midstory guides should be provided for vertical plastic pipe. Buried pipe running parallel to and deeper than the bottom of footings or foundation walls should be placed as illustrated in Figure 15-1.

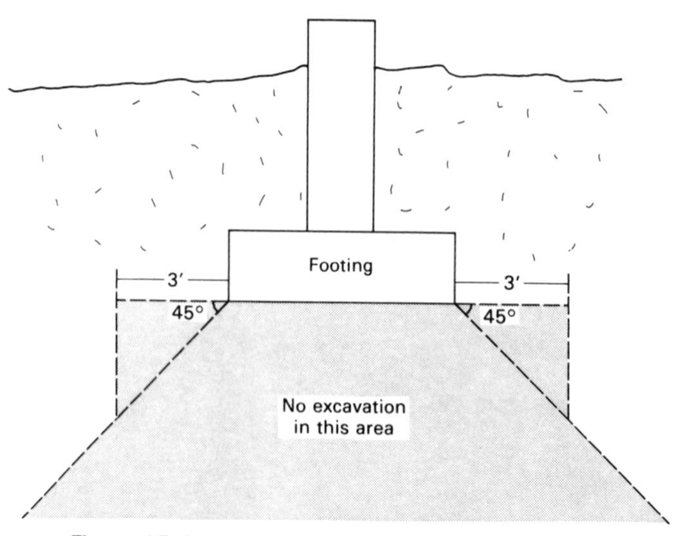

Figure 15-1 Location of buried pipes near footings.

Water Supply Systems

Basics. A building's water supply system is made up of the water service line, the water distribution lines, and the water heater(s). The *water service line* extends from the utility company's *water main* to the building. *Water distribution lines* supply cold and hot water to the building's plumbing fixtures and equipment. The water pressure at the point where the water supply line enters the building should be between 40–80 psig (276–552 kPa). If the pressure exceeds 80 psig (552 kPa), a pressure-reducing valve should be installed to reduce the pressure in the water distribution system.

A knocking or vibrating noise often occurs when the flow of water in a pipe is suddenly interrupted by closing a valve or faucet. This phenomenon is called *water hammer.* Water hammer can be prevented by placing an air chamber in the supply line near each valve or faucet. Although manufactured air chambers are available, common practice is to fabricate an air chamber by extending and capping the water supply line above the point where it joins the fixture branch line (see Figure 15-2). This type of air chamber should consist of at least a 12 in. (30 cm) length of 1/2 in. (13 mm) or larger pipe.

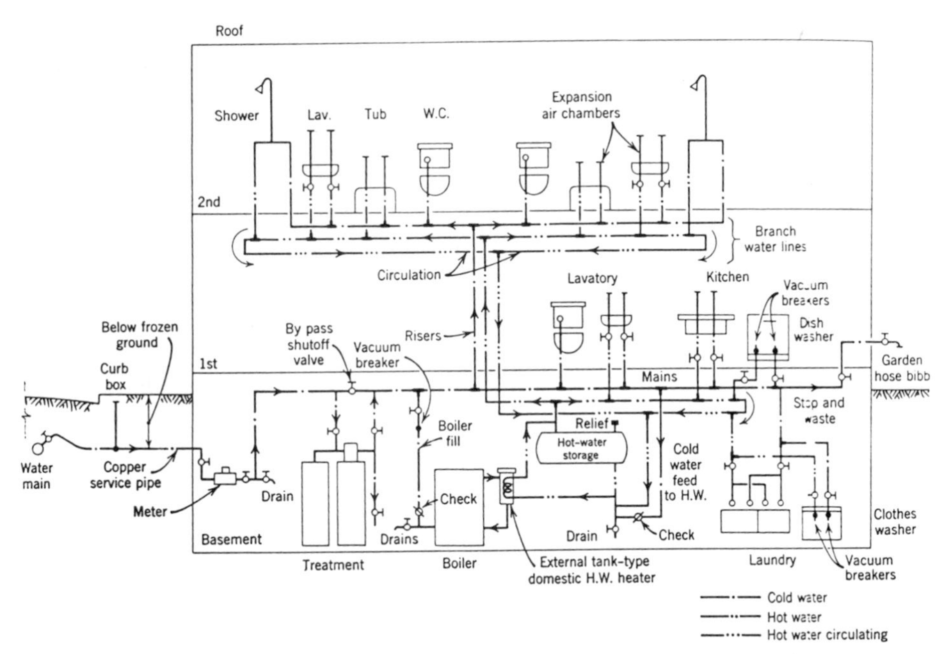

Figure 15-2 Typical building water distribution system (from *Building Technology: Mechanical & Electrical Systems* by William J. McGuinness and Benjamin Stein, © 1977 by John Wiley & Sons, Inc. Reproduced by permission).

Pipe Materials. The pipe materials available for water service and water distribution lines include:

1. Copper pipe or tube
2. Galvanized steel
3. Plastic pipe, including:
   ABS (acrylonitrile-butadiene-styrene)*

*Use for cold water lines only.

TABLE 15-1
Minimum Size of Water Distribution Pipe

Pipe function	Minimum nominal pipe size in.	Minimum nominal pipe size mm
Service line	3/4	19
Fixture branch lines:		
Dishwasher, lavatory, or tank-type toilet	1/4	6
All others	3/8	10

CPVC (chlorinated polyvinyl chloride)
PB (polybutylene)
PE (polyethylene)*
PVC (polyvinyl chloride)*

Copper pipe is available as Type K, Type L, and Type M. Type K is heavyweight, Type L is medium weight, and Type M is lightweight. Copper pipe is joined by sweating it to fittings. *Sweating* is the process of heating the joint while adding solder to the joint. Steel pipe is joined by screwing the threaded connections together. Plastic pipe is joined by glueing the pipe and fittings together or by using special clamps. No joints should be allowed in buried PB plastic pipe. The minimum working pressure rating should be 160 psig (1103 kPa) at 73°F (23°C) for cold water pipe and 100 psig (690 kPa) at 180°F (82°C) for hot water pipe.

Pipe Size. The minimum pipe size required for water service and distribution lines is determined by the required water flow expressed in gallons per minute (gpm), liters per minute (l/m), or water supply fixture units (w.s.f.u.). Plumbing codes normally contain specific requirements for sizing such lines. The CABO code (Reference 1), for example, presents both a simplified method and an engineering analysis method for sizing water supply lines. To use the simplified method, the water pressure at the water service must be at least 40 psig (276 kPa), and the elevation of the highest plumbing fixture must not be more than 25 ft. (7.6 m) above the service valve.

The basic procedure for selecting pipe size is to work upstream (toward the main) from the most distant fixture, calculating a cumulative total demand for each pipe. However, to account for the fact that it is unlikely that every fixture or outlet will be used simultaneously, a demand factor is applied to the sum of the individual demands for a group of fixtures. For example, the expected (or probable) demand for one bathroom (consisting of tub with shower, lavatory, and toilet) is only 4.3 w.s.f.u. versus 5.2 w.s.f.u. obtained by adding the individual demands of these fixtures. Likewise, the expected demand for one bathroom is 4.3 w.s.f.u. while the expected demand for a two bathroom group is only 5.2 w.s.f.u.

After determining the expected demand (w.s.f.u. or gal./min.) for each pipe, the minimum pipe size for each line is determined from appropriate tables or charts. Since pipe friction factors vary with the type and condition of the pipe, be sure to use the chart or table for the pipe material being used. In any case, the pipe size should not be less than the values shown in Table 15-1. Also, the water pressure at any outlet should not be less than 8 psig (55 kPa).

System Layout. A diagram of a portion of a typical water distribution system for a dwelling is shown in Figure 15-2. Notice that the 3/4 in. (19 mm) cold

*Use for cold water lines only.

water line extends from the service line to the hot water heater and continues on until branch lines are encountered. Likewise, a 3/4 in. (19 mm) hot water line extends from the water heater to the first branch line.

Water Heaters. A supply of hot water is required for every dwelling and virtually all other buildings. For dwelling and other small buildings, hot water is normally supplied by a *water heater.* The usual water heater consists of a storage tank containing a heating element(s). Energy for the heater may be supplied by electricity, gas, or oil. Common sizes of domestic water heaters range from 30 gal. (114 l) to over 120 gal. (454 l). The minimum size of the water heater is determined by the anticipated hot water demand (or family size for a dwelling unit). For larger buildings, separate hot water boilers and storage tanks are often used.

Hot water heaters, boilers, and storage tanks must be equipped with a pressure relief valve and a temperature relief valve, or a combination pressure/temperature relief valve. Pressure relief valves must open at 150 psig (1034 kPa) or less and temperature relief values at 210°F (99°C) or less. No cutoff or check valve may be installed between the heater or tank and its relief valve.

A typical installation of an electric hot water heater is illustrated in Figure 15-3. Notice the installation of the pressure/temperature relief valve and its discharge line, as well as the cutoff valve for the cold water supply line.

Drainage, Waste, and Vent Systems

Basics. A building's drainage, waste, and vent (DWV) system consists of the drain lines serving plumbing fixtures and equipment and connecting them to the

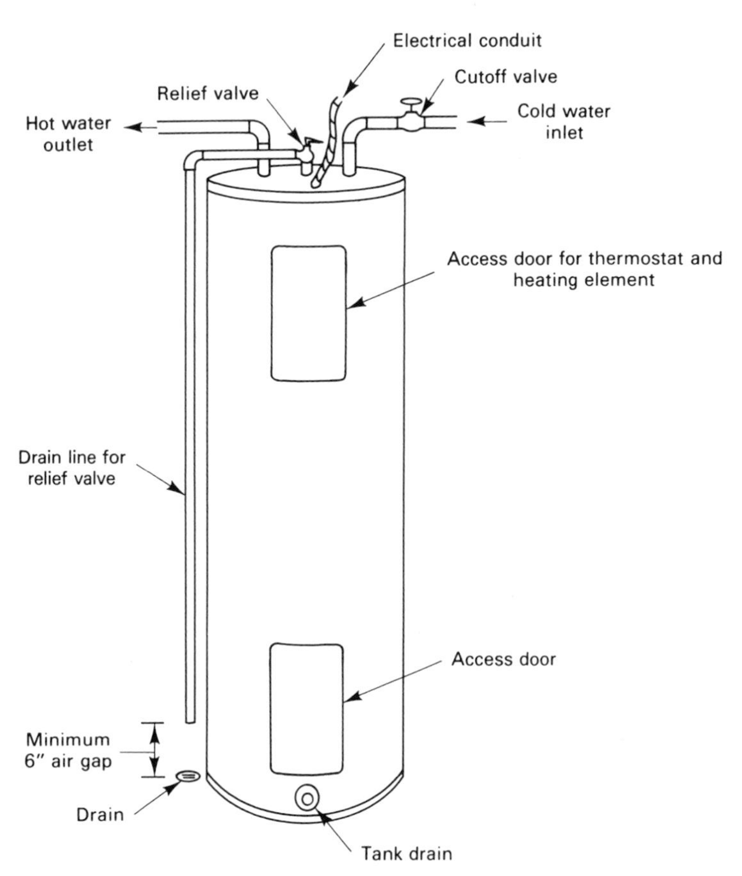

Figure 15-3 Typical installation of electric hot water heater.

public (or private) sewer system, as well as the vent system serving these lines. Types of drainage lines include soil, waste, and storm drains. *Soil pipes* handle sewage that contains human waste (fecal matter). *Waste pipes* handle liquid wastes that do not contain fecal matter; that is, the effluent from sinks, lavatories, showers, etc. *Storm drains* handle the runoff from rainfall. Many plumbing codes prohibit the introduction of storm water into a building's drain system and require a separate storm drain system. Thus, the combined drain system illustrated in Figure 15-4 may not be allowed.

Horizontal drain lines for a building include building sewers, building drains, branch drains, and fixture drains. A *building sewer* or *house sewer* extends from the building drain to the public sewer line (see Figure 15-4). The main horizontal drain line within a building is the *building drain* or *house drain*. A *fixture branch* or *horizontal branch drain* extends from a plumbing fixture drain to the intersection of the branch with another drain or stack.

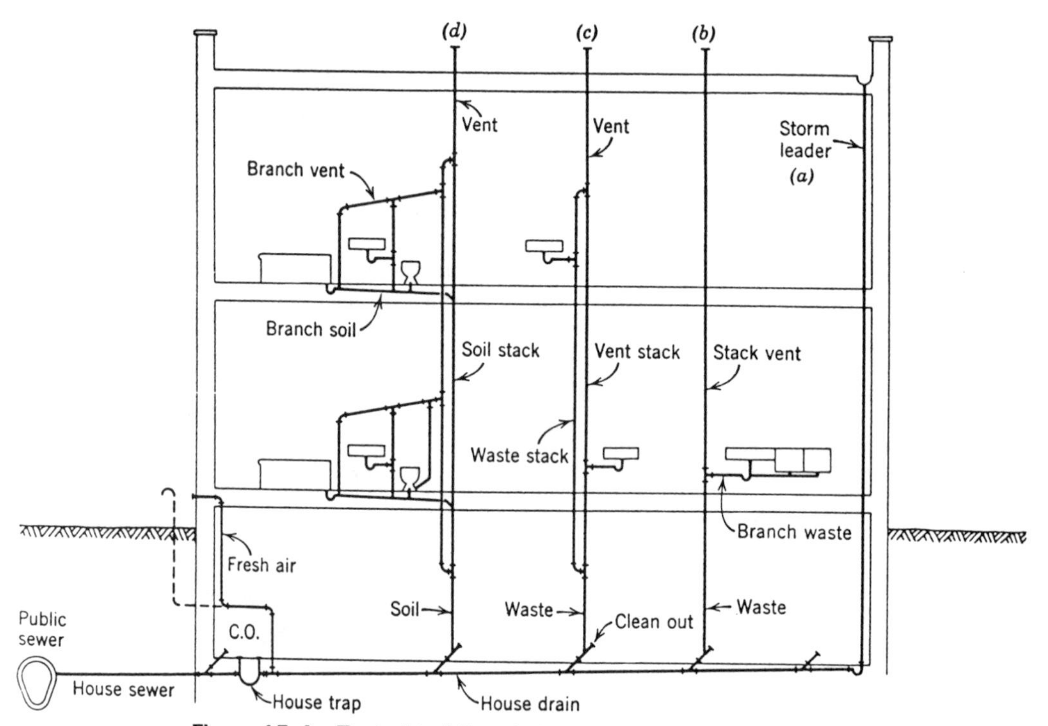

Figure 15-4 Typical building drainage system (from *Building Technology: Mechanical & Electrical Systems* by William J. McGuinness and Benjamin Stein, © 1977 by John Wiley & Sons, Inc. Reproduced by permission).

Most vertical DWV lines are known as *stacks*. They include soil stacks, waste stacks, and vent stacks. A *soil stack* handles liquid waste containing fecal matter; a *waste stack* handles liquid waste that does not contain fecal matter; and a *vent stack* is a vertical pipe that serves as a vent for the DWV system and extends through the roof. A vent stack which is located above the highest horizontal drain and is an extension of a soil stack or waste stack is a *stack vent*. A vent pipe that connects the drainage system with a stack vent or vent stack is as *branch vent*. A *wet vent* is a soil or a waste pipe that also functions as a vent. A *dry vent* is a vent pipe that does not carry liquid waste.

A plumbing *trap* prevents the flow of DWV system gases into a building. As illustrated in Figure 15-5, a trap basically consists of a liquid-filled U-shaped section

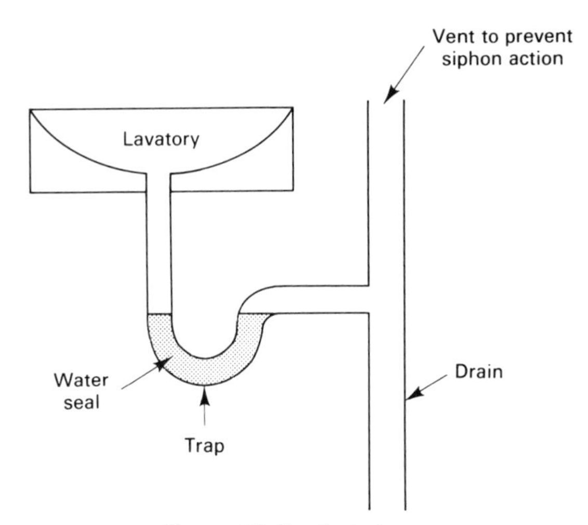

Figure 15-5 A drain trap.

of drain. A *trap seal* of 2–4 in. (5–10 cm) of liquid should be provided as illustrated. Some plumbing fixtures such as toilets contain an integral trap. However, most traps are separate devices inserted in a drain line. The vent system serves to prevent traps from being drained by siphon action. It also prevents a buildup of pressure in the drain system that could force gases back through the trap. In addition to fixture traps, a *house trap* and fresh air inlet may be required on the house drain line as illustrated in Figure 15-4. If a storm drain is connected to the building drain as illustrated in Figure 15-4, a trap must be installed at the junction of these two lines as shown.

Cleanouts for the drain system must be provided in accessible locations as required by code. Some common required locations for cleanouts include:

1. At each change of direction greater than 45° in a drain line.
2. At the base of each soil or waste stack.
3. In horizontal drain lines at a designed maximum spacing.
4. At the junction of the building drain and building sewer.

Materials. The pipe materials available for DWV systems include:

- Cast iron pipe
- Concrete pipe*
- Copper tube, Type DWV
- Galvanized steel pipe**
- Plastic pipe, including:
  ABS (acrylonitrile-butadiene-styrene)
  PVC (polyvinyl chloride)
- Vitrified clay pipe*

Pipe joint connections must be made using a method approved for the type of pipe involved. DWV pipe should not be burned, drilled, tapped, or welded.

*Building sewer only.

**Above-grade only.

Pipe Size. The minimum pipe size required for DWV lines is determined by the expected flow expressed in drainage fixture units (d.f.u.). Plumbing codes contain specific requirements for sizing drain and vent lines. The basic procedure for selecting pipe size is similar to that used for sizing water distribution pipe. Work downstream (toward the sewer) from the highest floor and most distant fixture calculating the cumulative expected load for each pipe. After determining the expected load, determine the minimum pipe size from appropriate tables or charts. The minimum size of building drain and building sewer lines is also governed by the slope of the line, which can range from 1/8 to 1/2 in. per foot (10 to 42 mm per m). Common minimum pipe size requirements are shown in Table 15-2.

TABLE 15-2
Suggested Minimum Size of DWV Pipe

	Minimum nominal pipe size	
Pipe function	*in.*	*mm*
Below-grade line	1 1/2	38
Single-stack soil stack	3	76
Building sewer	4	102

All horizontal drain pipe should run at a uniform slope of not less than 1/4 in./ft. (21 mm/m) for 3 in. (7.6 cm) or smaller pipe, and 1/8 in./ft. (10 mm/m) for 4 in. (10 cm) or larger pipe. Buried pipe placed parallel to and below the bottom elevation of foundations or footings should be placed as shown in Figure 15-1. Most codes prohibit the installation of exposed DWV lines outside the building unless the minimum design temperature is above 32°F (0°C).

Installation

The installation of plumbing that can be installed prior to the installation of drywall and plumbing fixtures is known as plumbing *rough in, roughing-in, rough plumbing,* or *roughing*. Figure 15-6 shows the plumbing rough in for the laundry room of a single-family dwelling. On the right, notice the copper water distribution lines and faucets (hose bibbs) for connecting a washing machine, as well as the plastic drain line with trap for the washing machine drain hose. On the left are shown the copper water supply lines and plastic drain line for the later installation of a laundry sink. The fixture drain lines connect with a plastic waste stack and stack vent in the center.

After the installation of drywall, plumbing fixtures are installed and connected. Next, the building water supply system is connected to the water meter, and the building sewer is connected to the public sewer. The plumbing system is now ready for final testing. Before final inspection, check to ensure that all vent stacks penetrating the roof have been properly sealed to the roofing to make a watertight joint.

15-3 HEATING, VENTILATING, AND AIR CONDITIONING

Fundamentals

The mechanical systsems used to create a comfortable climate-controlled environment within a building are commonly referred to as HVAC (heating, ventilating, and air conditioning) systems. The requirements for occupant comfort as well as

Figure 15-6 Plumbing rough in.

the energy efficiency of buildings are discussed in Chapter 12. In this chapter we will look at the components of the HVAC system and their installation.

The principal elements of an HVAC system include heating equipment, cooling equipment, ventilating equipment, and their supply lines and ducts. Care must be taken in the use of the term "air conditioning." Technically, it means the control of the temperature and humidity of the air within a building to produce the desired level of occupant comfort. However, the term is also often used to describe only the cooling component of an HVAC system. A typical forced-air central air conditioning system for a house is illustrated in Figure 15-7. Notice that equipment for both heating and cooling is incorporated in the system.

Heating

The two principal types of heating systems are *hot water systems* and *forced air systems.*

Hot Water Systems. In a hot water heating system, a *boiler* is used to heat water. From the boiler, hot water flows through a piping system to heat-transfer devices installed in each room. These devices heat the air within the room. Either gravity flow or forced-water flow may be used to circulate the water within the heating system. After giving up heat in a heat transfer device, the cooler water returns to the boiler where it is reheated and recirculated. Other system components include expansion tanks, vents, pumps, thermostats, and a makeup water supply.

Available types of heat transfer devices include radiators, convectors, and unit heaters. A *radiator* or *radiant heater* primarily transfers heat to objects in the room by radiation. However, it also heats the air within the room by convection. The principal types of radiators include cast iron wall units and cast iron baseboard units. A *convector* uses a heat transfer tube enclosed in a cabinet or other enclosure to transfer heat to the surrounding air. Common types of convectors include baseboard units and cabinet units. A *unit heater* is a convector that contains a fan for

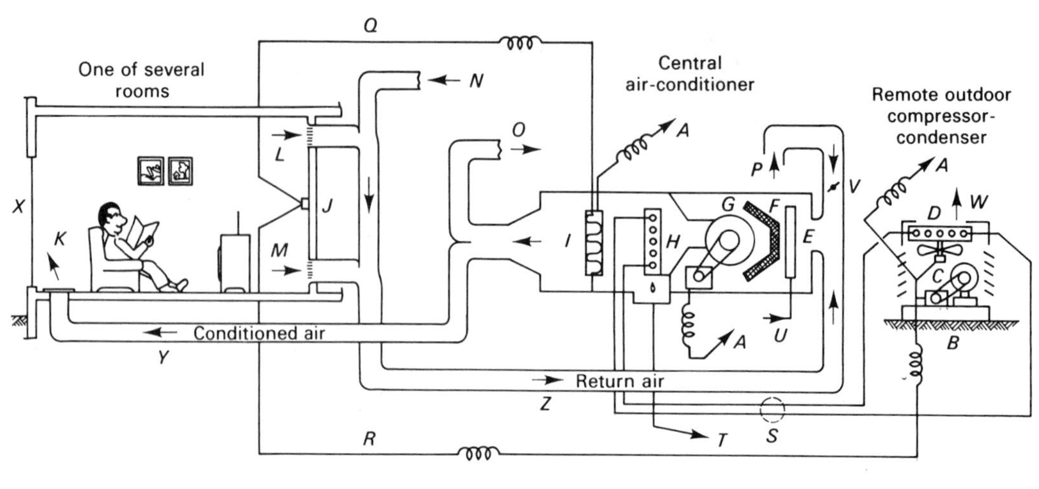

A	Wiring to electrical power panel
B	Compressor, compresses the freon gas
C	Blower, passes outdoor air over the condenser
D	Condenser, liquifies the freon gas
E	Humidifier, winter only
F	Filter, must be accessible for cleaning
G	Blower, circulates all air
H	Evaporator coil, freon evaporates (liquid to gas) absorbing heat from the air
I	Electric resistance heating element
J	Thermostat (summer-winter) best located on interior wall and near the return grills
K	Supply register, with deflecting vanes and volume damper
L/M	High and low return grills for reconditioning. *L* draws back warm air that accumulates there in summer. *M* returns cool air that settles there in winter.
N	Return air from other rooms
O	Supply air to other rooms
P	Outdoor fresh air, reduces odors
Q	Electric signal, controls resistance heating unit
R	Electric signal, controls compressor/condenser for cooling
S	Refrigerant tubing, to and from the evaporator coil, *H*
T	Plumbing drain, carries away the room moisture that condenses on the cold evaporator coil
U	Water supply to humidifier
V	Damper, regulates rate of fresh air intake
W	Hot air discharge from condenser, sometimes referred to as "heat rejection"
X	Glass, double (insulating) when budget permits
Y	Insulated supply duct
Z	Return duct

Figure 15-7 Air conditioning system for a house (from *Building Technology: Mechanical & Electrical Systems* by William J. McGuinness and Benjamin Stein, © 1977 by John Wiley & Sons, Inc. Reproduced by permission).

circulating the heated air. Unit heaters may be mounted on the floor, hung from the ceiling, or mounted in a recess in the ceiling, wall, or floor.

Forced-Air Systems. In a forced-air system, a fan is used to circulate air heated by a *furnace* (or the coil of a heat pump) to each room through large tubes called *ducts* (see Figure 15-7). Heated air is transferred from the furnace to each room by *supply ducts,* and *return ducts* bring cooler air back to the furnace for reheating. Ducts are commonly fabricated from sheet metal, but ducts of fiber glass lined with foil are coming into increasing use. Both rigid and flexible ducts are available. Ducts located in attics or other nonconditioned (unheated) space should be insulated to reduce energy loss. Mechanical codes commonly prescribe allowable duct materials, installation requirements, jointing methods, and required insulation.

The energy source for a furnace may be electricity, fuel oil, natural or liquified gas, or coal. Electric furnaces use an electric heating coil as a heat source, making such furnaces extremely simple. However, they are expensive to operate. Fuel-burning furnaces require a fuel supply system, a source of combustion air, and a venting system to remove combustion products. Again, codes normally prescribe minimum requirements for furnace installation, fuel systems, combustion air, and venting of combustion gases.

An electric *heat pump* is essentially a compressor-type cooling unit, which can operate in reverse so that it can supply either heating or cooling. As a heating unit, a heat pump is much more energy efficient than an electric furnace. Referring to Figure 15-7, a heat pump system would simply replace the inside and outside units of the cooling unit shown. When operated as a heater, the evaporator coil (*H*) now functions as a condenser to heat the conditioned air. The outside condenser (*D*) functions as an evaporator coil to chill the discharge air (*W*). The electric heating element (*I*) is normally retained in a heat pump system to supply additional heat during extremely cold weather.

Cooling Systems

The principal cooling system employed for small buildings is the forced-air cooling system. If the heating element shown in Figure 15-7 were removed, this system would be a forced-air cooling system. The components of a forced-air cooling system are similar to those of the forced-air heating system described earlier except that an *air handling unit* with an *evaporator coil* replaces the furnace. A forced-air heating/cooling system employing a heat pump also utilizes an air handling unit. However, other types of forced-air heating/cooling systems usually mount the cooling evaporator coil in a housing attached to the furnace. The furnace then contains the fan used to circulate the conditioned air.

Chilled water cooling systems are also available, but are seldom used in small buildings. A chilled water cooling system is similar to the hot water heating system previously described except that a *chiller* replaces the boiler. Chilled water is circulated to heat transfer units located in each room to cool room air.

Ventilation

The various requirements for building ventilation are described in Section 12-4. Duct systems (which include ducts, duct fittings, dampers, fans, plenums, and other accessories) are used to transmit and circulate air for ventilation purposes. The requirements for the supply of combustion air and for venting combustion gases are discussed below.

Combustion Air Supply. An adequate supply of combustion air must be available for fuel-burning furnaces and other appliances. For single dwelling units, this air supply is commonly taken from the air within the unit and replaced by infiltration. If the building is tightly sealed, infiltration may not be sufficient to provide adequate air for both combustion air and occupant health. In this case, a supply of outside air may be required (see Section 12-4). In any case, the use of a separate outside supply of combustion air to the appliance is usually energy efficient.

When fuel-burning appliances are located in a confined area such as a closet or utility room, vent openings into other space may be required. Typically, if the room housing the appliance has a volume of less than 50 cu. ft. (1.4 m^3) per 1000 BTU/h (293 W) of appliance input energy, an air supply vent is required. At least 2 sq. in. (12.9 cm^2) of vent area should be provided per 1000 BTU/h (293 W) of input energy, with a minimum vent area of 200 sq. in. (1290 cm^2).

Venting of Combustion Gases. The requirements for venting combustion gases to the outside atmosphere are normally specified by code. The major types of vents include:

1. Chimneys.
2. Type B Gas Vent: Used for most gas furnaces and other gas appliances.
3. Type BW Gas Vent: Used for designated gas wall furnaces.

4. Type L Vent: Used for oil-burning appliances and designated gas appliances.
5. Plastic Pipe: Used only for condensing appliances; must use plastic pipe approved for this purpose.

All vent systems must meet code requirements for size, materials, clearance, pitch, length, and access.

Sizing HVAC Systems

The selection of the proper size for heating and cooling equipment and the design of system ducts or pipes is beyond the scope of this chapter. The heat gain or loss in a building depends on the thermal resistance of the building envelope, the design interior and exterior temperatures, the amount of air infiltration, and the heat generated by building occupants and equipment. Many of these factors are discussed in Chapter 12. The end-of-chapter references provide HVAC system design procedures and example calculations.

Installation

The installation of HVAC ducts, supply lines, and accessories that can be installed prior to the installation of drywall is known as HVAC rough in. It is similar to electrical and plumbing rough in. The installation of a conditioned air supply duct in the attic of a house is shown in Figure 15-8. Notice the location of the electrical cable also visible in the photo. Care must be taken to minimize interference with electrical and plumbing lines when locating HVAC ductwork and accessories.

After the installation of drywall, registers or grilles are installed at the end of supply and return ducts. After all HVAC equipment has been installed and connected, the system is tested to ensure proper performance.

Figure 15-8 HVAC duct rough in above ceiling.

15-4 CONSTRUCTION PROCEDURES

Scheduling

The problems associated with the scheduling of plumbing and mechanical work are similar to those encountered in electrical construction (see Section 14-4). A suitable period of time must be available for the plumbing and mechanical contractors to efficiently accomplish their work without delaying project completion. At the same time, conflicts between the plumbing and mechanical workers and other trades must be minimized. The areas in which conflicts between trades are most likely to occur are similar to those listed in Section 14-4.

Construction Sequence

Plumbing. The principal steps in the installation of a building's plumbing system include:

1. Installation of buried water distribution lines and DWV lines within the building foundations.
2. Installation of buried plumbing lines outside the foundation lines.
3. Roughing in: installation of concealed pipe, fittings, traps, and accessories.
4. Installation of fixtures.
5. Connection of the building drain to the building sewer and of the building sewer to the public sewer.
6. Connection of the water supply line to the water meter.
7. Final testing of the plumbing system.

Mechanical. The principal steps in the installation of a building's HVAC system include:

1. Installation of buried ducts, supply lines, and conduit within the building foundation.
2. Installation of buried fuel supply lines outside the building foundation.
3. Roughing in: installation of concealed ducts, fittings, accessories, and pipes.
4. Installation of registers, grilles, and HVAC equipment.
5. Connection of all ducts, tubing, pipe, and other system components, as well as electrical lines.
6. Final testing of the HVAC system.

Quality Control

As with the electrical system, the builder is responsible for ensuring that the plumbing and mechanical systems are installed in accordance with the plans, specifications, applicable codes, and the standards of good workmanship.

Testing is an important element of quality control. Tests should be carried out by the specialty contractor, verified by the builder and an owner's representative, and approved by the local building inspector as required. Any test procedures detailed in project plans and specifications must be carried out as specified. The requirements of the plumbing code must be met in any case.

Plumbing. Plumbing codes usually require that the water supply system be tested under a water pressure of 100 psig (690 kPa) for 30 minutes without leakage. DWV systems should be tested twice: after rough plumbing and again after the installation of fixtures. Rough plumbing can be tested using either a water test or an air test. To run a water test, fill the system (or section being tested) with water to a height of 10 ft. (3 m) above the highest fitting. There should be no visual evidence of leakage after 15 minutes of testing. To run an air test, seal the system, apply a pressure of 5 psig (34 kPa), and remove the pressure source. There should be no loss of system pressure after 15 minutes. After fixtures have been installed and traps filled with water, test the system again by filling and draining all fixtures. There should be no visual evidence of leakage. A test for gas-tightness of the DWV system may also be required. When required, perform this test in accordance with the specified procedures.

Mechanical. The procedures for testing HVAC systems are specified by the system designer or equipment manufacturer or developed by the HVAC contractor. The local building official may prescribe additional tests.

Final Inspection. The final inspection should verify that all plumbing and mechanical fixtures and equipment conform to the specifications, are undamaged, are properly installed, and function as intended.

PROBLEMS

1. List four U.S. codes applying to building mechanical systems.
2. What temporary mechanical services are usually supplied during construction of a building?
3. What is "backflow" and why is it an item of concern in a water supply system?
4. Briefly describe "water hammer" and how it can be prevented.
5. What types of relief valves are required for boilers and hot water heaters?
6. Explain the difference between "soil" lines and "waste" lines.
7. Briefly explain the function and operation of a drain trap.
8. What unit of demand is used in sizing DWV lines?
9. Briefly explain the operation of a heat pump.
10. List the ventilation requirements for a building (see Section 12-4).

REFERENCES

1. Council of American Building Officials. *CABO One and Two Family Dwelling Code.* Falls, Church, VA.
2. Dagastino, Frank R. *Mechanical and Electrical Systems in Construction and Architecture.* Reston, VA: Reston, 1978.
3. Hettema, Robert M. *Mechanical and Electrical Building Construction.* Englewood Cliffs, NJ: Prentice-Hall, 1984.
4. Hornung, William J. *Estimating Building Construction.* 2nd ed. Englewood Cliffs, NJ: Prentice-Hall, 1986.
5. McGuiness, William J., and Benjamin Stein. *Building Technology: Mechanical and Electrical Systems.* New York: Wiley, 1977.
6. Nunnally, S. W. *Construction Methods and Management.* 2nd ed. Englewood Cliffs, NJ: Prentice-Hall, 1987.

chapter 16

Construction Tools and Equipment

16-1 HAND TOOLS

Hand tools have traditionally been the "tools of the trade" for the carpenter and other members of the building construction team. Some of the more common hand tools and their uses are briefly described in the following paragraphs. However, only instruction and extensive experience enables individuals to become skilled in the use of these tools. Powered versions of many of these tools are available and are described in Section 16-2. Many of these power tools have largely replaced their hand tool equivalent.

Measuring Tools

The measuring devices ordinarily used by the carpenter or builder are *rules* or *tapes*. Some common rules and tapes are illustrated in Figure 16-1. For many years, the *folding rule*, or *zig zag rule*, was the carpenter's principal measuring tool. Its usual length is 6 ft. (1.8 m). In recent years, the power-return pocket *tape rule* has largely replaced the folding rule. Tape rules are available with tape lengths of 8 ft. (2.4 m) to 25 ft. (7.6 m). Longer measuring tapes, or *long tapes*, use metal or plastic blades and are manually rewound. Their usual length is 50 ft. (15.2 m) or 100 ft. (30.5 m). They are primarily used for laying out building foundations, measuring long framing members, and locating walls and partitions.

Hand Saws

Some of the many available types of hand saws are illustrated in Figure 16-2. The traditional *hand saw* is available as a *rip saw*, a *crosscut saw*, or a *combination saw*, depending on the type of saw teeth employed. A saw's smoothness of cut and ability to cut a particular piece of wood is determined by the number of teeth per inch of saw and the set of the teeth. Teeth are *set* or bent outward to create a cut wider than

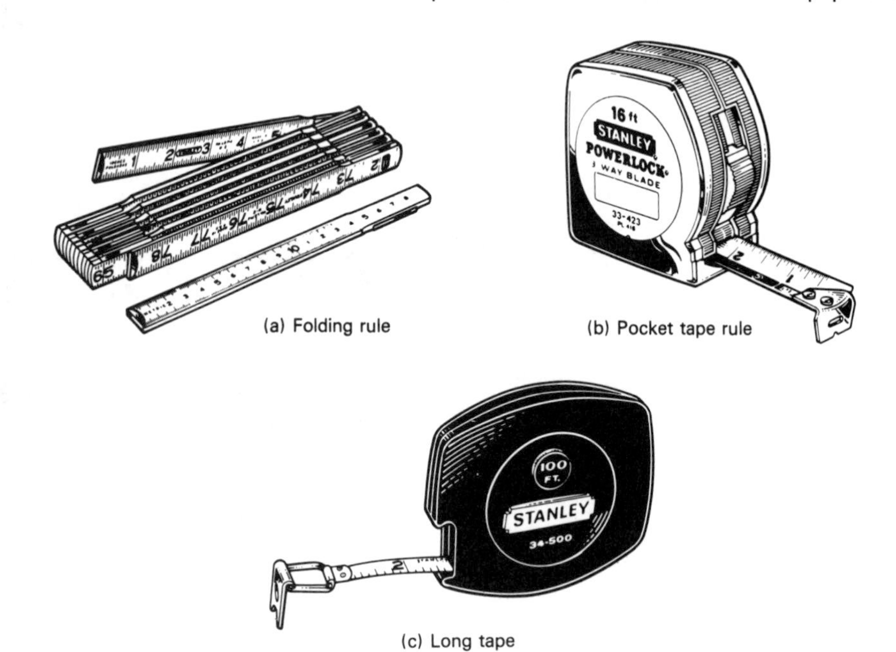

(a) Folding rule

(b) Pocket tape rule

(c) Long tape

Figure 16-1 Measuring tools (courtesy Stanley Tools).

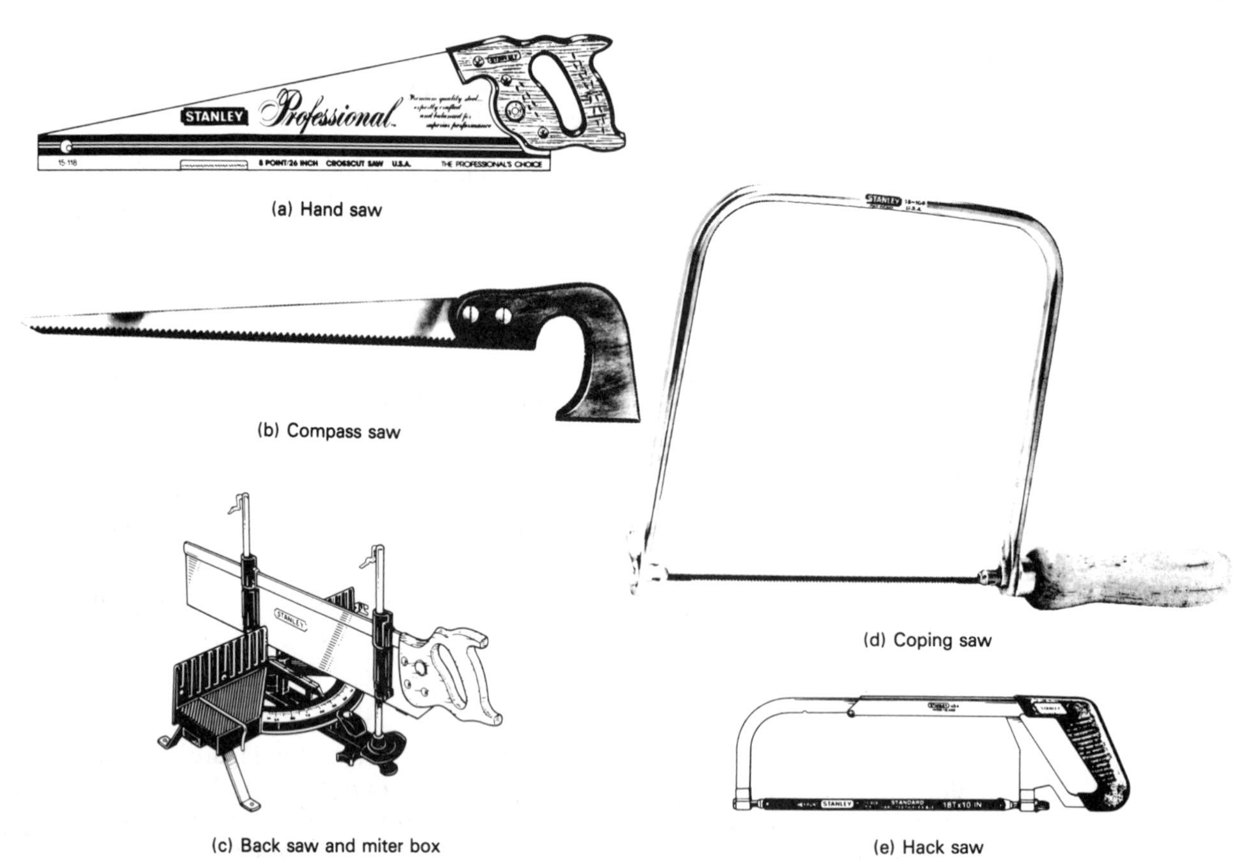

(a) Hand saw

(b) Compass saw

(c) Back saw and miter box

(d) Coping saw

(e) Hack saw

Figure 16-2 Hand saws (courtesy Stanley Tools).

the saw blade. This helps to prevent the saw from binding in the cut. The number of saw teeth per inch is usually expressed in "points per inch," which is actually one number larger than the number of teeth per inch. That is, a saw having 6 points per inch actually has 5 teeth per inch. The *rip saw* is designed to efficiently cut parallel to the wood grain (rip). It usually has 5 1/2 or 6 points per inch. The *cross-cut* saw is designed for fast cutting perpendicular to the wood grain (crosscut). It usually has 7 or 8 points per inch. *Combination saws* are suitable for either ripping or crosscutting. They often use 10 or more points per inch.

The *compass saw* or *keyhole saw* is used to cut small straight or curved openings in sheets of wood or gypsum board. A cut may be started by inserting the blade tip into a drilled hole. The *back saw* (also called a *tenon saw* or *miter saw*) uses a stiffener on its back or upper edge to keep the saw perfectly straight. It is used for making smooth, straight cuts, often in conjunction with a miter box. The *miter box* holds a piece of wood in position so that it may be cut at a precise angle. A simple fixed miter box uses slots incorporated into the box to permit cuts at an angle of 45° or 90°. An adjustable miter box places the saw in an adjustable guide frame so that accurate cuts may be made over a wide range of angles.

The *coping saw* is used to cut intricate designs in thin wood or light metal. The *hack saw* is used for cutting metal pipe, rods, bolts, and similar objects.

Hammers and Staplers

The *nail hammer* or *claw hammer* (Figure 16-3) is probably the carpenter's most basic tool. It is primarily used for driving and removing nails but can also be used to force framing members into alignment or to dismantle frame construction. The slotted curved end, or *claw,* is used to grip a nail and extract it as the hammer head is rotated across the wood surface. When extracting long nails, insert a small block of wood under the claw to raise the claw against the nail head and to provide greater leverage for extracting the nail. The *rip claw hammer* has a less curved claw than

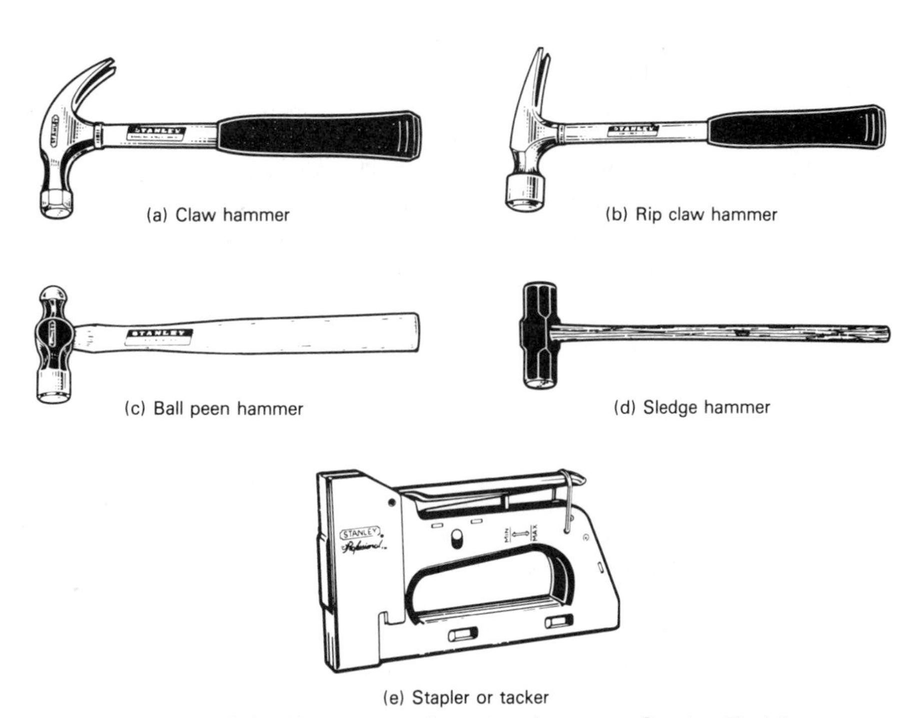

Figure 16-3 Hammers and staplers (courtesy Stanley Tools).

does a conventional claw hammer. This claw can be used as a ripping tool by driving the claw into the joint to be separated.

The *ball peen hammer* has one ball-shaped end (peen) and one conventional flat round face. The hammer is primarily used for riveting, chipping, and other metal work, although the flat face may be used for driving nails. The *sledge* or *sledge hammer* is a heavy hammer with a long handle. It is used for driving stakes and posts and for demolition work. A smaller, short-handle version of the sledge is also available and is used for driving masonry drills and for other heavy duty hammering.

Hand operated *staplers* or *tackers* are used to drive staples. Staples are coming into increasing use as fasteners for sheathing, decking, and other sheet material. Although power staplers are often used for such work, the hand stapler may also be used.

Squares and Levels

A *square* is a tool used for laying off or checking 90° angles. Some common forms of squares include the steel square, the try square, the bevel square, and the combination square, illustrated in Figure 16-4. The *steel square* or *framing square* is widely used for measuring and marking framing members. Its use for laying out rafters is described in Section 8-3. The *try square* is a small square with fixed blade. It is used for marking square cuts and for checking the squareness of board edges, faces, and ends. The *bevel square* has an adjustable blade, which can be locked at any desired angle. A *combination square* is a small square with a sliding blade. It can lay out 45° angles as well as 90° angles and is often equipped with a level vial.

Levels and plumb bobs are used to establish true horizontal and vertical planes. A *level* incorporates a vial and bubble to indicate when the edges of the instrument are aligned with the horizontal or vertical. Conventional levels are made of wood, metal, or plastic and are available in lengths from 9 in. (23 cm) to over 48 in. (122 cm). A *line level* is a small level that hooks onto a string line to indicate when a taut string line is level. A *plumb bob* is essentially a weight with a pointed end suspended by a string. It is used to locate a point directly below another point. The line suspending the plumb bob forms a *plumb* (vertical) line.

A *chalk line* is a string line coated with chalk. When stretched tautly between two points, pulled to one side, and released, it forms a straight line marked by chalk dust between the two points. The housing serving as a reel for the string line contains a reservoir for chalk dust, thus automatically coating the string line when it is reeled into the housing. The string line housing is often shaped so that it can also serve as as plumb bob.

Planes

Planes are used to smooth, level, or trim a wood surface (surfacing planes), create a groove or opening (grooving planes), or form a curved edge or molding (molding planes). The surfacing plane is most commonly used in building construction. It is available in several sizes and styles including block planes, all-purpose planes, and jointer planes illustrated in Figure 16-5. The *block plane* is a small plane about 7 in. (18 cm) long used for one-handed trimming. The *all-purpose plane,* a two-handed plane about 10 in. (25 cm) long, can be used for almost any type of surface planing. The *jointer plane,* about 22 in. (56 cm) long, is the largest of the surfacing planes. It is used for leveling and trueing large surfaces as well as for trimming the edges of doors and windows.

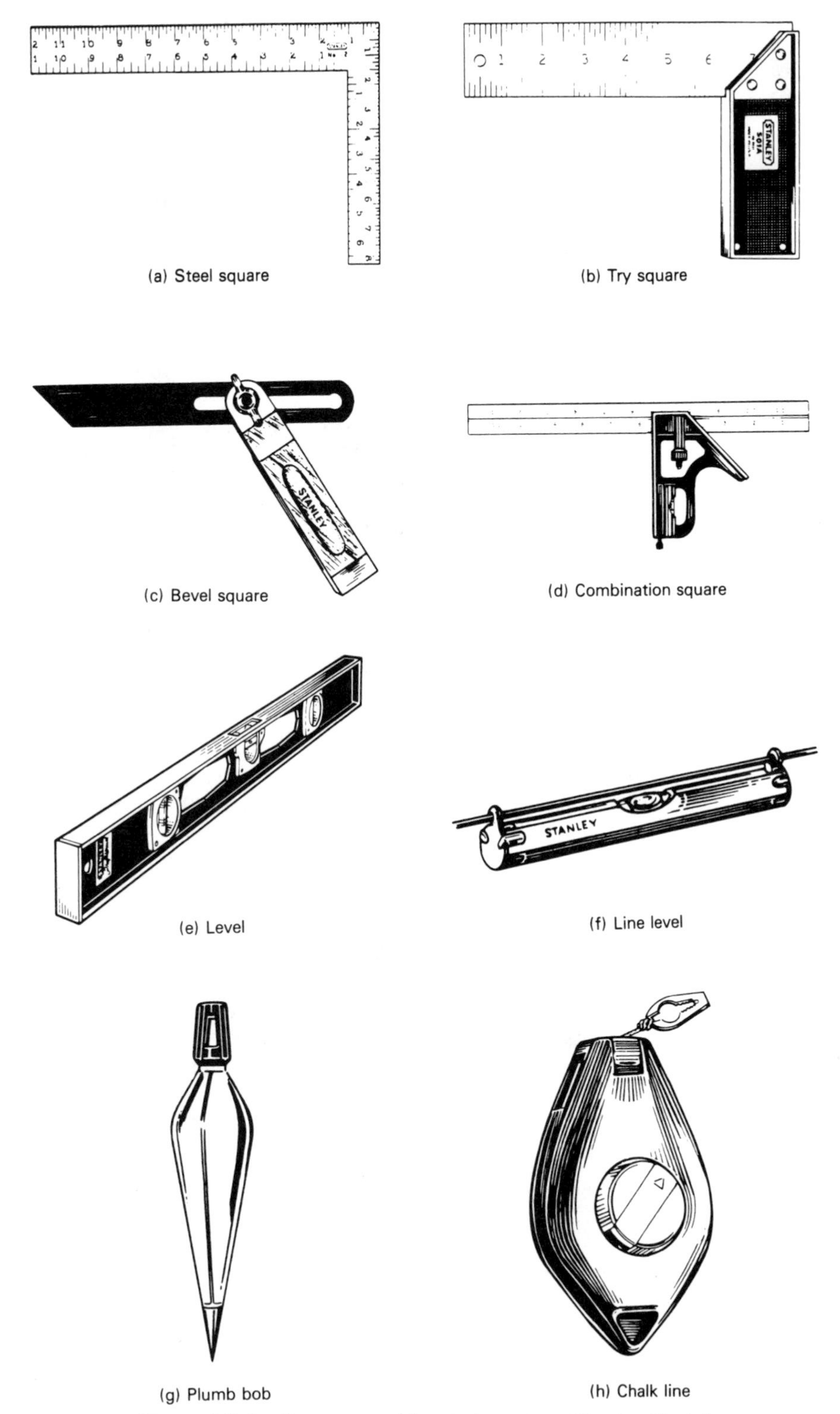

(a) Steel square
(b) Try square
(c) Bevel square
(d) Combination square
(e) Level
(f) Line level
(g) Plumb bob
(h) Chalk line

Figure 16-4 Squares and levels (courtesy Stanley Tools).

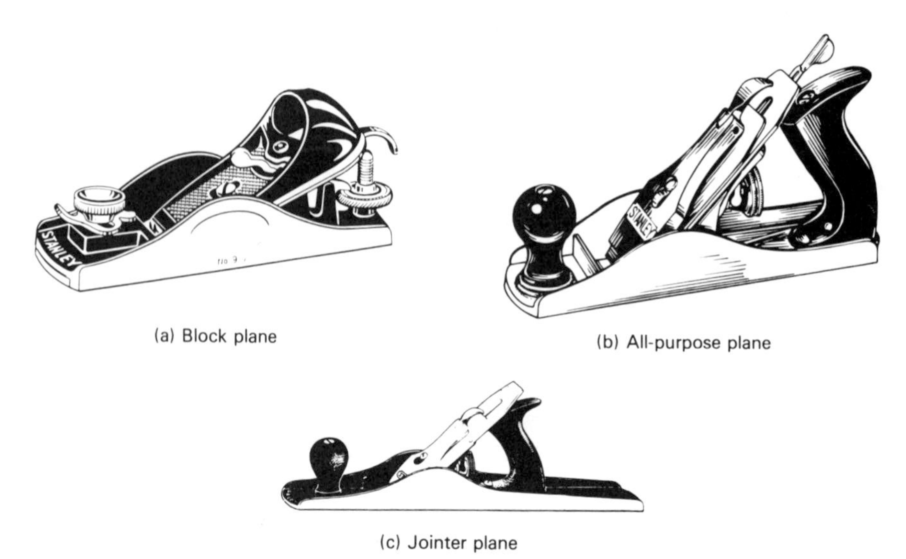

(a) Block plane (b) All-purpose plane

(c) Jointer plane

Figure 16-5 Planes (courtesy Stanley Tools).

Drills

The traditional carpenter's wood drill is known as a *bit brace* (Figure 16-6). It is used with an *auger bit* having a screw point and a square base (tang). The penetration of the bit into wood is accomplished by the screw point aided by pressure on the brace head. Auger bits are marked on their tang with a number indicating their size in 16ths of an inch (1.6 mm). Common auger bit sizes range from 1/4 in. (6 mm) to 1 in. (25 mm) and vary in length from 7 to 10 in. (18 to 25 cm). However, bits with extra length shanks are available. Adjustable *expansion bits* are available for boring holes ranging from 5/8 in. (16 mm) to 3 in. (75 mm) in diameter.

Hand drills and push drills are available but have been largely replaced by the power drill. The *hand drill* uses a crank to rotate a *twist drill bit,* which can be used to drill wood, metal, or plastic. Although commonly used with hand drills and power drills, they are available with a square tang for use in a bit brace. The handle of a *push drill* is pushed in and out to create an oscillating action for the special drill bits used.

Chisels

Common types of chisels (Figure 16-7) include wood chisels and cold chisels. *Wood chisels* are used for cuting grooves and notches in wood as well as for trimming and shaping. The wood chisel may be pushed with the hands for fine work but is more often struck with a hammer to create the cutting action. The *cold chisel* is designed for chipping and shearing metal when struck with a hammer.

Bars and Pullers

Bars and pullers (Figure 16-8) are used for heavy-duty nail extraction and for demolition of frame construction. Common types of bars include the wrecking bar and the rip bar. The *wrecking bar* (also called a *crow bar, pinch bar,* or *pry bar*) has a curved head with claw on one end and a chisel-shaped blade on the other end. The *rip bar* is smaller and shaped slightly differently than the wrecking bar. However, it is used for similar purposes.

Nail pullers extract nails more efficiently than do claw hammers. They are smaller than bars and are intended solely for extracting nails.

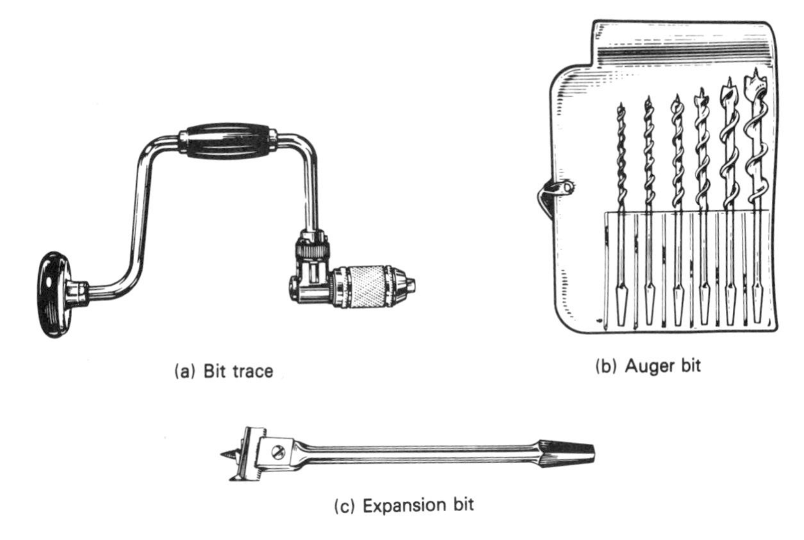

(a) Bit trace

(b) Auger bit

(c) Expansion bit

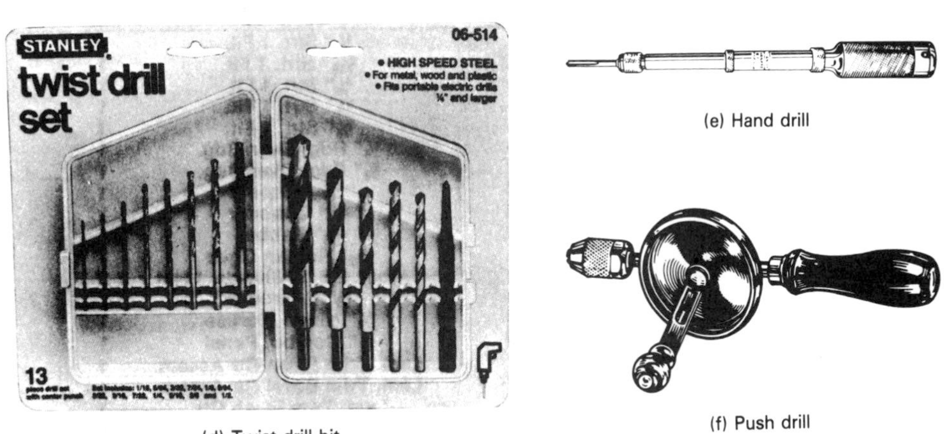

(d) Twist drill bit

(e) Hand drill

(f) Push drill

Figure 16-6 Drills (courtesy Stanley Tools).

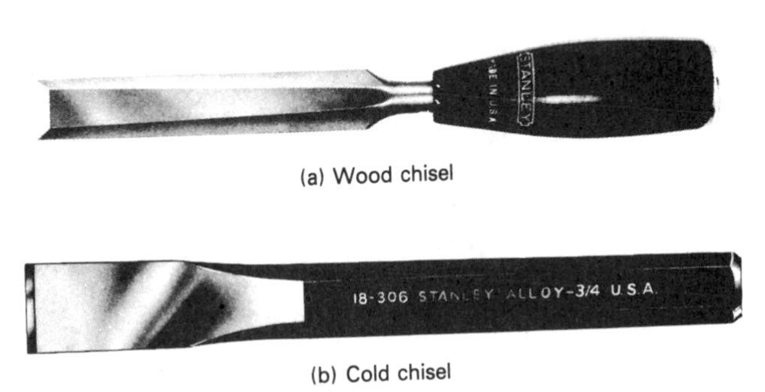

(a) Wood chisel

(b) Cold chisel

Figure 16-7 Chisels (courtesy Stanley Tools).

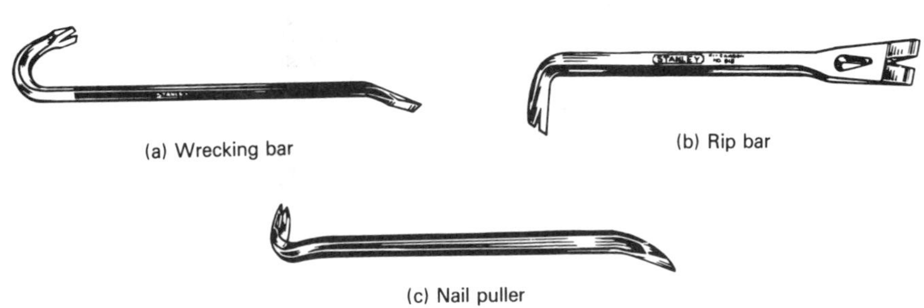

(a) Wrecking bar

(b) Rip bar

(c) Nail puller

Figure 16-8 Bars and pullers (courtesy Staney Tools).

Other Hand Tools

Other common hand tools often used by the builder are illustrated in Figure 16-9. These include pliers, wrenches, screwdrivers, nail sets, clamps, files, and rasps. Common *slip joint pliers* are versatile tools for holding objects as well as for turning nuts, bolts, and irregularly shaped objects. The jaws may be adjusted to two positions by turning one handle and moving it into the alternate position. Pliers having

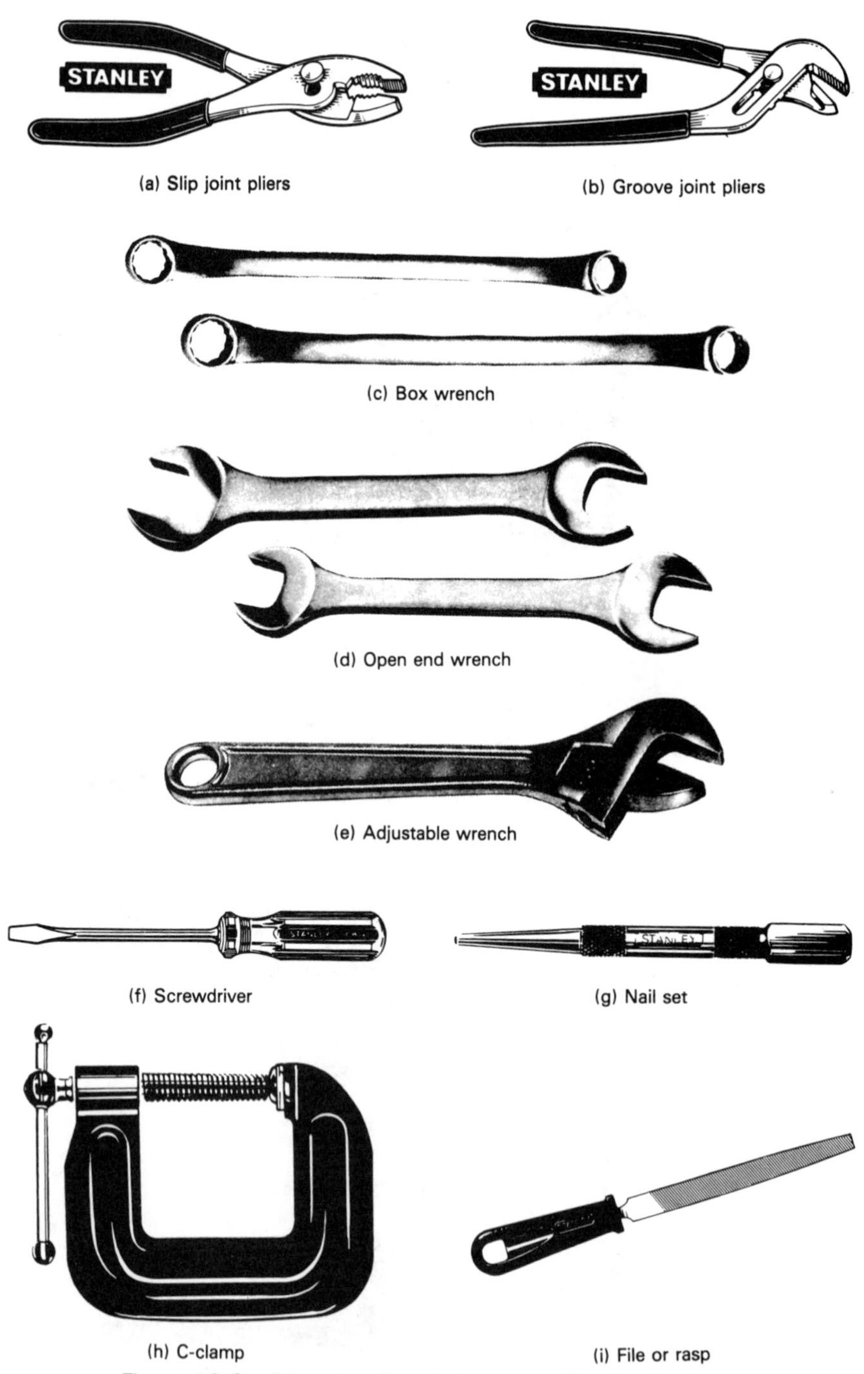

(a) Slip joint pliers
(b) Groove joint pliers
(c) Box wrench
(d) Open end wrench
(e) Adjustable wrench
(f) Screwdriver
(g) Nail set
(h) C-clamp
(i) File or rasp

Figure 16-9 Other hand tools (courtesy Stanley Tools).

the jaws placed at an angle to the handles are often called *groove joint pliers* or *waterpump pliers.* The jaw opening is adjustable to several positions using either a groove joint or a slip joint arrangement.

Wrenches, which come in a number of varieties, are also used for turning nuts, bolts, and other objects. The *box-end wrench* or *box wrench* has a closed end designed to fit hex head bolts and nuts. The *open end wrench* has an open slotted head designed to fit either square or hex head nuts and bolts. A *combination wrench* has one box end head and one open end head of the same size. An *adjustable wrench* has an adjustable open end head placed at an angle to the handle. A *pipe wrench* is a form of adjustable wrench designed to turn pipe or other smooth objects.

A *screwdriver* is designed to drive and remove screws. The two principal types are *slotted (or flat blade) screwdrivers* and *Phillips screwdrivers.* Slotted screwdrivers are designed to drive and remove screws having slotted heads. Phillips screwdrivers are designed to drive and remove Phillips head screws, which contain a cross-shaped recess. The Phillips head screw retains the screwdriver tip better than does a slotted head screw. It is also better suited to the use of a power screwdriver.

A *nail set* is a form of punch designed to drive finish nails below the wood surface. Nail sets come in several sizes to match nail size. They are used with a nail hammer.

Clamps, used for holding objects in position for drilling or glueing, come in a variety of types and sizes. The *C-clamp* illustrated in Figure 16-9 is probably the most widely used. Typical maximum opening size ranges from 1 in. (25 mm) to 12 in. (305 mm).

Files are used for sharpening tools and smoothing rough surfaces. A *rasp* is a file with coarse projections. Wood rasps are often used for removing material, enlarging openings, and for smoothing wood surfaces.

16-2 PORTABLE POWER TOOLS

Use of Power Tools

Powered versions of many of the hand tools described in the previous section are available. Power for these tools may be provided by electricity, compressed air, or gasoline engines. Electric power tools are most widely used because of their low cost and ease of handling. Pneumatic tools pose fewer safety hazards than do electric tools but require an on-site supply of compressed air. Pneumatic tools often used in building construction include staplers, nailers, and spray guns. Although many varieties of gasoline powered tools are available, only the gasoline powered chain saw is commonly used by the builder.

The hazard of electrical shock accompanies the use of electric power tools. Safety precautions should be observed in the use of such tools. Some electric tools are *double-insulated* to reduce the possibility of insulation failure within the tool. Such tools do not require electrical grounding. All other electric tools require the use of a grounded electrical outlet and a three-wire extension cord. Use only Underwriters Laboratory (UL) listed extension cords of adequate size and in good condition. If the cord is to be used outside or under damp conditions, ensure that it is approved for outdoor use. Such cords should carry the marking "WA" on the outside of the cord. Inspect extension cords regularly for damaged insulation, loose or exposed wires, and damaged connectors. Promptly repair or replace damaged cords. The minimum wire size for 120 volt tools rated at 10 amps or less should be No. 18 for lengths up to 50 ft. (15 m), No. 16 for lengths of 51–100 ft. (16–30 m), No. 14 for lengths of 101–150 ft. (31–45 m), and No. 12 for lengths of 151–200 ft. (46–61 m). For tools rated over 10 amps but not more than 15 amps, the minimum wire size should be No. 14 for lengths up to 50 ft. (15 m) and No. 12 for lengths of 51–175 ft. (16–53 m).

Safety goggles should be worn when operating any power cutting, hammering, sanding, or drilling tool. Do not remove tool guards or other safety devices when using a power tool. Be sure the tool has stopped turning and power is off before attempting to change or adjust blades or bits.

Drills and Screwdrivers

The portable *electric drill* (Figure 16-10) is among the most widely used builder's power tools. Variable speed, reversible drills may be used for a variety of tasks, including driving and removing screws. However, the *power screwdriver* is better suited to driving and removing screws.

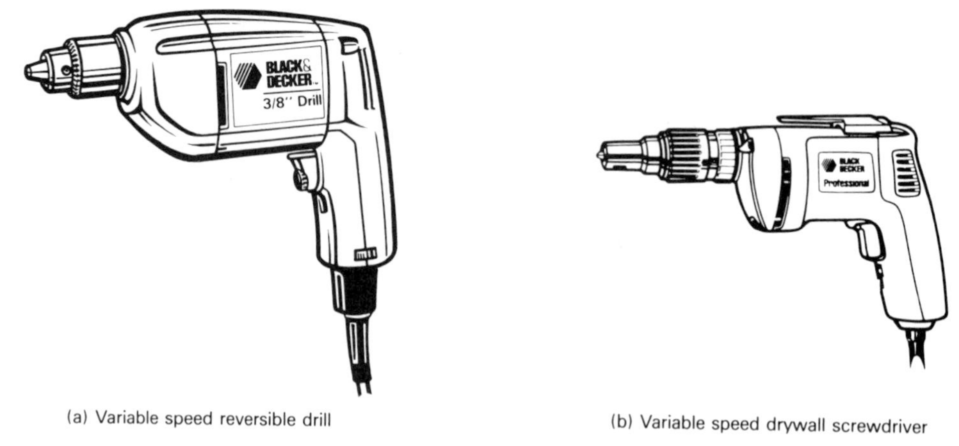

(a) Variable speed reversible drill

(b) Variable speed drywall screwdriver

Figure 16-10 Power drills and screwdrivers (courtesy of Black & Decker (U.S.), Inc.).

Saws

The electric *circular saw* (Figure 16-11) is another widely used building tool. The saw is primarily used for ripping and crosscutting wood framing, sheathing, and trim. The blade of the saw may also be tilted for making sloped cuts. Pocket or interior cuts may be started by resting the front edge of the saw's baseplate against the member to be cut and slowly lowering the blade into the member.

The *saber saw,* also known as a *jig saw* or *bayonet saw,* has a reciprocating blade mounted perpendicular to the body of the tool. It is a versatile cutting tool for making straight or curved cuts in wood, metal, or plastic. To make an interior or pocket cut, drill a hole slightly larger than the saw blade, insert the blade into the hole, and start the saw. Always use a blade suitable for the type of material being cut and the nature of the cut. Do not force the blade into the work while sawing. Some saber saws incorporate a scrolling mechanism, which allows the blade to be rotated in relation to the tool body, facilitating the making of intricate cuts.

The *reciprocating saw* is similar to a saber saw except that the blade is mounted parallel to the body of the tool. It is a heavy duty saw that can be used to cut pipe and other metals as well as wood, plastic, and gypsum board. Portable electric *hack saws* are also available for cutting pipe and other metal shapes.

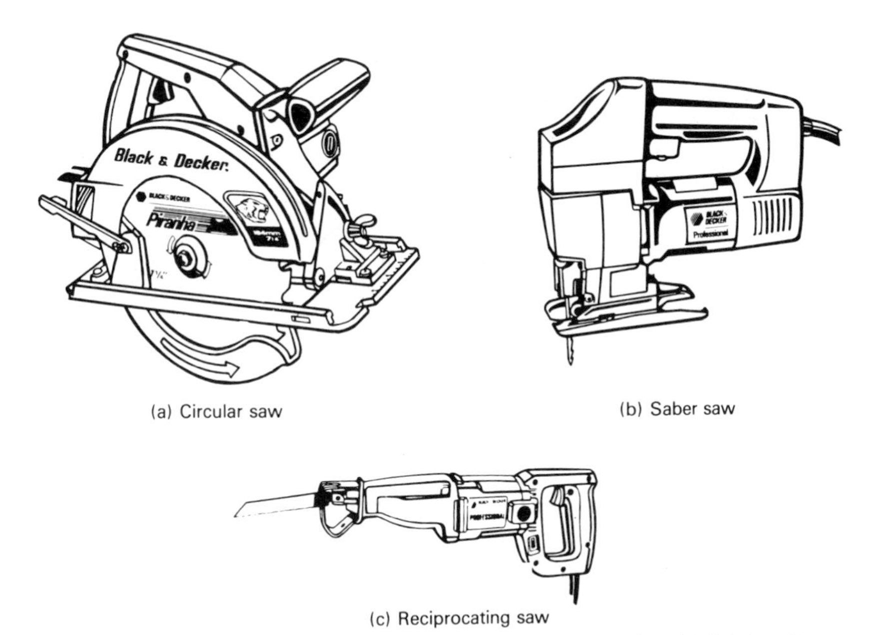

(a) Circular saw

(b) Saber saw

(c) Reciprocating saw

Figure 16-11 Power saws [courtesy of Black & Decker (U.S.), Inc.].

Sanders

Portable power sanders include belt sanders, finishing sanders, and rotary sanders, as illustrated in Figure 16-12. *Belt sanders* use an endless sanding belt to sand a relatively large surface at one time, and can remove material rapidly when equipped with a coarse sanding belt.

Finishing or *pad sanders* are used for producing a smooth wood finish before painting or staining. They usually have one or more handles as illustrated in Figure 16-12 and use an orbital or elliptical sanding action. Dual-motion sanders that can provide either an orbital or straight-line sanding action are also available. *Palm sanders* are small pad sanders that are held and guided by the palm of one hand. They also employ an orbital sanding action.

Rotary or *disc sanders* are equipped with a revolving head to which a sandpaper disc is attached. They are designed for fast sanding of wood or metal. When using one of these sanders, take care to avoid creating deep circular grooves in the surface being sanded. Rotary sanders may also be equipped with a polishing pad for use as a polishing machine.

Routers and Planes

The *router* (Figure 16-13) uses a rotary cutting bit to create a mortice (a depression such as that required for a door hinge or lock), groove, or rounded edge. A wide variety of cutting and shaping bits are available to accomplish these tasks.

The portable electric *planer* is a powered version of the hand plane. It is used for quickly trimming, squaring, or smoothing the edges of door and window frames.

Nailers, Staplers, and Power Hammers

Nailers, staplers, and power hammers (Figure 16-14) are used for rapid fastening of framing, sheathing, roofing, and other materials. *Nailers* replace the nail hammer

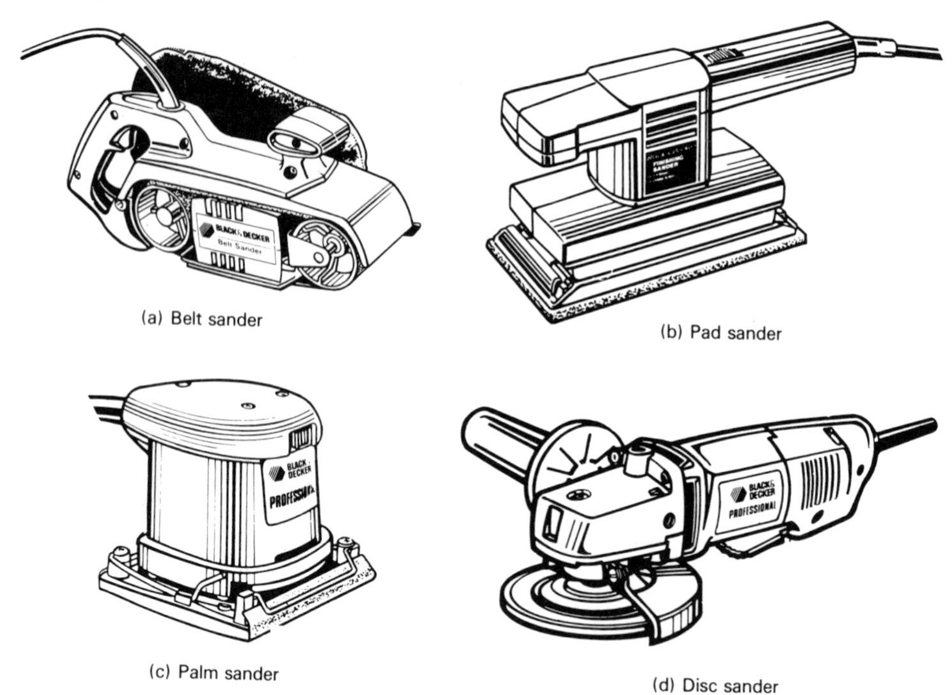

(a) Belt sander

(b) Pad sander

(c) Palm sander

(d) Disc sander

Figure 16-12 Power sanders [courtesy of Black & Decker (U.S.), Inc.].

for driving brads, common nails, or roofing nails. *Roofing nailers* can drive roofing nails up to 1 3/4 in. (44 mm) long. *Framing nailers* can drive nails as large as 16d for general purpose nailing. Power *staplers* are used for fastening sheathing, decking, and other sheet materials. *Power hammers* use an explosive cartridge and special fastener to attach framing or furring to concrete or masonry by driving the fastener through the member into the concrete or masonry.

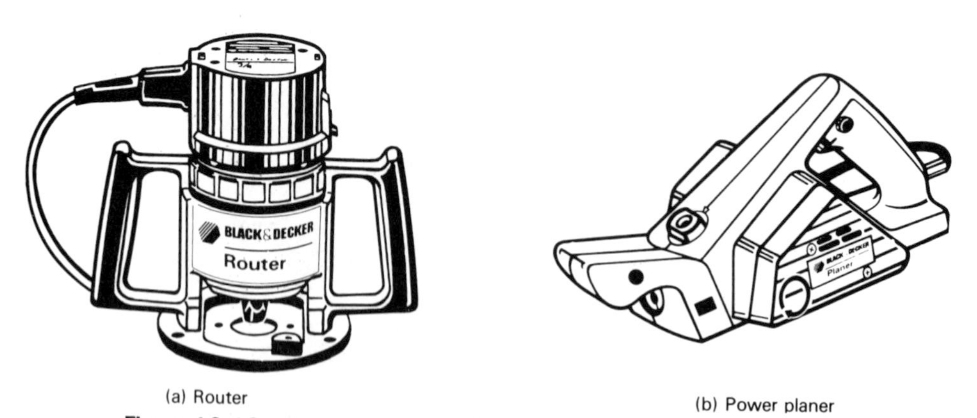

(a) Router

(b) Power planer

Figure 16-13 Power routers and planers [courtesy of Black & Decker (U.S.), Inc.]

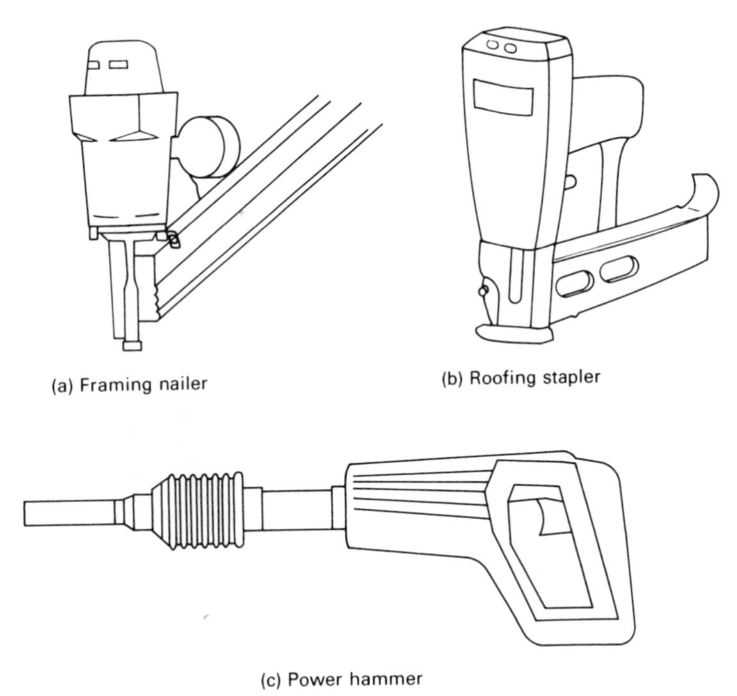

(a) Framing nailer (b) Roofing stapler

(c) Power hammer

Figure 16-14 Nailers, staplers, and power hammers.

16-3 CONSTRUCTION EQUIPMENT

Introduction

The large machines manufactured for use in the construction process are classified as *construction equipment.* Most construction equipment is designed for use in the earthmoving process. *Earthmoving* involves excavating, loading, transporting, dumping, grading, and compacting soil and rock. While the amount of earthmoving involved in building construction is usually small, excavation is required for foundations and basements, fill may be needed to bring the site to the required grade, soil must be compacted, and the area surrounding the building must be sloped in accordance with the site plan. Other construction equipment is often needed for lifting and loading materials and building components, and for concrete construction.

Although a detailed discussion of construction equipment is beyond the scope of this book, this section provides a brief description of those items of construction equipment that the builder is most likely to encounter. See References 2 and 3 for additional information on this topic.

Loaders and Excavators

Loaders, also called *front-end loaders* or *bucket loaders,* are versatile pieces of equipment capable of excavating soil or broken rock, loading material into trucks or other haul units, transporting material, and shaping banks or grades. *Wheel loaders* (Figure 16-15) are most often used by builders because of their maneuverability and ability to move between job sites. Although very large wheel loaders are available for use in earthmoving, typical loaders used by builders use buckets with a capacity ranging from 1.5–3.0 cu. yd. (1.1–2.3 m^3). A variety of attachments such as lift

(a) Loader/backhoe

(b) Track loader

(c) Hydraulic excavator (backhoe)

Figure 16-15 Loaders and excavators [(a) courtesy JCB, Inc., (b) courtesy Caterpillar, Inc., (c) courtesy Caterpillar, Inc.].

forks, augers, sweepers, and backhoes are available for loaders. Loaders designed for quick mounting and removal of attachments are sometimes called *tool carriers. Track loaders* are used when soil conditions are poor or steep grades are involved. Perhaps the most common type of loader used by the builder is the *loader/backhoe,* which combines a wheel loader with a small backhoe.

Excavators are used for loosening, removing, and loading soil or rock. The excavator most likely to be encountered in building construction is the *backhoe.* Hydraulic backhoes are also called *hydraulic excavators.* The backhoe, which digs by pulling the bucket back toward the machine, is most effective for digging below grade, and it is widely used for digging basements, footings, trenches, and similar excavations.

Cranes and Lifts

The *crane* is used for lifting, lowering, and transporting material and equipment. The *mobile crane,* or *truck crane,* illustrated in Figure 16-16 is often used by builders because of its mobility on and between job sites. The two major hazards involved in crane operations are the crane overturning and the presence of high voltage lines.

(a) Hydraulic truck crane

(b) Rough terrain fork lift

Figure 16-16 Cranes and lifts [(a) courtesy Bucyrus-Erie Co., (b) courtesy Koehring Construction Equipment].

Care must be taken to ensure that the crane load does not exceed the crane's rated capacity under the conditions involved. Likewise, always ensure that the required clearance is maintained between the crane (including its load) and any high voltage lines. As always, observe all other construction safety regulations.

Forklifts are also used for lifting and transporting materials and building components. Forklifts designed for use at a construction site are also called *rough terrain forklifts* and *shooting boom lifts.*

(a) Walk-behind plate compactor

(b) Rammer

(c) Walk-behind roller compactor

Figure 16-17 Compactors [(a) courtesy Dynapac Mfg., Inc., (b) courtesy Wacker Corp., (c) courtesy Dynapac Mfg., Inc.].

Compactors

Soil compaction is often required for the soil supporting foundations and floor slabs, as well as for trench and foundation wall backfill. Of the available types of compactors, those most often used in building construction include vibratory plate compactors, rammers, and small rollers.

The *vibratory plate compactor* (Figure 16-17) achieves soil compaction by a combination of vibration and static weight. It is widely used for compacting soil around foundations and other confined areas.

The *rammer* or *tamper* achieves compaction primarily by impact on the soil. However, vibratory rammers are available, which have an operating frequency high enough to produce vibratory compaction as well as impact compaction.

Small self-propelled *smooth wheel rollers* and *tamping foot rollers* are also used in building construction. In many cases, roller wheels are interchangeable so that a roller can be used as either a smooth wheel roller or a tamping foot roller. Some machines can also operate in either a static or vibratory mode.

PROBLEMS

1. Explain the difference between a crosscut saw and a rip saw.
2. What is a framing square?
3. Briefly describe the use of a chalk line.
4. What type of hand drill is used to turn an auger bit?
5. List some of the major safety precautions that should be observed when using electric power tools.
6. What is a nailer?
7. How should a pocket cut be started in a wood board using an electric circular saw?
8. Briefly describe a loader/backhoe and its uses.
9. What is a hydraulic excavator?
10. Briefly identify and explain the two major safety hazards involved in the use of a crane.

REFERENCES

1. Baudendistel, Robert F. *Modern Carpentry.* Englewood Cliffs, NJ: Prentice-Hall, 1986.
2. Nunnally, S. W. *Construction Methods and Management.* 2nd ed. Englewood Cliffs, NJ: Prentice-Hall, 1987.
3. Nunnally, S. W. *Managing Construction Equipment.* Englewood Cliffs, NJ: Prentice-Hall, 1977.
4. Stanley-Proto Industrial Tools. *Stanley Tool Guide.* Covington, GA, 1968.

appendix

Metric Conversion Factors

Multiply English unit	*By*	*To obtain metric unit*
foot	0.3048	meter (m)
square feet	0.0929	square meters (m^2)
cubic feet	0.0283	cubic meters (m^3)
cubic feet/min./ft.	1.549	liters/second/meter (l/s/m)
cubic yards	0.7646	cubic meters (m^3)
footcandle	10.76	lux (lx)
gallon	3.785	liter (l)
inch	2.540	centimeter (cm)
pound	0.4536	kilogram (kg)
pounds per foot	1.488	kilograms per meter (kg/m)
pounds per square foot	47.88	pascal (Pa)
pounds per square inch	6.895	kilopascals (kPa)
pounds per cubic foot	16.02	kilograms per cubic meter (kg/m^3)
pounds per cubic yard	0.5933	kilograms per cubic meter (kg/m^3)
thermal resistance (R)	0.1761	sq meter degree/watt (m^2K/W)
thermal transmission (U)	5.678	watt/sq meter degree (W/m^2K)
ton (2000 lbs.)	0.9072	metric ton (t)

Index